UNIVERSITY OF STRATHCLYDE
30125 00756957 6

D1761088

ANDERSONIAN LIBRARY
WITHDRAWN
FROM
LIBRARY
STOCK
UNIVERSITY OF STRATHCLYDE

Food Flavorings

Third Edition

Edited by

Philip R. Ashurst

Dr. P.R. Ashurst & Associates

Kingstone, Hereford, United Kingdom

AN ASPEN PUBLICATION®

Aspen Publishers, Inc.

Gaithersburg, Maryland

1999

The author has made every effort to ensure the accuracy of the information herein. However, appropriate information sources should be consulted, especially for new or unfamiliar procedures. It is the responsibility of every practitioner to evaluate the appropriateness of a particular opinion in the context of actual clinical situations and with due considerations to new developments. The author, editors, and the publisher cannot be held responsible for any typographical or other errors found in this book.

Aspen Publishers, Inc., is not affiliated with the American Society of Parenteral and Enteral Nutrition.

Library of Congress Cataloging-in-Publication Data
Food flavorings / edited by Philip R. Ashurst. — 3rd ed.
p. cm.
ISBN 0-8342-1621-3
1. Flavoring essences. I. Ashurst, P. R.
TP450.F66 1999
664′.5—dc21 99-27422
CIP

Orders: (800) 638-8437
Customer Service: (800) 234-1660

Editorial Services: Ruth Bloom
Library of Congress Catalog Card Number: 99-27422
ISBN: 0-8342-1621-3

Printed in the United States of America
1 2 3 4 5

Table of Contents

Contributors

Philip Ashurst, Ph.D.
Dr. P.R. Ashurst & Associates
Kingstone
Hereford
United Kingdom

D.G. Ashwood
Burtons Gold Medal Biscuits
Quality House
Blackpool
United Kingdom

D. Bahri
Givaudan Aromen GmbH
Dortmund
Germany

Belayet H. Choudhury
Takasago International Corporation (USA)
Rockeigh, New Jersey

D.C.F. Church
D.C. Flavours Ltd.
Clacton-on-Sea
Essex
United Kingdom

H. Kuentzel
Givaudan Research Company
Dubendorf
Switzerland

D.V. Lawrence
Flavex Ltd.
Kingstone
Hereford
United Kingdom

Charles H. Manley, Ph.D.
Takasago International Corporation (U.S.A.)
Rockeigh, New Jersey

Günter Matheis
Dragoco Gerberding & Co. AG
Holzminden
Germany

A.C. Mathews
Net Consultancy
Drake House
Cinderford
United Kingdom

Peter Mazeiko
Takasago International Corporation (USA)
Rockeigh, New Jersey

David A. Moyler
Fuerst Day Lawson Ltd.
St. Katherines Way
London
United Kingdom

Roger N. Penn, Ph.D.
Director
Tobacco Business Unit
MANE
Bar-sur-Loup
France

Barry Taylor
Danisco Flavours
Dennington Industrial Estate
Wellingborough
Northhampton
United Kingdom

Geoff White
The Edlong Company
Martlesham Heath
Ipswich
Suffolk
United Kingdom

Suzanne White
The Edlong Company
Martlesham Heath
Ipswich
Suffolk
United Kingdom

John Wright
Bush Boake Allen Inc.
Montvale, New Jersey

Preface

It is almost a decade since the first edition of this volume was produced and nearly 5 years from the second. Despite the many organizational changes that have taken place in both the flavor and publishing businesses, it is again gratifying to find that the demand for a third edition remains.

This is in no small part due to the fact that the flavor industry continues to flourish despite the takeovers and amalgamations that increase the size of the major manufacturers. These activities tend inevitably to create a fallout of skilled personnel, a proportion of whom restart in a small way and start the cycle all over again.

The industry generally becomes more oriented to high technology operations, although these generally impact more on the control of manufacture, sales and finance than on the actual creation of flavors themselves. The heart of a good flavor remains the simple blending of high-quality ingredients to produce a creation that is, to its user, more than the sum of its parts.

The industry continues to supply traditional demands for flavors in the food industry—soft drinks, baking and confectionery—and at the same time to meet the new challenges of ready-prepared meals and other new developments in products and processing.

For the first time, however, this third edition acknowledges the important contribution made by flavors to areas other than food. Flavors play a vital part in the formulation, acceptability and, therefore, efficacy of most oral medicines—both over-the-counter (OTC) and ethical preparations. A chapter on the use of flavors in pharmaceutical applications deals with this subject.

Similarly, a chapter reflects the major use of flavors in tobacco products. Whatever the level of social acceptability of smoking, the subject is of great significance with the flavor industry.

Other new chapters deal with the important topics of flavor modifiers and the whole realm of flavor quality control. Of the remaining contributions, some are unchanged and others updated and amended.

Overall, it is hoped that readers will again find this to be a useful work. Its contributors are widely experienced and I am extremely grateful to them for taking time from busy schedules to prepare and edit manuscripts.

Any errors and omissions are those of the editor and I accept responsibility for them. I hope, however, they will not detract too much from the important contribution this book has made and will continue to make to those seeking knowledge of the flavor industry.

I would, finally, like to acknowledge the help of my colleague, Sue Bate, for her invaluable help and support in preparing this volume.

Philip R. Ashurst

Chapter 1

Essential Oils

John Wright

1.1 INTRODUCTION

Most foods derive their characteristic flavor from chemicals that are present at levels ranging from parts per billion to parts per million. On the broad canvas of nature, some plant species evolved with far higher levels of flavor chemicals than others. Dried clove buds, for example, contain 12% eugenol. Such herbs and spices have been used from very early times to flavor other foods. With the discovery of distillation, it became possible to separate the flavor chemical mixture from the botanical material, and essential oils as commodities were born.

Essential oils and material extracted by solvent from herbs and spices found ready use in the rapid evolution of the soft drinks and confectionery industries. They formed the backbone of the raw materials used by the early flavor industry. Even today, essential oils form a significant part of the flavorist's repertoire and training normally covers the subject in great detail.

1.2 THE PRODUCTION OF ESSENTIAL OILS

1.2.1 *Steam Distillation*

Essential oils are produced by a variety of methods. Steam distillation is the most widely used (Figure 1–1). The steam is normally generated in a separate boiler and then blown through the botanical material in a still. The basic principle behind the distillation of two heterogeneous liquids, such as water and an essential oil, is that each exerts its own vapor pressure as if the other component were absent. When the combined vapor pressures equal the surrounding pressure, the mixture will boil. Essential oil components, with boiling points normally ranging up to 300° C, will thus evaporate at a temperature close to the boiling point of water. The steam and essential oil are condensed and separated. Essential oils

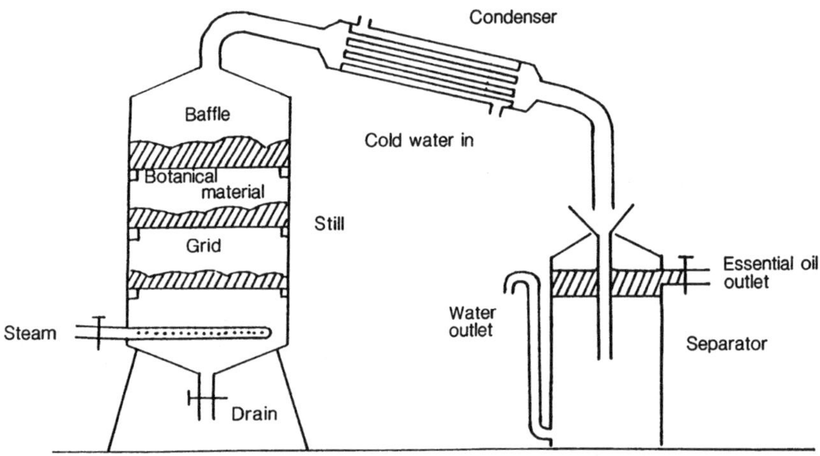

Figure 1–1 Diagram of typical steam distillation plant.

produced in this way are frequently different from the original oil in the botanical in a number of respects. Chemicals that are not volatile in steam, for example 2-phenyl ethanol in rose oil, are mainly left behind in the still. Many of these non-volatile components are responsible for taste rather than odor effects. Some very volatile chemicals may be lost in the distillation, and the process itself may induce chemical changes such as oxidation or hydrolysis.

1.2.2 *Water Distillation*

Water distillation is similar in many respects to steam distillation except that the botanical material is in contact with the boiling water. "Still odor" is a frequent problem with this type of distillation particularly if the still is heated directly. The burnt character known as "still odor" will gradually reduce on storage of the oils.

Steam and water distillation combines some features of both methods. The botanical material is separated from boiling water in the lower part of the still by a grid. If the still is heated carefully this method reduces the danger of "still odor."

Hydrodiffusion is a variation on the normal steam distillation process where the steam enters from the top of the vessel and the oil and water mixture is condensed from the bottom. This method reduces distillation time and is particularly suitable for distilling seeds.

Water distillation is the method of choice for fine powders and flowers. Steam fails to distill these materials effectively because it forms channels in powder and clumps flowers. For all other botanical materials, steam distillation, which offers the choice of high or low pressure steam, is the method of choice. Water and steam distillation offers many of the advantages of steam distillation but restricts the choice to low pressure steam. Steam distillation causes less hydrolysis of oil components, is much quicker and gives a better yield because less high boiling and water soluble components remain in the still. Steam distillation also removes the need to return the distillation waters to the still (cohobation).

In all these types of distillation the essential oil and water vapors are condensed in a condenser and collected in an oil separator. Most essential oils are lighter than water and will form a layer at the top of the separator. Much more water is distilled over than essential oil, so it is vital to remove the excess water continuously. Some oils are heaver than water and the function of the separator must be reversed.

1.2.3 *Distillation Methods*

Dry distillation is only suitable for a small range of essential oils and is often used to distill the oil from exudates such as balsams. Destructive distillation involves the formation of chemicals not present in the original botanical material. A typical example is cade oil, which is produced by the destructive distillation of juniper wood. The natural status and safety in use of such oils are open to considerable question.

1.2.4 *Expression of Oils*

Citrus oils are often isolated from the peel by expression ("cold pressing"). This process involves the abrasion of peel and the removal of the oil in the form of an aqueous emulsion that is subsequently separated in a centrifuge. Expressed citrus oils have superior odor characteristics compared with distilled oils, because of the absence of heat during processing and the presence of components that would not be volatile in steam. They are also more stable to oxidation because of the presence of natural antioxidants, such as tocopherols, which are not volatile in steam. The lack of heat damage to the oil is also significant.

1.2.5 *Extraction (see also Chapter 2)*

Extractions of the oils with supercritical liquid carbon dioxide provides the advantages of a cold process and the incorporation of some of the non-volatile components. It is expensive in terms of plant and, in some cases, results in an unusual balance of extracted oil components. Carbon dioxide extracts resemble essential oils more closely than oleoresins and, when well balanced, can offer unique raw materials for flavor creation.

1.3 FURTHER PROCESSING OF ESSENTIAL OILS

Raw essential oils produced by any of these methods may be processed further in a number of ways. Simple redistillation is often carried out to clean the oil. Cassia oil, for example, is often contaminated with impurities such as iron in its raw state. Redistillation, without any separation of fractions, solves the problem.

Rectification takes this process one step further and involves the selection of desirable fractions. Some of the first fractions of peppermint oil have undesirable harsh and vegetal odors. The later fractions have heavy cloying odors. Selection of the 80–95% of oil between these extremes results in a cleaner, sweeter smelling peppermint oil.

1.3.1 *Rectification*

The process of rectification can be extended to cover the removal of a substantial part of the terpene hydrocarbons in an oil. These chemicals are generally more volatile than the rest of the oil and can easily be separated by distillation under vacuum. Depending on the quantity of terpene hydrocarbons originally present, the residues in the still may be as little as 3% of the original volume of raw oil. The disadvantages of this process are the effects of heat and the loss of desirable components with the hydrocarbons. The profile of the oil will be changed and the degree of flavor concentration will not match the physical concentration.

Terpeneless oils are produced by distilling, under vacuum, the volatile fractions of a concentrated oil from the still residues. The main disadvantage is the loss of desirable components in their residues. Concentrated and terpeneless oils reduce the problems of oxidation of terpene hydrocarbons, particularly in citrus oils. Terpeneless oils provide clear aqueous solutions and are especially useful for flavoring soft drinks. Sesquiterpeneless oils carry the fractionation one step further and remove the sesquiterpene hydrocarbons. They offer increased strength, stability and solubility but decreased odor fidelity because of missing components.

Fractionation can be carried out for other reasons than concentration. In some instances the objective is to isolate a desirable individual component, for example, linalol from rosewood oil. Alternately the intention may be to exclude an undesirable component from the oil, for example, pulegone from peppermint oil.

The objective of obtaining selected parts of essential oils can be achieved in other ways. Chromatography offers a relatively cold process and great selectivity, often at a high price. Carbon dioxide extraction of oils also offers a cold process but less selectivity. Both these processes can result in concentrated oils with an unusual balance of components. Extraction of the oil with a solvent, typically a mixture of an alcohol and water, results in a well-balanced flavor. This process is very widely used in the manufacture of soft drink flavors.

1.3.2 *Washed Oils*

"Washed oils" produced in this way retain most of the character of the original oils. The solvent mixture may be varied to remove most or all of the hydrocarbons, depending on the degree of water solubility required. Some desirable oxygenated components are lost in the oil fraction, the "terpenes," particularly when a low level of hydrocarbons is required in the "washed oil." The inefficiency of this extraction process may mean that as little as 60% of the desirable flavor in the oil is finally extracted. "Washed oils" are not often used outside soft drinks and limited dairy applications because of the high dose rates that would be required. Countercurrent liquid/liquid extraction, followed by removal of solvent, produces a high quality terpeneless oil with many of the flavor characteristics of a "washed oil." It provides the added bonuses of more efficient extraction and much greater flavor concentration.

The terpene hydrocarbon fractions which are the byproducts of these processes are also used in the flavor industry. Quality, and usefulness, will vary considerably. Citrus terpenes from the production of "washed oils" often retain much of the character of the original oils. They can be washed to remove residual solvent, dried, and used in the production of inexpensive nature identical flavor oils for confectionery use. Most terpenes can be used to dilute genuine oils and this is one of the most difficult types of adulteration to detect. The only change in the analysis of the oil may be in the ratios of some of the components. Distilled terpenes, particularly citrus terpenes, oxidize more readily than the original oil. Adulteration of oils with terpenes results in reduced stability as well as an inferior profile.

Some delicate raw materials are processed in a different way for use in flavors and fragrances. Flowers, such as jasmin, are first extracted with hexane. The solvent is then removed under vacuum resulting in a semi-solid mass known as a "concrete." This raw material is used in fragrances but is not normally soluble enough for flavor use. Extraction of the concrete with ethanol, followed by solvent removal produces an "absolute" that mainly consists of the volatile components of the original flowers. Absolutes are obviously not strictly essential oils but they perform a similar function and are powerful raw materials in the flavorist's arsenal.

1.3.3 *Oil Quality*

Much of the quality control effort of flavor companies is dedicated to the detection of adulteration in essential oils and other natural raw materials. Traditional standards, often specified in official pharmacopeia, are usually out of date. Most physical and simple chemical characteristics are easily altered to meet a specification. Gas chromatography (GC) presents a much more formidable challenge. It is very difficult to reformulate an oil in the minute detail that is revealed by this technique. One traditional method of quality control also works well. A trained nose can recognize many faults and foreign components in essential oils.

All essential oils are subject to deterioration, mainly caused by oxidation, polymerization and hydrolysis, on storage. They should be stored dry, in full airtight containers, under nitrogen if possible, and in cold dark conditions.

1.4 THE USES OF ESSENTIAL OILS

The uses of essential oils in flavors can be grouped into two broad categories. The main use for most oils is to give their own characteristic flavor to an end-product. The flavor may be simple or part of a blend with other essential oils. In some cases the character may be enhanced by the addition of nature identical raw materials. The addition of ethyl butyrate (fruity, juicy) and *cis*-3-hexanal (green, leaf) gives a typically orange juice character to orange peel oil.

The most challenging use of essential oils is in the creation of natural flavorings. Many natural flavorings cannot be produced effectively from the named product because of problems of availability, price, concentration, heat stability and variability. Essential oils sometimes contain key characteristics of other flavors and can be used, often in trace quantities, in natural flavors to enhance those characteristics. Coriander oil contains linalol, which is an important component of the natural aroma of apricots, and coriander oil is frequently used to good effect in natural apricot flavors.

The main disadvantage of this use of essential oils is that other, less desirable, components of the oil can detract from the overall character of the flavor. This problem can be reduced by the use of isolated fractions from the oil. Eugenol, derived from clove leaf oil, is almost universally used in natural banana flavors.

A small, but growing, category of essential oils find their main use in natural flavorings. The most important example is davana oil (*Artemesia pallens,* family Compositae). Only about two tons of this oil are produced annually in India but it plays a major part in the juicy character of many natural fruit flavors, especially raspberry.

1.5 THE COMPOSITION OF ESSENTIAL OILS

The following essential oils have been selected from many hundreds. They represent the major commercial oils that are used in flavorings. Only the most important information on composition is given. Chemical formulae of the key components are shown in Figure 1–2. The quantities are typical but individual oils will often vary widely in practice.

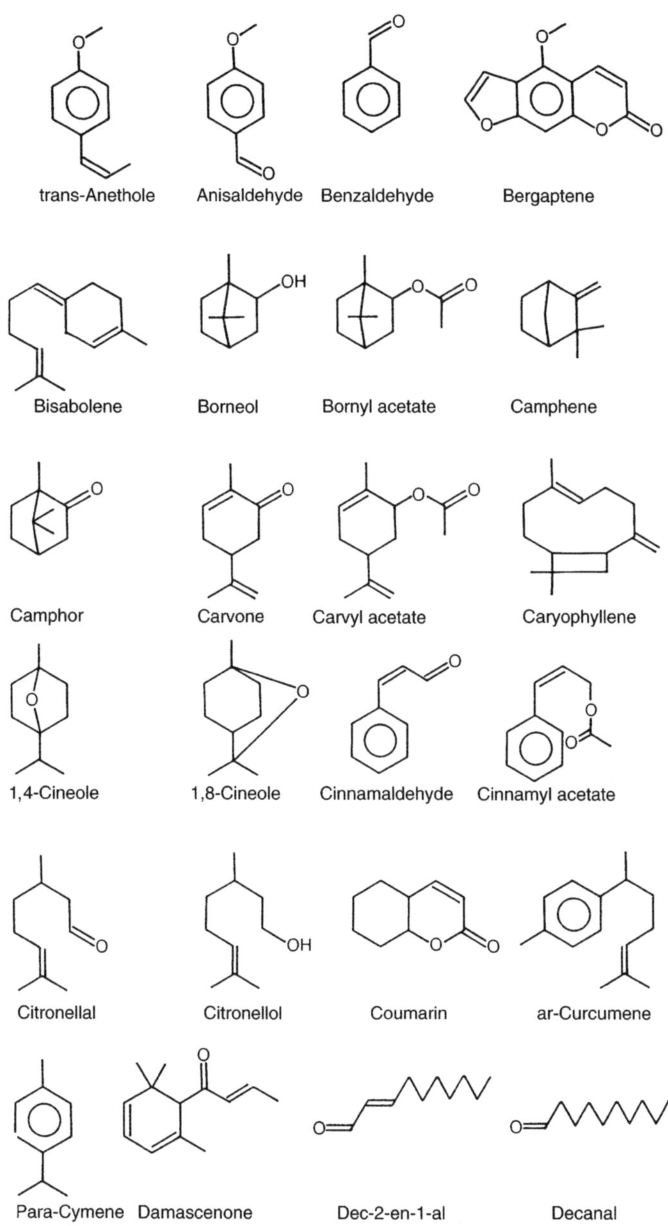

continues

Figure 1–2 Chemical formulae of the key components of principal essential oils.

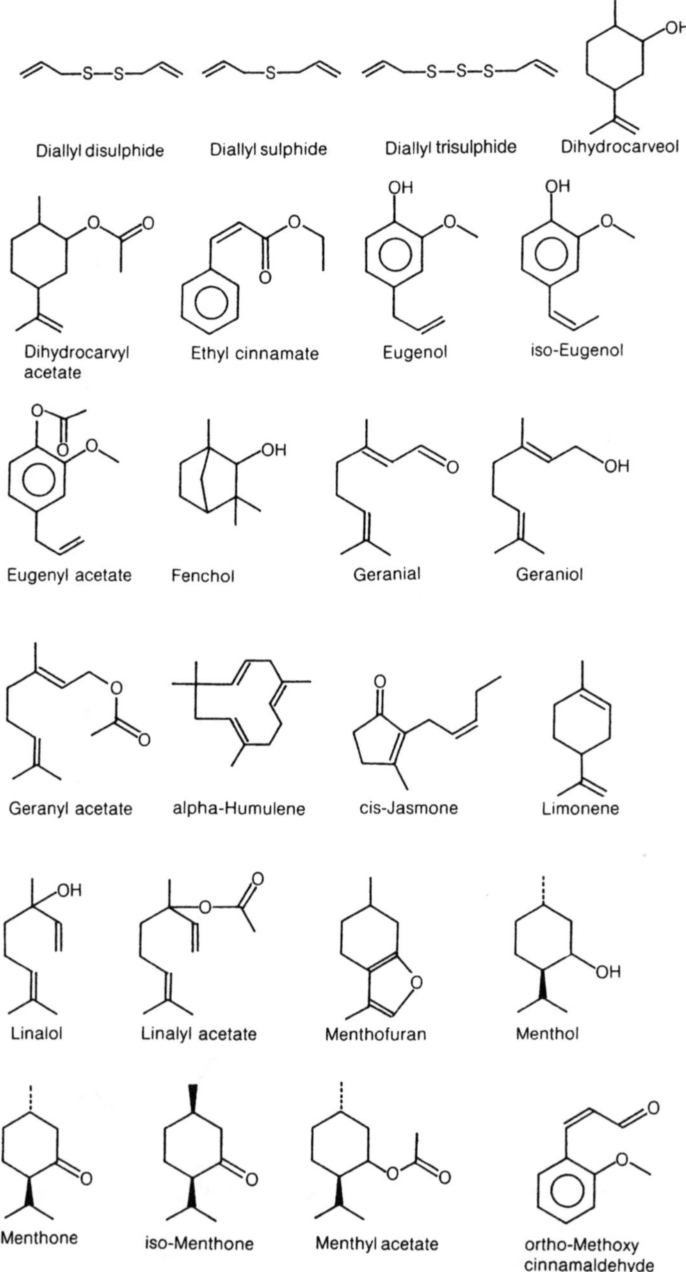

Figure 1–2 continued

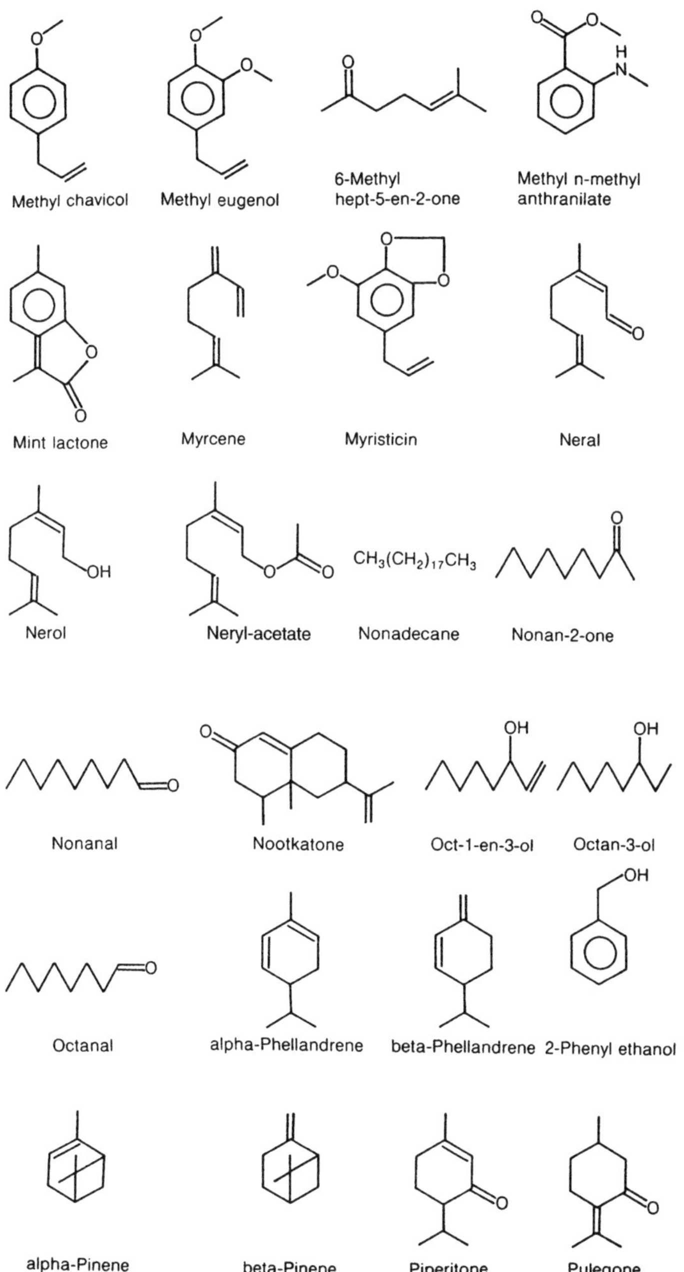

Methyl chavicol
Methyl eugenol
Mint lactone
Myrcene
Myristicin
Neral
Nerol
Neryl-acetate
Nonadecane
Nonan-2-one
Nonanal
Nootkatone
Oct-1-en-3-ol
Octan-3-ol
Octanal
alpha-Phellandrene
beta-Phellandrene
2-Phenyl ethanol
alpha-Pinene
beta-Pinene
Piperitone
Pulegone

continues

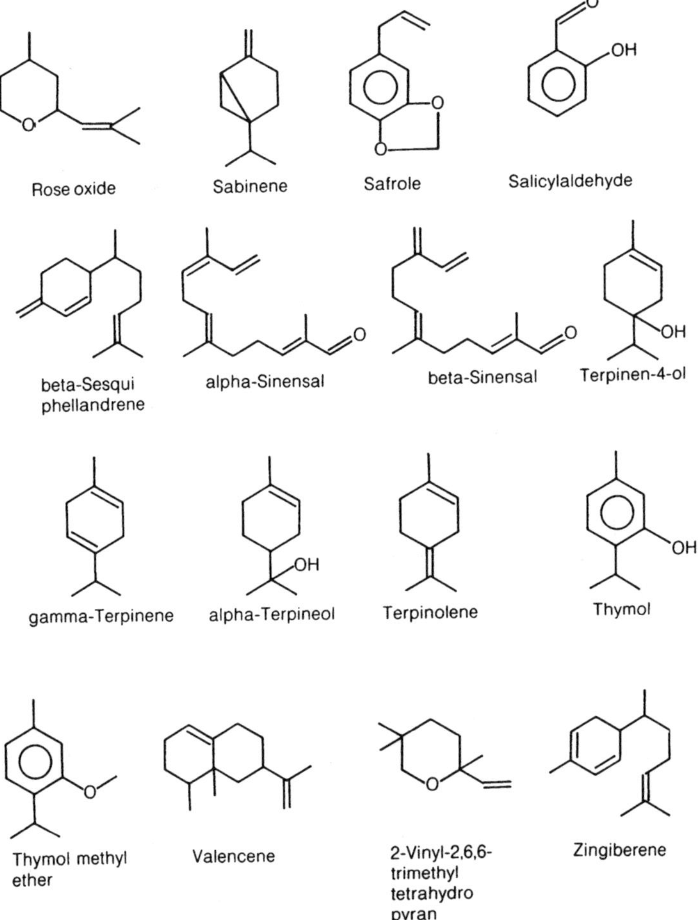

Figure 1–2 continued

The oils belong to the following plant families:

Compositae	davana
Cupressaceae	juniperberry
Geraniaceae	geranium
Gramineae	lemongrass
Grossulariaceae	blackcurrant buds
Iridaceae	orris
Labiatae	cornmint, peppermint, rosemary, spearmint, sweet basil, sweet marjoram, thyme

Lauraceae	cassia, cinnamon, litsea cubeba
Liliaceae	garlic, onion
Magnoliaceae	star anise
Myristicaceae	nutmeg
Myrtaceae	clove, eucalyptus
Oleaceae	jasmine
Rosaceae	bitter almond, rose
Rutaceae	bergamot, bitter orange, boronia, buchu, grapefruit, lemon, lime, petitgrain, sweet orange, tangerine
Umbelliferae	anise seed, coriander, cumin, dill, sweet fennel
Violaceae	violet leaf
Zingiberaceae	cardamom, ginger

1.5.1 *Anise Seed Oil*

Pimpinella anisum. Annual production of this oil has dwindled to around 8 tons, mainly from Spain, because of competition from star anise oil. The seeds yield 2.5% of oil on steam distillation.

The major components of the oil are typically:

95%	*trans*-anethole (strong, sweet, anise)
2%	methyl chavicol (strong, sweet, tarragon)

Anise seed oil is used in alcoholic drinks, seasoning and confectionery applications and in small quantities in natural berry flavors. There are no legal restrictions on the use of anise seed oil in flavorings. It is FEMA GRAS (2094) and Council of Europe listed.

1.5.2 *Bergamot Oil*

Citrus aurantium. Most of the annual total of 200 tons of bergamot oil is produced in Italy and the Ivory Coast. Other producers are Guinea, Brazil, Argentina, Spain and the former U.S.S.R. (BSI, BS 5750/ISO9000). Peel yields 0.5% cold pressed oil. Only small quantities of distilled oil are produced. Demand is stable. The major market is France (50%) followed by the Netherlands (17%) and the United States (15%) (BSI, BS 4778).

The major quantitative components of bergamot oil are typically:

35%	linalyl acetate (fruity, floral, lavender)
30%	limonene (weak, light, citrus)
15%	linalol (light, lavender)
7%	beta-pinene (light, pine)
6%	gamma-terpinene (light, citrus, herbaceous)

2%	geranial (lemon)
2%	neral (lemon)

Other qualitatively important components are:

1%	geraniol (sweet, floral, rose)
0.4%	neryl acetate (fruity, floral, rose)
0.2%	geranyl acetate (fruity, floral, rose)
0.2%	bergaptene (very weak odor)

Bergaptene is a skin sensitizer and some bergamot oil is distilled to produce oils free from bergaptene and terpenes. The level of linalol is generally higher in African than Italian oils. Adulteration of bergamot oil is sometimes carried out with synthetic linalol and linalyl acetate, together with orange and lime terpenes. It is easily detected on odor and by gas chromatography.

The major flavor use of bergamot oil is in "Earl Grey" type tea flavors, where it is normally the major component. It is also used as a minor component in citrus soft drink flavors and some natural fruit flavors, especially apricot. There are no legal restrictions on the use of bergamot oil in flavorings. It is FEMA GRAS (2153) and Council of Europe listed.

1.5.3 *Bitter Almond Oil*

Prunus amygdalus. Interest in the United States in natural cherry flavors has fueled the resurrection of this oil. Production in the U.S., France and Italy is increasing rapidly and is now around 100 tons. Bitter almonds are rarely used because the yield of oil is only 0.6%. Apricot kernels yield 1.2% of oil and are the preferred starting material.

The major components of the oil are typically

97.5%	benzaldehyde (bitter almond)
2%	hydrogen cyanide (bitter almond)

Hydrogen cyanide is highly toxic and is removed chemically before the oil is used in flavors.

Bitter almond oil was very often adulterated with synthetic benzaldehyde but such adulteration is readily detectable by isotopic ratio measurements such as the $^{13}C/^{12}C$ ratio. The main use to the oil is to give a bitter almond or cherry note to natural flavors. There are no legal restrictions on the use of bitter almond oil in flavorings. It is FEMA GRAS (2046) and Council of Europe listed.

1.5.4 *Bitter Orange Oil*

Citrus auranthium. Around 30 tons are produced annually, mainly in the West Indies. The yield of cold pressed oil from the fruit is 0.4%.

The major components of the oil are typically:

93%	limonene (weak, light, citrus)
0.2%	decanal (strong, waxy, peel)
0.2%	linalol (light, lavender)

Bitter orange oil is easy to adulterate with sweet orange oil. A small proportion of bitter orange oil can be added to sweet orange flavors to give a very attractive effect. It is widely used in tonic drink flavors and can also be used in small quantities in a wide range of other natural flavors, especially apricot. There are no legal restrictions on the use of bitter orange oil in flavorings. It is FEMA GRAS (2823) and Council of Europe listed.

1.5.5 *Blackcurrant Buds Absolute*

Ribes nigrum. The principal producer is now Tasmania, but some extracts are still produced in France. Blackcurrant buds (from prunings) yield 3% concrete on extraction with hexane. The concrete yields 80% absolute. The absolute contains 13% of volatile oil, which is not available commercially.

The major quantitative components of the oil are typically:

14%	delta-3-carene (herbal, medicinal)
10%	caryophyllene (spicy, woody)
4%	terpinen-4-ol (strong, herbaceous, nutmeg)

The key minor component is:

0.2%	4-methoxy-2-methyl-2-mercaptobutanone (very strong, catty, blackcurrant)

Adulteration of this raw material is not unusual, but can be easily detected. Blackcurrant buds concrete is relatively soluble in flavor solvents at low levels and may be used in preference to the absolute because it retains more of the catty character. The concrete and absolute are widely used in natural blackcurrant flavors, but also uniquely useful in other natural fruit flavors which have a catty note, such as grapefruit and peach. There are no legal restrictions on the use of blackcurrant buds absolute in flavorings. It is FEMA GRAS (2346) and Council of Europe listed.

1.5.6 *Boronia Absolute*

Boronia megastigma. Small quantities of absolute are produced, mainly in Tasmania. Flowers yield 0.7% concrete on extraction with hexane. The concrete yields 60% absolute or 20% of a volatile oil, which is not available commercially.

The major components of the oil are typically:

38%	beta-ionone (floral, violet)
4%	methyl jasmonate (strong, floral, jasmin)

Boronia absolute is frequently adulterated, but it is easy to detect by gas chromatography. The very attractive and unique character of this absolute fits well into many natural fruit flavors. It is especially effective in natural raspberry flavors. There are no legal restrictions on the use of boronia absolute in flavorings. It is FEMA GRAS (2167) and Council of Europe listed.

1.5.7 *Buchu Leaf Oil*

Barosma betulina and *crenulata*. Annual production of buchu leaf oil is around 1 ton, mainly from South Africa. The leaves yield 2.2% of oil on steam distillation. Demand is increasing in the major markets in Europe.

The major components of buchu oil are typically:

20%	diosphenol (weak, minty)
18%	menthone (harsh, herbal, mint)
0.5%	mentha-8-thiol-3-one (sulfury, catty)

Buchu oil is not easily adulterated. The oil is a very effective source of the catty note which is vital in a wide range of natural flavors, especially blackcurrant, grapefruit, peach and passion fruit. If buchu oil is used at excessively high levels in blackcurrant flavors the minty note can be intrusive. There are no legal restrictions on the use of buchu leaf oil in flavorings. It is FEMA GRAS (2169) and Council of Europe listed.

1.5.8 *Cardamom Oil*

Elettaria cardamomum. Around 8 tons are produced annually, mainly in Guatemala, India and Sri Lanka. The seeds yield 6% oil by steam distillation.

The major components of the oil are typically:

37%	terpinyl acetate (fruity, herbal, citrus)
34%	1,8-cineole (fresh, eucalyptus)

The oil is fairly easy to adulterate and often smells more like eucalyptus than cardamom. The genuine oil is very useful in seasoning blends, and, at low levels, in tea flavors. There are no legal restrictions on the use of cardamom oil in flavorings. It is FEMA GRAS (2241) and Council of Europe listed.

1.5.9 *Cassia Oil*

Cinnamonum cassia. Virtually all of the 500 tons of cassia oil produced annually originate in China. Very small quantities are produced in Taiwan,

Indonesia and Vietnam. Leaves, twigs and sometimes inferior bark yield 0.3% oil by water distillation. Demand is increasing steadily despite unpredictable production levels in China. The main markets are the United States (56%), Japan (25%) and Western Europe (11%).

The major quantitative components of the oil are typically:

85%	cinnamaldehyde (spicy, warm, cinnamon)
11%	*o*-methoxy cinnamaldehyde (musty, spicy)
6%	cinnamyl acetate (sweet, balsamic)

Other qualitatively important components are:

1%	benzaldehyde (bitter almond)
0.4%	ethyl cinnamate (balsamic, fruity)
0.2%	salicylaldehyde (pungent, phenolic)
0.2%	coumarin (sweet, hay)

Coumarin is suspected of being toxic. Cinnamaldehyde is the most important contributor to the characteristic odor of cassia but *o*-methoxy cinnamaldehyde is mainly responsible for the unique note which distinguishes cassia from cinnamon oil. Cassia oil is often imported in a crude state and requires redistillation to improve the odor and remove metallic impurities. Adulteration of the oil with cinnamaldehyde is practiced but can be easily detected by gas chromatography.

In many respects cassia was originally a "poor man's cinnamon." In the flavor industry, the oil makes a unique contribution in its own right. It is a major part of the traditional flavor of cola drinks. It is also used in confectionery, sometimes in conjunction with capsicum oleoresin. Use of cassia oil in other natural flavors is restricted to cherry, vanilla and some nut flavors. There are no legal constraints on the use of cassia oil in flavors. It is FEMA GRAS (2258) and Council of Europe listed.

1.5.10 *Cinnamon Oil*

Cinnamonum zeylanicum. Two oils are produced from the cinnamon tree. Cinnamon leaf oil is mainly produced in Sri Lanka (90 tons) together with a further 8 tons produced in India and 1 ton in Seychelles. Only about 5 tons of cinnamon bark oil are produced annually, mainly in Sri Lanka (2 tons). Leaves yield 1% oil and bark yields 0.5% oil. Demand for the oils is declining. The main markets are Western Europe (50%) and the United States (33%).

The major quantitative components of the oils are typically:

(a) Cinnamon leaf oil

80%	eugenol (strong, warm, clove)
6%	caryophyllene (spicy, woody)
3%	cinnamaldehyde (spicy, warm, cinnamon)

2%	iso-eugenol (sweet, carnation)
2%	linalol (light lavender)
2%	cinnamyl acetate (sweet, balsamic)

(b) Cinnamon bark oil

76%	cinnamaldehyde (spicy, warm, cinnamon)
5%	cinnamyl acetate (sweet, balsamic)
4%	eugenol (strong, warm, clove)
3%	caryophyllene (spicy, woody)
2%	linalol (light, lavender)

Other qualitatively important components of the bark oil are:

0.7%	alpha-terpineol (sweet, floral, lilac)
0.7%	coumarin (sweet, hay)
0.6%	1,8-cineole (fresh, eucalyptus)
0.4%	terpinen-4-ol (strong, herbaceous, nutmeg)

Coumarin, as previously noted, is under suspicion of being toxic. Eugenol is the main odor component in the leaf oil but the bark oil has a complex odor with significant contributions from all the components listed above.

Cinnamon leaf oil is used in seasoning blends as an alternative to clove oil, which it resembles on odor. It is also sometimes used to adulterate the bark oil, in conjunction with cinnamaldehyde. The bark oil is used in high quality seasoning blends and occasionally in natural flavors. There are no legal constraints on the use of cinnamon leaf and bark oils. They are FEMA GRAS (2292) leaf (2291) bark and Council of Europe listed.

1.5.11 *Clove Oil*

Eugenia caryophyllata. World production of clove leaf oil is around 2,000 tons, with Madagascar (900 tons), Indonesia (850 tons), Tanzania (200 tons), Sri Lanka and Brazil being the main producers. Only 100 tons of clove stem oil are produced, with Tanzania, Madagascar and Indonesia being the main producing countries. Fifty tons of clove bud oil are produced annually, mainly in Madagascar. Leaves and twigs yield 2% leaf oil, the stems attached to buds and flowers yield 5% stem oil and the buds yield 15% bud oil. Demand for clove oils is static with the main markets currently North America and Western Europe.

The major components of clove oil are typically:

81%	eugenol (strong, warm, clove)
15%	caryophyllene (spicy, woody)
2%	alpha-humulene (woody)
0.5%	eugenyl acetate (warm, spicy)

Quantities of the same components in stem and bud oils are typically:

Stem	Bud	
93%	82%	eugenol
3%	7%	caryophyllene
0.3%	1%	alpha-humulene
2%	7%	eugenyl acetate

The bud oil has by far the finest odor character of the three oils but is also the most expensive. Stem oil is used as a substitute for bud oil. Leaf oil is little more than a source of natural eugenol. Adulteration of bud oil by stem, leaf oils, eugenol and stem oil terpenes is carried out but can be detected by gas chromatography.

Clove oils are used in seasoning blends but also have an interesting part to play in other natural flavors. They form an essential part of the character of banana and a useful background note in blackberry, cherry and smoke flavors. There are no legal constraints on the use of clove oils. They are FEMA GRAS (2323) bud, (2325) leaf and (2328) stem and Council of Europe listed.

1.5.12 *Coriander Oil*

Coriandrum sativum. The annual production of coriander seed oil is about 700 tons, virtually all from the former U.S.S.R. Other minor producers are Yugoslavia, India, Egypt, Romania, South Africa and Poland. Production of coriander herb oil in France, the former U.S.S.R. and Egypt is very limited. Coriander seeds yield 0.9% oil on steam distillation. The fresh herb yields only 0.02% oil. Demand is increasing slightly. The major market is the United States.

The major components of coriander seed oil are typically:

74%	linalol (light, lavender)
6%	gamma-terpinene (light, citrus, herbaceous)
5%	camphor (fresh, camphoraceous)
3%	alpha-pinene (light, pine)
2%	*para*-cymene (light, citrus)
2%	limonene (weak, light, citrus)
2%	geranyl acetate (fruity, floral, rose)

The major component of coriander leaf oil is

10%	dec-2-enal (strong, orange marmalade)

Coriander seed oil can be adulterated with synthetic linalol but this is readily detectable by gas chromatography.

The seed oil is used in a very wide variety of flavor applications. It is part of the traditional flavoring of a number of alcoholic drinks, especially gin. It is widely used in meat seasonings and curry blends. It provides a very attractive natural source of linalol in natural fruit flavors, particularly apricot. The herb oil is very widely used in South Asian seasoning blends, but also provides a unique citrus character in natural flavors. There are no legal restrictions on the use of coriander oil in flavorings. It is FEMA GRAS (2334) and Council of Europe listed.

1.5.13 *Cornmint Oil*

Mentha arvensis. About 7,100 tons of cornmint oil (sometimes incorrectly called Chinese peppermint oil) are produced annually. It is almost all converted into menthol (2,800 tons) and dementholized oil (4,300 tons). China accounts for around 65% of the world production and India accounts for most of the remainder. Dried plants yield 2.5% oil by steam distillation. Cheap synthetic menthol has reduced the demand for cornmint oil into the main markets in the United States, Western Europe and Japan.

The major quantitative components of the dementholized oil are typically:

35%	laevo-menthol (cooling, light, mint)
30%	laevo-menthone (harsh, herbal, mint)
8%	iso-menthone (harsh, herbal, mint)
5%	limonene (weak, light, citrus)
3%	laevo-menthyl acetate (light, cedar, mint)
3%	piperitone (herbal, mint)
1%	octan-3-ol (herbal, oily)

Cornmint oil contains about 1% of pulegone (pennyroyal, mint odor) which is suspected of being toxic. The raw oils are rectified to remove some of the front and back fractions. Careful blending of fractions can reduce the characteristically harsh odor of cornmint oil but it still remains much less attractive than peppermint oil. Adulteration of cornmint oil is not a commercially attractive proposition.

Most cornmint oils are used to give a cheap peppermint flavor to a wide range of applications, often blended with true peppermint oil. It is more frequently used in blended flavors than peppermint oil because of its price advantage. Cornmint oil is Council of Europe listed with provisional limits for pulegone levels ranging from 25 ppm in food to 350 ppm in mint confectionery.

1.5.14 *Cumin Seed Oil*

Cuminium cyminum. Most of the annual total of 12 tons is produced in Iran, Spain and Egypt. Cumin seeds yield 3% of oil on steam distillation. Demand is steady in the major markets in Europe.

The major component of cumin seed oil is typically:

33%	cuminic aldehyde (sweet, spicy, cumin)

Cumin seed oil is not easy to adulterate because some important components are not available synthetically. The main use of the oil is in seasoning blends, especially curry, and also in some natural citrus and other fruit flavors. There are no legal restrictions on the use of cumin seed oil in flavorings. It is FEMA GRAS (2343) and Council of Europe listed.

1.5.15 *Davana Oil*

Artemisia pallens. Two tons of davana oil are produced each year in India. Davana grass yields 0.4% of oil on steam distillation. Demand is increasing in the major markets in Europe and the United States.

The major component of davana oil is typically:

40%	davanone (sweet, berry)

Davana oil is frequently adulterated but it is very difficult to adulterate the oil without changing its distinctive odor. The unusual fruity/berry character of davana oil is not widely known outside the flavor industry. It is widely used in natural flavors to give a berry note, especially to raspberry flavors. There are no legal restrictions on the use of davana oil in flavorings. It is FEMA GRAS (2359) and Council of Europe listed.

1.5.16 *Dill Oil*

Anethum graveolens. The annual world production of dill weed oil is around 140 tons. The major producers are the United States (70 tons), Hungary (35 tons) and Bulgaria (20 tons), followed by the former U.S.S.R. and Egypt. Annual production of dill seed oil is only 2.5 tons from the former U.S.S.R., Hungary, Poland, Bulgaria and Egypt. The weed oil is steam distilled from the whole plant at a yield of 0.7%. The seeds yield 3.5% oil. The major market for dill oil is the United States.

The major components of dill weed oil are typically:

35%	dextro-carvone (warm, spearmint, caraway)
25%	alpha-phellandrene (light, fresh, peppery)
25%	limonene (weak, light, citrus)

Dill weed oil can be adulterated with distilled orange terpenes, but this can be readily detected by gas chromatography.

The main use for dill oil is in seasoning blends, particularly for use in pickles. There are no legal restrictions on the use of dill oils in flavorings. They are FEMA GRAS (2383) and Council of Europe listed.

1.5.17 *Eucalyptus Oil*

Eucalyptus globus. Eucalyptus oil production now totals 5,000 tons. The majority is produced in China from camphor oil fractions, with a steadily decreasing proportion of true eucalyptus oil coming from Portugal, South Africa and Spain. An additional 400 tons of a "eucalyptus" oil from camphor oil fractions are produced annually in China. Eucalyptus leaves yield 1.5% on steam distillation. The demand for cineole type eucalyptus oils is increasing steadily. The major markets are Western Europe (60%) and the United States (20%).

The major components of eucalyptus oil are typically:

75%	1,8-cineole (fresh, eucalyptus)
10%	alpha-pinene (light, pine)
2%	*para*-cymene (light, citrus)
2%	limonene (light, weak, citrus)

Eucalyptus oils are often sold by their cineole (eucalyptol) content.

The major use for eucalyptus oil is in blends to give a fresh bright, slightly medicinal note particularly in conjunction with peppermint and aniseed oils. It may also be used in small quantities in other natural flavors such as blackcurrant. There are no legal restrictions on the use of eucalyptus oil in flavorings. It is FEMA GRAS (2466) and Council of Europe listed.

1.5.18 *Garlic Oil*

Allium sativum. Annual production of garlic oil is around 40 tons. China is the main producer (30 tons), followed by Mexico. Garlic bulbs yield 0.1% oil on steam distillation. Demand is steady in the major markets in France and Spain.

The major components of garlic oil are typically:

30%	diallyl disulfide (strong, garlic)
30%	diallyl trisulfide (strong, heavy, garlic)
15%	diallyl sulfide (strong, fresh, garlic)

Garlic oil can be adulterated with nature identical raw materials but this can be detected by gas chromatography.

The main use of the oil is in seasoning blends but it can also be used, in very small quantities, in other natural flavors to give a sulfur note. There are no legal restrictions on the use of garlic oil in flavorings. It is FEMA GRAS (2503) and Council of Europe listed.

1.5.19 *Geranium Oil*

Pelargonium graveolens. Around 200 tons of geranium oil are produced annually in China and Egypt. The flowers yield 0.1% of oil by steam distillation or 0.2% of concrete by extraction with hexane. The concrete yields 65% absolute.

The major components of the oil are typically:

32%	citronellol (fresh, floral, rose)
12%	geraniol (sweet, floral, rose)
6%	iso-menthone (harsh, herbal, mint)

Geranium oil may be used as an inexpensive alternative to rose oil in Turkish Delight confectionery, but the mint note is intrusive at higher levels of use. It is also useful in natural fruit flavors, especially blackcurrant, where the mint note does not present a problem. There are no legal restrictions on the use of geranium oil in flavorings. It is FEMA GRAS (2508) and Council of Europe listed.

1.5.20 *Ginger Oil*

Zingiber officinale. The annual production of ginger oil is close to 155 tons. The major producers are China and India with much smaller quantities from Sri Lanka, Jamaica, Australia and South Africa. Significant quantities are produced in Europe and the United States from imported roots. Dried roots yield 2% oil on steam distillation. Demand is increasing slowly. The major markets are the United States, Western Europe and Japan.

The major quantitative components of ginger oil are typically:

35%	zingiberene (warm, woody)
10%	AR-curcumene (woody)
10%	beta-sesquiphellandrene (woody)
8%	bisabolene (woody, balsamic)
6%	camphene (camphoraceous, light)
3%	beta-phellandrene (light, fresh, peppery)
2%	1,8-cineole (fresh, eucalyptus)

Other qualitatively important components are:

0.5%	bornyl acetate (camphoraceous, earthy, fruity)
0.5%	linalol (light, lavender)
0.3%	geranial (lemon)
0.2%	neral (lemon)
0.2%	nonan-2-one (pungent, blue cheese)
0.1%	decanal (strong, waxy, peel)

Oils from some sources, especially Australia, contain much higher levels of geranial and neral. Terpeneless ginger oil is produced to improve solubility in soft drinks, but this product is frequently a complicated recombination of a number of different fractions from a distillation. It varies widely in strength and character. Ginger oil is not often adulterated.

Large quantities of the oil are used in soft drinks. It is also used in spice blends for bakery and confectionery. The oil lacks the hot taste character of the oleoresin and the two raw materials are often blended together. Small quantities of ginger oil may be used in natural flavors such as raspberry. There are no legal restrictions on the use of ginger oil in flavorings. It is FEMA GRAS (2522) and Council of Europe listed.

1.5.21 *Grapefruit Oil*

Citrus paradisi. The total world production of grapefruit oil is about 700 tons. The major producers are the United States and Israel, followed by Argentina and Brazil. Peel yields 0.4% cold pressed oil. Demand is steady. The major market is the United States.

The major quantitative components of grapefruit oil are typically:

90%	limonene (weak, light, citrus)
2%	myrcene (light, unripe mango)

Other qualitatively important components are:

0.5%	octanal (strong, fresh, peel)
0.4%	decanal (strong, waxy, peel)
0.3%	linalol (light, lavender)
0.2%	nootkatone (strong, peel, grapefruit)
0.1%	citronellal (strong, citrus, green)
0.1%	neral (lemon)
0.1%	geranial (lemon)
0.02%	beta-sinensal (strong, orange marmalade)

Grapefruit oil often smells disappointingly like orange oil, rather than grapefruit. Nootkatone is frequently taken as an indicator of quality but it only represents a part of the recognizable character. Small quantities of grapefruit essence oil are recovered during juice concentration. This oil has a more typical fresh juice character but is relatively unstable to oxidation. Concentrated and terpeneless oils, together with extracts, are produced to solve the twin problems of oxidation and solubility. They are frequently not very reminiscent of grapefruit in odor. Grapefruit oil can be adulterated with orange terpenes and, as the genuine oil has more than a passing resemblance to orange, it is easier to detect by gas chromatography than odor.

The main use of grapefruit oil in flavors is to impart a grapefruit character to a wide range of application. It is sometimes mixed with other citrus flavors but is not often used outside this field. Much more natural tasting grapefruit flavors can be achieved by adding nature identical raw materials (catty, green and juicy notes) to oil, and most of the flavors produced are of this type. There are no legal restrictions on the use of grapefruit oil in flavorings. It is FEMA GRAS (2530) and Council of Europe listed.

1.5.22 *Jasmine Concrete and Absolute*

Jasminum officinale. About 10 tons of concrete and absolute are produced each year in Morocco, Egypt and Italy. The flowers yield 0.3% of concrete on solvent extraction, normally by hexane. The concrete yields 50% of absolute on extraction by ethanol. The absolute is normally used in flavors because of its superior solubility.

Demand is increasing in the major markets in Europe and the United States.

The major components of jasmine "oil" (16% of the concrete) are typically:

11%	benzyl acetate (fruity, floral)
3%	linalol (light, floral, lavender)
1.4%	*cis*-jasmone (herbal)
0.9%	methyl jasmonate (strong, floral, jasmine)
0.5%	indole (heavy, animalic)

Jasmine absolute is frequently adulterated, often with a very high degree of sophistication. Gas chromatography analysis is capable of detecting any significant dilution of the genuine material. The rich blend of useful characters in the absolute make it a valuable addition to many natural fruit flavors, particularly strawberry and raspberry. There are no legal restrictions on the use of jasmine absolute in flavorings. It is FEMA GRAS (2598) and Council of Europe listed.

1.5.23 *Juniperberry Oil*

Juniperus communis. Most of the annual production of 12 tons originates from Croatia. Juniperberries yield 1.2% of oil on steam distillation.

The major component of the oil is typically:

34%	alpha-pinene (light, pine)

Most of the oil is used, in conjunction with coriander and other oils, in the production of gin flavors. There are no legal restrictions on the use of juniperberry oil in flavorings. It is FEMA GRAS (2604) and Council of Europe listed.

1.5.24 *Lemongrass Oil*

Cymbopogon flexuosus and *citratus. Cymbopogon flexuosus* is known as East Indian lemongrass and is indigenous to South Africa. *Cymbopogon citratus* is mainly cultivated in Central and South America and is known as West Indian lemongrass. Only 10 tons are still produced. India and China are the major producers, followed by Guatemala, Brazil, Sri Lanka, Haiti and the former U.S.S.R. Dried grass yields 0.5% oil on steam distillation. Demand continues to decline because of the increasing dominance of *Litsea cubeba* as a source of natural citral. The main markets are Western Europe (30%), the United States (20%) and the former Soviet Union (20%).

The major quantitative components of East Indian lemongrass oil are typically:

40%	geranial (lemon)
30%	neral (lemon)
7%	geraniol (fruity, floral, rose)
4%	geranyl acetate (fruity, floral, rose)
2%	limonene (light, weak, citrus)

Other qualitatively important components are:

1.4%	caryophyllene (spicy, woody)
1.1%	6-methyl hept-5-en-3-one (pungent, fruity, herbaceous)
0.9%	linalol (light, lavender)
0.5%	citronellal (strong, citrus, green)

Before the advent of *Litsea cubeba,* lemongrass oil was the main source of natural citral. The other components of the oil also have a strongly characteristic effect which is evident even in citral fractionated from lemongrass oil. Adulteration with synthetic citral can be detected by gas chromatography.

Lemongrass citral still has some use in lemon flavors. Some traditional lemonade flavors derived part of their recognizable character from lemongrass citral and this can be recaptured by using a little lemongrass oil together with synthetic citral. There are no legal constraints on the use of lemongrass oil. It is FEMA GRAS (2624) and Council of Europe listed.

1.5.25 *Lemon Oil*

Citrus limon. About 3,600 tons of lemon oil are produced annually. Argentina, the United States and Italy are the major producers. Other producers are Brazil, Ivory Coast, Greece, Spain, Israel, Cyprus, Australia, Peru, Guinea, Indonesia, Venezuela and Chile. Lemon peel contains 0.4% oil and most production is by cold pressing. Cheaper, but inferior, distilled oil is mainly used in the production of terpeneless oil. Small quantities of lemon juice oil are produced during juice concentration. Although different varieties of tree are planted, the main reasons for the differences in character of oils from different areas are the dissimilarities of terrain, climate and production methods. Sicilian oil has the best odor characteristics. Demand for lemon oil is only increasing slowly. The main markets are Western Europe (40%), the United States (35%) and Japan (8%).

The major quantitative components of cold pressed lemon peel oil are typically (see also Appendix I):

63%	limonene (weak, light, citrus)
12%	beta-pinene (light, pine)
9%	gamma-terpinene (light, citrus, herbaceous)

Other qualitatively important components are:

1.5%	geranial (lemon)
1.0%	neral (lemon)
0.5%	neryl acetate (fruity, floral, rose)
0.4%	geranyl acetate (fruity, floral, rose)
0.2%	citronellal (strong, citrus, green)
0.2%	linalol (light, lavender)
0.1%	nonanal (strong, peel)

The citral (neral and geranial) level in United States oil is lower than normal and these oils have an individual, rather atypical odor. Winter oils are better than those produced in summer and generally command about 15% higher prices. Production methods also affect the quality of the oil. Sfumatrice oils have a fine fresh lemon character and change in color from yellow/green to yellow/orange as the fruit ripens. They are on average 15% more expensive than pelatrice oils, which have a poorer, grassy odor and change in color from dark green/yellow to yellow brown through the season. The terpene hydrocarbons which constitute the bulk of the oil are insoluble in water and susceptible to oxidation. Extraction, concentration and deterpenation are carried out to produce stable, soluble lemon flavors for soft drinks. Terpeneless lemon oils give a reasonably balanced lemon character. Adulteration of cold pressed lemon oil with distilled oil and terpenes is detectable by gas chromatography.

Lemon oils are used in a wide variety of applications. Juice oil is not widely used because it does not have the attractive fresh character of orange juice oil. Nature identical components can be added to duplicate the elusive note of fresh juice. Lemon oil is also used in other flavors, such as butterscotch, pineapple and banana. There are no legal constraints on the use of lemon oil in flavorings. It is FEMA GRAS (2625) and Council of Europe listed.

1.5.26 *Lime Oil*

Citrus aurantifolia. The annual production of distilled lime oil is around 1,000 tons. Mexico, Peru and Haiti are the major producers, followed by Brazil, Cuba, Ivory Coast, the Dominican Republic, Guatemala, Jamaica, Ghana, Swaziland and China. Only 160 tons of cold pressed oil are produced each year in Brazil (90 tons), the United States (40 tons), Mexico (25 tons) and Jamaica. Ten tons of oil are produced annually from a related variety, Persian lime (*Citrus latifolia*) in Brazil. Lime peel yields 0.1% of oil. Demand for lime oil continues to increase. The major market is the United States (67%) followed by the United Kingdom (10%).

The major quantitative components of distilled lime oil are typically:

52%	limonene (light, weak, citrus)
8%	gamma-terpinene (light, citrus, herbaceous)
7%	alpha-terpineol (sweet, floral, lilac)
5%	terpinolene (fresh, pine)
5%	*para*-cymene (light, citrus)
3%	1,4-cineole (fresh, eucalyptus)
2%	1,8-cineole (fresh, eucalyptus)
2%	beta-pinene (light, pine)

Other qualitatively important components are:

1%	bisabolene (woody, balsamic)
0.7%	fenchol (camphoraceous, earthy, lime)
0.7%	terpinen-4-ol (strong, herbaceous, nutmeg)
0.5%	borneol (camphoraceous, earthy, pine)
0.4%	2-vinyl-2,6,6-trimethyl tetrahydropyran (light, lime)

The major quantitative components of cold pressed lime oil are:

45%	limonene (light, weak, citrus)
14%	beta-pinene (light, pine)
8%	gamma-terpinene (light, citrus, herbaceous)
3%	geranial (lemon)
2%	neral (lemon)

The distilled oil is chemically very different from the cold pressed oil because it is distilled over acid. Much of the citral (neral and geranial) is lost in the process and with it the raw/fresh lemon character which typifies lemon peel. This character is replaced in the distilled oil by the lilac/pine/camphoraceous/earthy complex character that is normally recognized as lime flavor. Terpeneless lime oil is produced to improve solubility in soft drinks. Adulteration of lime oil is normally carried out by adding synthetic terpineol, terpinolene and other components to lime terpenes. It can often be easily identified by odor and can also be detected by gas chromatography.

Distilled lime oil is a major component of flavors for cola drinks. It is also used in traditional clear lemon lime soft drinks. Cold pressed oil is used in some more recent lemon lime drinks. Lime is little used in other natural flavors. There are no legal constraints on the use of lime oils in flavorings. Lime oil is FEMA GRAS (2631) and Council of Europe listed.

1.5.27 *Litsea Cubeba Oil*

Litsea cubeba. About 1,100 tons of oil are produced in China. The fruit yields about 3.2% oil on steam distillation. Demand for the oil is increasing rapidly. The main market is China itself, followed by the United States, Western Europe and Japan.

The major quantitative components of the oil are typically:

41%	geranial (lemon)
34%	neral (lemon)
8%	limonene (weak, light, citrus)
4%	6-methyl hept-5-en-2-one (pungent, fruity, herbaceous)
3%	myrcene (light, unripe mango)
2%	linalyl acetate (fruity, floral, lavender)
2%	linalol (light, lavender)

Other qualitatively important components are:

1%	geraniol (sweet, floral, rose)
1%	nerol (sweet, floral, rose)
0.5%	caryophyllene (spicy, woody)
0.5%	citronellal (strong, citrus, green)

The dominant citral (neral and geranial) odor is considerably modified by the other components of the oil, particularly methyl heptenone and citronellal.

The oil is frequently fractionated to increase the citral content above 95% and improve the purity of the lemon odor. It is possible to treat the oil with bisulfite to form bisulfite addition compounds of the aldehydes. The original aldehydes can later be regenerated, giving a high citral content and an attractive odor. This process is chemical and the purified oil should not be described as "natural." Adulteration with synthetic citral can be detected by gas chromatography.

Citral redistilled from *Litsea cubeba* oil is widely used in lemon flavors. It demonstrates good stability in end-products and many of the minor impurities reflect minor components of lemon oil itself. It can also be used in some natural fruit flavors, especially banana. *Litsea cubeba* oil is Council of Europe listed.

1.5.28 *Nutmeg Oil*

Myristica fragrans. Annual production of nutmeg oil is around 300 tons. The major producers are Indonesia and Sri Lanka. Some European oil is distilled from Grenadian nutmegs. Nutmegs yield 11% oil on steam distillation. Mace, the coating of the nutmeg, yields 12% oil but very little is produced. Demand for nutmeg oil is static. By far the main market is the United States (75%), followed by Western Europe.

The major components of nutmeg oil are typically:

22%	sabinene (light, peppery, herbaceous)
21%	alpha-pinene (light, pine)
12%	beta-pinene (light, pine)
10%	myristicin (warm, woody, balsamic)
8%	terpinen-4-ol (strong, herbaceous, nutmeg)
4%	gamma-terpinene (light, citrus, herbaceous)
3%	myrcene (light, unripe mango)
3%	limonene (weak, light, citrus)
3%	1,8-cineole (fresh, eucalyptus)
2%	safrole (warm, sweet, sassafras)

Safrole and myristicin are suspected of being toxic. Myristicin is normally present at a much lower level in oil distilled from West Indian nutmegs, but unfortunately this is not a normal item of commerce. Terpeneless nutmeg oil is produced to improve solubility for use in soft drinks. The oil can be adulterated by the addition of terpenes and nature identical raw materials, but this is easily detected on odor and by gas chromatography.

Nutmeg oil is a major component of cola flavors, and this use accounts for most of the world-wide production. It is also used in meat seasonings and in spice mixtures for bakery products. Small quantities can be used in natural fruit flavors, where it imparts depth and richness. Nutmeg oil is FEMA GRAS (2793) listed together with mace oil FEMA GRAS (2653). Both oils are Council of Europe listed with provisional limits for safrole levels ranging from less than 1 ppm in food to 15 ppm in products "containing mace or nutmeg."

1.5.29 *Onion Oil*

Allium cepa. About 3 tons are produced annually, mainly in Egypt. The yield of oil from fresh onions is 0.015% by steam distillation.

The major components are typically:

22%	dipropyl trisulfide (strong, heavy, onion)
20%	dipropyl disulfide (strong, onion)
5%	dipropyl tetrasulfide (strong, stale, onion)

The distillation process has a considerable influence on the composition of onion oil. Longer distillation times increase the proportion of tri- and tetrasulfides, with a correspondingly more cooked and staler odor. The oil is mainly used in seasoning blends, but is also useful in natural fruit flavors (at very low levels). There are no legal restrictions on the use of onion oil in flavorings. It is FEMA GRAS (2817) and Council of Europe listed.

1.5.30 *Orris Oil Concrete*

Iris germanica. Small quantities are produced in Italy, France and North Africa. The yield of steam distilled "oil concrete," so named simply because it solidifies on cooling, is 0.2% from dried and aged roots. Absolutes are produced from this oil to increase the concentration of alpha-irone.

The major component of the oil is typically:

15%	alpha-irone (strong, violet)

Adulteration is not uncommon, but can be detected by gas chromatography. Absolutes are used in preference to the oil concrete in natural flavors to give a clean violet note. This is especially useful in raspberry and other berry flavors. There are no legal restrictions on the use of orris oil concrete in flavorings. It is FEMA GRAS (2829) and Council of Europe listed.

1.5.31 *Peppermint Oil*

Mentha piperita. The majority of the 4,500 tons of peppermint oil produced annually is from the United States. Eighty-four percent of the United States oil is produced in the far west states, which use irrigation to obtain a high yield per acre. Sixteen percent is produced in the midwest. An increasing quantity of peppermint oil is now being produced in India, with small quantities coming from a number of European countries. The oils are produced by steam distillation at a yield of 0.4%. Demand for peppermint oil is only increasing slowly. The main market is the United States.

The major quantitative components are typically:

50%	levo-menthol (cooling, light, mint)
20%	levo-menthone (harsh, herbal, mint)
7%	levo-menthyl acetate (light, cedar, mint)
5%	1,8-cineole (fresh, eucalyptus)
4%	menthofuran (sweet, hay, mint)
3%	limonene (weak, light, citrus)
3%	iso-menthone (harsh, herbal, mint)

Other qualitatively important components are:

0.4%	octan-3-ol (herbal, oily)
0.1%	oct-1-en-3-ol (raw, mushroom)
0.03%	mint lactone (creamy, sweet, mint)

Peppermint oil contains about 1% of pulegone (pennyroyal, mint odor) which is suspected of being toxic. Menthofuran is the main analytical difference between peppermint oil and the cheaper, but similar, cornmint oil, but it is not a reliable indicator of quality. Raw oils are usually rectified to remove some of the "front" and "back" fractions. The nature of this processing varies considerably and can contribute to a recognizable house style. Oils with heavier cuts are sometimes described as "terpeneless" and can be used for cordials or liqueurs where a clear end-product is required. Complicated re-blending of individual fractions from peppermint oil can achieve selective results, reduction in the level of pulegone or softening of the odor profile.

Adulteration of peppermint oil with cornmint oil and terpenes occurs and can be easier to detect on odor than by analysis. Stretching with nature identical components is much more unusual and is rarely effective.

The main use of peppermint oil is to give a peppermint flavor to a wide range of applications. It is frequently blended with spearmint oils and also with many other common flavorings, including eucalyptus oil, methyl salicylate and anethole. Small quantities can be used to give subtle effects in natural fruit flavors. Peppermint oil is FEMA GRAS (2848) and Council of Europe listed with provisional limits for pulegone levels ranging from 25 ppm in food to 350 ppm in mint confectionery.

1.5.32 *Petitgrain Oil*

Citrus auranthium. Paraguay produces 150 tons of oil each year. The yield of oil from bitter orange twigs and leaves is 0.2% by steam distillation.

The major quantitative components of the oil are typically:

35% linalyl acetate (fruity, floral, lavender)
30% linalol (light, lavender)

A key minor component is:

0.002% 2-methoxy 3-isobutyl pyrazine (intense, green, bell pepper)

The lavender character of the oil fits into a number of natural fruit flavors, but the bell pepper note is especially useful, giving a realistic green note to flavors such as blackcurrant. There are no legal restrictions on the use of petitgrain oil in flavorings. It is FEMA GRAS (2855) and Council of Europe listed.

1.5.33 *Rose Oil*

Rosa damescena. Total production of rose oil and concrete from *Rosa damascena* and *Rosa centifolia* (Rose de Mai) is around 15 tons. The major producers are Turkey (4 tons), Bulgaria (3 tons) and the former U.S.S.R. (2 tons) (all *damascena*) and Morocco (3 tons) (*centifolia*). Other producers are India, Saudia Arabia, France, South Africa and Egypt. Flowers yield 0.02% oil (sometimes called otto) on steam distillation or 0.2% concrete on extraction with hexane. The concrete yields about 50% absolute. Demand is increasing. The major markets are the United States and Western Europe.

The major quantitative components of rose oil are typically:

50%	citronellol (fresh, floral, rose)
18%	geraniol (sweet, floral, rose)
10%	nonadecane (odorless)
9%	nerol (sweet, floral, rose)
2%	methyl eugenol (warm, musty)
2%	geranyl acetate (fruity, floral, rose)

Other qualitatively important components are:

1%	eugenol (strong, warm, clove)
1%	2-phenyl ethanol (sweet, honey, rose)
0.6%	rose oxide (warm, herbaceous)
0.5%	linalol (light, lavender)
0.04%	damascenone (strong, berry, damson)

The volatile fractions of concretes and absolutes contain a much higher level (up to 60%) of 2-phenyl ethanol. The relatively low level in distilled oil is caused by the fact that 2-phenyl ethanol is not steam volatile. Nonadecane and other stearoptenes in the distilled oil contribute very little to the odor of the oil but they solidify in the cold and cause solubility problems. Oils with these components removed are called stearopteneless. Rose oil is frequently adulterated with nature identical components such as citronellol and geraniol. This is usually obvious on odor and can be detected by gas chromatography.

The oil is used alone to flavor Turkish Delight and many other products in Asia. It is a very useful raw material in other natural flavors. Small quantities can be used to very good effect in many fruit flavors, especially raspberry. There are no legal restrictions on the use of rose oil in flavorings. It is FEMA GRAS (2989) and Council of Europe listed.

1.5.34 *Rosemary Oil*

Rosmarinus officinalis. About 250 tons of rosemary oil are produced each year. The major producer is Spain (130 tons), followed by Morocco (60 tons), Tunisia (50 tons), India, the former U.S.S.R., Yugoslavia, Portugal and Turkey. Fresh plants yield 0.5% steam distilled oil. Demand for rosemary oil continues to decline steadily. The major market is Western Europe.

The major components of rosemary oil are typically:

20%	alpha-pinene (light, pine)
20%	1,8-cineole (fresh, eucalyptus)

18%	camphor (fresh, camphoraceous)
7%	camphene (camphoraceous, light)
6%	beta-pinene (light, pine)
5%	borneol (camphoraceous, earthy, pine)
5%	myrcene (light, unripe mango)
3%	bornyl acetate (camphoraceous, earthy, fruity)
2%	alpha-terpineol (sweet, floral, lilac)

Moroccan and Tunisian oils normally contain about 40% of 1,8-cineole, with correspondingly lower levels of alpha-pinene, camphor and camphene. Camphor and eucalyptus fractions are used to adulterate rosemary oil and can easily be detected by gas chromatography.

Rosemary oil is used in seasoning blends and in flavor oil blends with a medicinal connotation. It is not often used in other natural flavors. There are no restrictions on the use of rosemary oil in flavorings. It is FEMA GRAS (2992) and Council of Europe listed.

1.5.35 *Spearmint Oil*

Mentha spicata. The majority of the total world production of 1500 tons is produced in the United States. Eighty-five percent of the United States oil is produced in the west, the remainder originating from the midwest. Other producers are China, Italy, Brazil, Japan, France and South Africa. Just over 50% of the oil produced in the United States is from *Mentha spicata* and is called "native" oil. The less hardy, but superior in terms of odor quality, *Mentha cardiaca* cross produces an oil called "Scotch." Production of this variety has increased in recent years. Western Europe is the largest market with demand increasing very slowly.

The major quantitative components are typically:

70%	levo-carvone (warm, spearmint)
13%	limonene (weak, light, citrus)
2%	myrcene (light, unripe mango)
2%	1,8-cineole (fresh, eucalyptus)
2%	carvyl acetate (sweet, fresh, spearmint)

Other qualitatively important components are:

1%	dihydrocarvyl acetate (sweet, fresh, spearmint)
1%	dihydrocarveol (woody, mint)
1%	octan-3-ol (herbal, oily)
0.04%	*cis*-jasmone (warm, herbal)

The raw oil is rectified to reduce the harsh sulfurous front fractions and sometimes to reduce the limonene producing a "terpeneless" oil. The odor character of spearmint oil frequently improves considerably with age. Adulteration with levo-carvone and blending from different sources occurs but can be detected on odor and by analysis.

The main use of spearmint oil is to give a spearmint flavor to chewing gum and oral hygiene products. It is often blended with peppermint and other flavors. There are no restrictions on the use of spearmint oil in flavors. It is FEMA GRAS (3032) and Council of Europe listed.

1.5.36 *Star Anise Oil*

Illicium verum. Most of the world annual production of 400 tons originates from China. Dried fruit yields 8.5% oil on steam distillation. This oil should not be confused with the chemically similar, but botanically unrelated, anise seed oil, which is produced from *Pimpinella anisum*, an umbelliferous plant. Only 8 tons of anise seed oil are still produced each year. Demand for star anise oil continues to decline. France is the main market.

The major components of star anise oil are typically:

87%	*trans*-anethole (strong, sweet, anise)
8%	limonene (weak, light, citrus)

Other qualitatively important components are:

1%	anisaldehyde (sweet, floral, hawthorn)
0.8%	linalol (light, lavender)
0.5%	methyl chavicol (strong, sweet, tarragon)

Redistilled oils, containing less limonene, are prepared for use in clear drinks. The oil can be adulterated with anethole but this can be detected by gas chromatography.

Star anise oil is used in large quantities in alcoholic drinks. It is also a popular flavoring in confectionery, particularly with a medicinal connotation, and in oral hygiene applications. Use in other natural flavors is restricted to small quantities in flavors such as cherry. There are no legal restrictions on the use of star anise oil in flavorings. It is FEMA GRAS (2096) and Council of Europe listed.

1.5.37 *Sweet Basil Oil*

Ocimum basilicum. Several tons of oil are produced annually, mainly in France. The yield of oil from the fresh plants is 0.1% by steam distillation.

The major components of the oil are typically:

45%	linalol (light, lavender)
25%	methyl chavicol (strong, sweet, tarragon)

Basil oil from the Comores contains more methyl chavicol and less linalol than sweet basil oil. Sweet basil oil is used in seasoning blends, but is also useful in small quantities in other natural flavors. There are no legal restrictions on the use of sweet basil oil in flavorings. It is FEMA GRAS (2119) and Council of Europe listed.

1.5.38 *Sweet Fennel Oil*

Foeniculum vulgare var. Dulce. Spain accounts for most of the 25 tons of fennel oil produced annually. The seeds yield 3% of oil by steam distillation.

The major components of the oil are typically:

70%	*trans*-anethole (strong, sweet, anise)
9%	limonene (weak, light, citrus)

Sweet fennel oil is mainly used in terpeneless form in alcoholic drinks. Small quantities are also used in seasonings and other natural flavors. There are no legal restrictions on the use of sweet fennel oil in flavorings. It is FEMA GRAS (2483) and Council of Europe listed.

1.5.39 *Sweet Marjoram Oil*

Origanum majorana. Thirty tons of oil are produced annually, mainly in Morocco. Fresh plants yield 0.4% oil by steam distillation.

The major components of the oil are typically:

31%	terpinen-4-ol (strong, herbaceous, nutmeg)
15%	gamma-terpinene (light, citrus, herbaceous)
8%	alpha-terpinene (light, citrus, herbal)
6%	*para*-cymene (light, citrus, resinous)
3%	linalol (light, lavender)

Sweet marjoram oil is mainly used in seasoning blends, usually in conjunction with a solid extract. It is also useful as a natural source of the characteristically nutmeg terpinen-4-ol which is free from safrole and myristicin. There are no legal restrictions on the use of sweet marjoram oil in flavorings. It is FEMA GRAS (2663) and Council of Europe listed.

1.5.40 *Sweet Orange Oil*

Citrus sinenis. Over 26,000 tons of orange oil are produced each year. The major producers are Brazil and the United States. Other producers are Israel, Italy, Australia, Argentina, Morocco, Spain, Zimbabwe, Cyprus, Greece, Guinea, the former U.S.S.R., South Africa, Indonesia and Belize. The majority of the oils are cold pressed from the peel of the fruit. Only small quantities are still produced by steam distillation. A different type of oil is produced during the concentration of orange juice. Orange juice or essence oil is the oil phase of the volatiles recovered during the process. Typical yields from the fruit are 0.28% of peel oil and 0.008% of juice oil (Deming, 1992). Different varieties of tree are planted to ripen at different times of year, hence the terms early, mid and late season. Late season oils are said to be the best and are usually produced from the variety "Valencia." Demand for orange oil is increasing steadily. The main markets are the United States (30%), Western Europe (30%) and Japan (20%).

The major quantitative components of cold pressed sweet orange oil are typically (see also Appendix I):

94%	limonene (weak, light, citrus)
2%	myrcene (light, unripe mango)

Other qualitatively important components are:

0.5%	linalol (light, lavender)
0.5%	octanal (strong, fresh, peel)
0.4%	decanal (strong, waxy, peel)
0.1%	citronellal (strong, citrus, green)
0.1%	neral (lemon)
0.1%	geranial (lemon)
0.05%	valencene (orange juice)
0.02%	beta-sinensal (strong, orange marmalade)
0.01%	alpha-sinensal (strong, orange marmalade)

The total aldehydes in orange oil have traditionally been taken as a quality indicator. Brazilian oils often have fairly low aldehyde contents. At the other end of the scale Zimbabwean oils may contain double the average level of aldehydes. Aldehyde content is an important factor in determining the yield of terpeneless oils but it is not a reliable guide to quality. Minor components, particularly the sinensals, make a major contribution.

Some countries produce oils with different odor characteristics which command high prices. Sicilian and Spanish oils are the most important examples. Orange juice oil has a fresher, juicy odor. It contains much more valencene (up to 2%) together with traces of ethyl butyrate, hexanal and other components which are responsible for the fresh odor of orange juice. The major component, *d*-limonene, has little odor value and is susceptible to oxidation giving rise to unpleasant off notes. It is also insoluble in water at normal use levels. Some orange oil is processed to reduce the *d*-limonene content. Concentrated and terpeneless oils produced by distillation have "flatter," less fresh odors because octanal and other volatile components are removed with the limonene. Octanal can be recovered and added back to improve odor quality. Extraction and chromatography produce better results. Orange terpenes are the byproduct of all these processes and find a ready market as a solvent for flavor and fragrances.

Adulteration of orange oil is an infrequent practice because the market price of the oils from the major producers is so low. Higher priced oils are sometimes diluted with oils from cheaper sources. This practice can be very difficult to detect.

Orange oil gives an acceptable orange flavor in many applications. A proportion of juice oil may be added to give a fresh effect. In nature identical flavors orange oil will often provide a base. Components found in orange juice could be included to duplicate the fresh juice taste. Other attractive top-notes which can be subtly emphasized in orange oil based flavors include violet, lemon and vanilla. Orange oil is also used in other flavors, particularly apricot, peach, mango and pineapple. There are no legal constraints on the use of orange oil in flavors. It is FEMA GRAS (2821) and Council of Europe listed.

1.5.41 *Tangerine Oil*

Citrus reticulata. World production of tangerine oil is about 300 tons. The main producers are Brazil (250 tons) and the United States (45 tons), followed by the former U.S.S.R., South Africa and Spain. Production of the closely related mandarin oil is 120 tons, from Italy (50 tons), China (40 tons), Argentina (10 tons), Brazil (10 tons), the Ivory Coast, the United States and Spain. Cold pressed peel yields 0.5% oil. Demand is increasing steadily in the main markets of the United States, Western Europe and Japan.

Mandarin oil has a strong aromatic/musty odor and justifiably commands a much higher price than tangerine oil, which is more reminiscent of orange.

The major quantitative components of tangerine oil are typically:

93%	limonene (weak, light, citrus)
2%	gamma-terpinene (light, citrus, herbaceous)
2%	myrcene (light, unripe mango)

Other qualitatively important components are:

0.7%	linalol (light, lavender)
0.4%	decanal (strong, waxy, peel)
0.3%	octanal (strong, fresh, peel)
0.1%	thymol methyl ether (sweet, thyme, herbaceous)
0.05%	alpha-sinensal (strong, orange marmalade)
0.03%	thymol (sweet, phenolic, thyme)

The major quantitative components of mandarin oil are typically:

72%	limonene (light, weak, citrus)
18%	gamma-terpinene (light, citrus, herbaceous)

Other qualitatively important components are:

0.8%	methyl *N*-methyl anthranilate (heavy, musty, orange blossom)
0.5%	linalol (light, lavender)
0.2%	decanal (strong, waxy, peel)
0.1%	thymol (sweet, phenolic, thyme)
0.05%	alpha-sinensal (strong, orange marmalade)

Concentrated and terpeneless oils are produced, together with extracts, to give flavors used in clear soft drinks. Much of the character of the original oil is retained in terpeneless oils. Adulteration is normally carried out by adding synthetics, such as methyl *N*-methyl anthranilate, to sweet orange oil. It is usually easy to detect by gas chromatography.

Mandarin and tangerine oils are widely used in soft drink flavors and confectionery both alone and in conjunction with orange oils. They also find good use in other natural fruit flavors, especially mango and apricot. There are no legal restrictions on the use of either oil in flavorings. They are FEMA GRAS (3041) tangerine, (2657) mandarin and Council of Europe listed.

1.5.42 *Thyme Oil*

Thymus vulgaris. Crude (red) and redistilled (white) oils are produced. Twenty-five tons are produced annually in Spain from *Thymus vulgaris* and *Thymus zygis*. An additional 10 tons of so called Origanum oil are also produced in Spain from *Thymus capitatus* and *Thymus serpyllum*. Carvacrol predominates over thymol in these oils but they have similar odors. Demand for thyme oil is steady in the major markets in Europe.

The major components of thyme oil are typically:

50%	thymol (sweet, phenolic, thyme)
15%	*para*-cymene (light, citrus, resinous)
11%	gamma-terpinene (light, citrus, herbal)

Thyme oil is very easy to adulterate. The major use of the oil is in seasoning blends, but it finds some use in natural flavors. There are no legal restrictions on the use of thyme oil in flavors. It is FEMA GRAS (3064) and Council of Europe listed.

1.5.43 *Violet Leaf Absolute*

Viola odorata. Produced in moderate quantities in France, violet leaves yield 0.1% of concrete on extraction with hexane. The concrete yields 40% absolute or 8% volatile oil (which is not normally available commercially).

The major component of the oil is typically:

14% *trans*-2-*cis*-6 nonadienal (strong, green, cucumber)

Violet leaf absolute is expensive and is occasionally adulterated with spinach absolute. Violet leaf absolute is an extremely useful cucumber, green character in a wide range of natural flavors. It is especially useful in melon and cucumber flavors but also finds wide use in many less obvious fruit flavors such as banana and strawberry. There are no legal restrictions on the use of violet leaf absolute in flavorings. It is FEMA GRAS (3110) and Council of Europe listed.

REFERENCES

British Deming Association, *A2–Deming's 14 Points for Management*, BDA (1989).

BSI, BS 4778, *Quality Vocabulary*, BSI Publications, Milton Keynes.

BSI, BS 5750/IS09000 all parts, BSI Publications, Milton Keynes.

W. Edwards Deming, *Out of the Crisis*, Massachusetts Institute of Technology (1992).

A. Feignbaum, *Total Quality Control,* McGraw Hill (1983).

P. Gosby, *Quality is Free*, McGraw Hill (1979).

Ishihara's Test for Colour Blindness, Kane and Hara (1990).

J. Juran, *Quality Control Handbook,* McGraw Hill (1995).

K. Likawa, *What is Total Quality Control?–The Japanese Way,* Prentice Hall (1985).

J. Oahland, *Total Quality Management,* Heinemann (1989).

F. Price, *Right First Time,* Gower (1985).

F. Price, *Right Every Time,* Gower (1990).

Rules and Guidance for Pharmaceutical Manufacturers, H.M.S.O., London (1993).

Chapter 2

Oleoresins, Tinctures and Extracts

David A. Moyler

2.1 INTRODUCTION

2.1.1 *General Comments*

When preparing or choosing a natural ingredient for incorporation into a food or beverage product, some initial thought about its form or the isolation method used to obtain the flavor principles can save considerable time at later stages of the development of consumer products. This simple principle applies whether preparing vegetables in the kitchen before making a sauce or choosing one of a spectrum of ginger extracts for making a flavoring suitable for ginger ale. This chapter illustrates the principles of extraction in the food industry from a practical, not a theoretical, viewpoint. Illustrative examples of actual problems are explained.

The form of the finished product will influence the selection process. For instance, a clear carbonated ginger ale requires a water soluble ginger extract whereas a cloudy carbonated "brewed style" ginger beer does not. For the clear beverage, a soluble ginger CO_2 extract would be recommended (not the outdated steam distilled and oleoresin ginger mixture that precipitates a sticky resinous deposit that has to be filtered out). A cloudy ginger beverage can be made by boiling ginger root in water to give flavor and cloud together.

2.1.2 *Costs*

Financial constraints of formulating products are not just those of raw materials. It is pointless using cheaper raw materials if the processing or labor cost in rendering them to a stable soluble form in the product escalates the total cost to one higher than that of a selective soluble extract that could be incorporated by simple mixing. Cost efficiency and the accurate material and process costings that make up that total are the key to successful product formulating in today's competitive world.

2.1.3 *Raw Materials and Processes*

If simple chopped or powdered dry plant materials can be used to give the required flavor to a slowly cooked traditional curry dish, then use them. However, if an instant convenience curry sauce is required for some customers, then extracts of plant materials are preferred, because they quickly and consistently disperse into the sauce.

When the instant convenience of a carbonated cola beverage is required, many processing and extraction technologies are needed to prepare the vegetable and spice ingredients into a drinkable form, e.g., roasting, grinding and extraction of kola nuts; cold expression and subsequent solubilization of citrus fruit oils; grinding and distillation of spices such as cassia bark, nutmeg kernels, ginger root, peppercorns, etc. All this activity is before even thinking about the caramelization of sugar, addition of acid, water quality and that essential sparkle provided by carbon dioxide.

These technologies are so much a part of our lives that their costs of manufacture are accepted. To clarify the differences between forms of natural botanical extracts, Table 2–1 lists some of the terminology used within the industry.

2.2 PLANT MATERIALS

2.2.1 *Origin*

To ensure consistency in a formulated product containing natural extracts, it is important to specify the country of origin of a plant material ingredient as well as its botanical name. For instance, nutmeg kernels (*Myristica fragrans*) from the East Indies (E.I.) have a different organoleptic profile to those from the West Indies (W.I.). The E.I. nutmeg has a higher myristicin level (5% compared to 4%), which is stronger for flavor use (Analytical Methods Committee, 1984), whereas nutmeg from the W.I. is low in safrole (0.2% compared to 3%), which is considered desirable for safety reasons (RIFM, 1982). Sri Lankan nutmegs are different again (Analytical Methods Committee, 1987).

Such examples of variations in plants, soil, climate and growing conditions are commonplace in the essential oils and extract industry. The fact that a supplier can give details of botany, origin and crop time is usually a sign that the supplier is thoroughly conversant with the material and it is less likely to be blended (with other origin naturals or different plant materials or even nature identical chemicals).

Plant materials from different origins are also valued on quality and some, which are considered from a superior origin, are at a premium price in the market place. Purchasers should always be on their guard against inferior qualities being passed off as something they are not. Periods of market shortages are the most usual time for the "relaxation of standards." However, most suppliers with a high reputation will supply material from a specific origin on request, or offer a

Table 2–1 Some terminology used in the preparation of oleoresins, tinctures and extracts (Shaath & Griffin, 1988)

Type	*Definition*	*Example*
Concrete	Extract of previously live plant tissue; contains all hydrocarbons, soluble matter and is usually solid, waxy substance	All flowers (jasmine, orange flower, rose, carnation, etc.)
Absolute	An alcohol extract of a concrete which eliminates waxes, terpene and sesquiterpene hydrocarbons	All flowers (jasmine, clove, cubeb, cumin, lavender, chamomile, etc.)
Balsam	A natural raw material which exudes from a tree or plant; they have a high content of benzoic acid, benzoates and cinnamates	Peru, tolu, canada, fir needle, etc.
Resin	They are either natural or prepared; natural resins are exudations from trees or plants, and are formed in nature by oxidation of terpenes; prepared resins are oleoresins from which the essential oils have been removed	Orris, olibanum, mastic, cypress and flowers
Oleoresin	Natural oleoresins are exudations from plants; prepared oleoresins are liquid extracts of plants yielding the oleoresin upon evaporation	Gurjun, lovage, onion, pepper, melilotus, etc.
Extract	Concentrated products obtained from solvent treatment of a natural product	St. John's bread, rhatany root, mate, fenugreek, etc.

"blend" to meet a specification within the price constraints applicable to today's competitive products.

2.2.2 *Crop to Crop Variations*

Sometimes, despite all the skill that a grower and processer can apply, Mother Nature provides plants which, when extracted, are outside the previously encountered range of characteristics. This does not usually present a serious problem because some blending with previous carry-over crops or oils from nearby growing areas can take place. However, if oils are being marketed as being 100% natural and from a named specific origin, due allowance should be made for a possible wider variation than would otherwise be encountered as due to "seasonal variations."

One example is cinnamon bark oil which contains a large proportion of cinnamaldehyde. The leaf oil, obtained from the same tree, is principally eugenol. The U.S. Essential Oil Association has published a specification for the bark oil as 55–75% cinnamaldehyde.

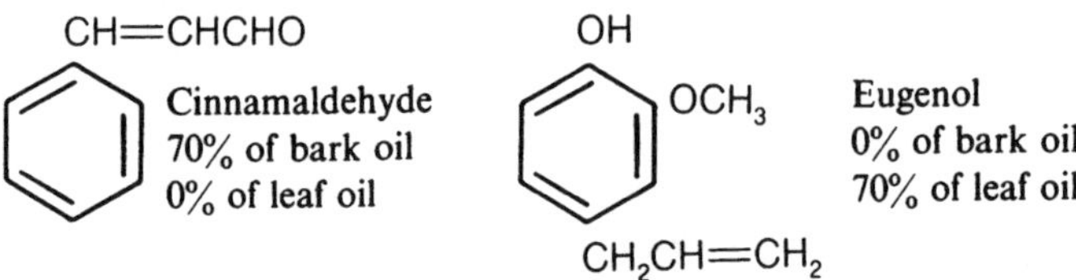

By careful raw material selection and careful low temperature processing, it is possible routinely to obtain bark oils with a cinnamaldehyde content in excess of 80%.

The bark oil from Sri Lanka (a major source) is frequently "standardized" to a minimum content of cinnamaldehyde, namely two grades at 60% and 30%. Cinnamon leaf oil is used for this purpose. This dilution serves to even out crop to crop and botanical variations. In any product application where the oil comes into contact with the skin, eugenol acts as a "quencher" for the irritation reaction that some people have toward cinnamaldehyde. Indeed I.F.R.A. (International Fragrance Research Association), in its published guidelines on perfumery, recommends the addition of eugenol, glycol, or phenyl ethyl alcohol at 1:1 ratio with cinnamon bark oil to desensitize the potential irritation.

It should be borne in mind that the leaf oil is one twentieth the price of the bark oil from cinnamon. Any prospective purchaser should be aware of this fact, making a selection of oil not only based on quality but value of natural cinnamaldehyde.

2.2.3 *Storage*

Plant materials, once prepared for storage and transportation by drying, curing, aging, etc., should be kept under appropriate conditions. For instance, sacks of peppercorns that are stored under damp conditions, without free circulation of air, develop a "musty" note that remains in the oil even after extraction. Some plant materials require several months of curing and maturation and are graded on the quality of this preparation (e.g., vanilla beans; details of their processing are discussed in Section 2.2.7).

2.2.4 *Yield*

Some plant materials, particularly seeds (e.g., coriander, celery, nutmeg), contain fixed oils (triglycerides of fatty acids) as well as essential oils. As extraction solvents have differing polarities and hence affinities for the components of plant materials other than essential oil, it is important to know the yield as well as the solvent. For example, some commercial CO_2 extracts of celery seed are alcohol soluble whereas others are not. Because alcohol solubility is often used as an analytical criterion for the assessment of essential oils, alcohol insolubility would classify such extracts as oleoresins (C.A.L., 1986; Moyler, *Developments,* 1989).

The yield can vary significantly with the solvent employed. For instance, the acetone extract of ginger root (*Zingiber officinale*) yields 7% whereas CO_2 yields only 3%. CO_2 extracts all of the essential oil in an undegraded form, together with the gingerols and shogaols responsible for the pungency (Moyler & Heath, 1988; Lawrence, 1985; McHale et al., 1989) because they have a molecular weight below 400, the cut off for liquid CO_2 extraction. Solutions of CO_2 extracted ginger are water soluble. Acetone, being more polar than CO_2, will extract resins that increase the weight of extract but not its organoleptic quality. However, for application to baked goods or cooked products, the extra resins of the solvent extract act as a fixative and help prevent other components from volatilizing during the manufacture of the food product, so are desirable (Heath, 1982). Solutions of acetone extracted ginger oleoresin are *not* water soluble.

The yield and composition of extracts (Moyler et al., 1992; Moyler, *Proceedings,* 1989; Calame & Steiner, 1987; Naik, Lentz, & Maheshwari, 1989; Naik, Maheshwari, & Gupta, 1989) vary with solvent polarity (Table 2–2).

2.2.5 *Degradation*

Some methods of preparing extracts cause degradation or changes of components. The time honored method of steam distillation causes many such degradations, such as formation of terpineol isomers, *p*-cymene from gamma-terpinene, loss of citral, loss of esters, etc. (Pickett et al., 1975). This is because the conditions of wet steam and elevated temperatures that are employed cause bond rearrangements. It should not be forgotten that even ethanol causes changes, especially if free acids are present in the plant material and esterification can take place (Liddle & de Smedt, 1981). Such changes are even considered desirable for some applications such as alcoholic beverages (Rose, 1977).

Formation of acetals, Schiff's bases, Strecker degradation compounds, Maillard reactions and other artifacts can also take place during processing, but not necessarily to the detriment of the final use of the extract in food. Many cooked food flavors are dependent on these types of reaction taking place during cooking. However, for true fresh natural tastes, processing without heating is more desirable.

Table 2–2 Comparison of extraction yields by % different techniques

Botanical	*Part*	*Source*	*Origin*	*Steam dist*	*L. CO_2*[†]	*S. CO_2*	*Solvent (specified)*
Almond	Seed	*Prunus amygdalus*	Italy	0.5	3.5	—	20 (expression)
Ambrette	Seed	*Hibiscus abelmoschus*	Africa	0.2–0.6	1.5	—	—
Angelica	Seed	*Angelica archangelica*	N. Europe	0.3–0.8	3	—	—
Aniseed	Seed	*Anisum pimpanellum*	N. Europe	2.1–2.8	4.3	7[a]	15 (ethanol)
Anise star	Seed	*Illicium verum*	China	8–9	10	—	28 (ethanol)
Arnica	Root	*Arnica montana*					
Basil	Leaf	*Ocimum basilicum*	N. Europe	0.3–0.8	—	1.3[b]	1.5 (ethanol)
Bay	Leaf	*Pimenta racemosa*	W. Indies	1–2	1.0	—	—
Buchu	Leaf	*Barosma betulina*	S. Africa	1–2.8	2.0	—	—
Calamus	Root	*Acorus calamus*	India				
Calendula	Flower	*Calendula officinalis*					
Caraway	Seed	*Carum carvi*	N. Europe	3–6	3.7	5[d]	20 (ethanol)
Cardamom	Seed	*Elletaria cardamomum*	Guatemala	4–6	4	5.8[b]	10 (ethanol)
Carob	Fruit	*Ceratonia siliqua*	Africa	<0.01	0.1	—	40 (water-ethanol)
Carrot	Root	*Daucas carota*	Russia	—	2.7	—	—
Carrot	Seed	*Daucas carota*	European	0.2–0.5	1.8	3.3	3.3 (ethanol)
Cassia	Bark	*Cinnamonum cassia*	China	0.2–0.4	0.6	—	5 (dichloromethane)
Cassis	Bud	*Ribes nigrum*	Europe	<0.01	0.7	—	3.7 (acetone)
Celeriac	Root	*Celeriac* sp.	Russia	—	7	—	—
Celery	Seed	*Apium graveolens*	India	2.5–3.0	3	—	13 (ethanol)
Chamomile	Flower	*Anthemis nobilis*	Russia	0.3–1.0	2.9	—	—
Chamomile	Flower	*Matricaria chamomilla*	Germany	0.3–1.0	0.5	1.4[d]	—
Chilli	Fruit	*Capsicum annum*	India	<0.1	4.9	—	10 (acetone) 30 (60% ethanol)
Cinnamon	Bark	*Cinnamonum zeylanicum*	Sri Lanka	0.5–0.8	—	1.4[a]	4 (dichloromethane)
Clove bud	Flower	*Syzygium aromaticum*	Madagascar	15–17	16	20[c]	20 (ethanol)

continues

Cocoa (defatted)	Seed	*Theobroma cacao*	Africa	<0.01	0.5	—	1	(ethanol)
Coffee	Seed	*Coffea arabica*	Arabia	<0.01	5	—	30	(ethanol-water)
Coriander	Seed	*Coriandrum sativum*	Romania	0.5–1.0	1.3	3[d]	20	(ethanol)
Cubeb	Fruit	*Piper cubeba*	E. Indies	10–16	13	—	12	(ethanol)
Cumin	Seed	*Cuminium cyminum*	India	2.3–3.6	4.5	—	12	(ethanol)
Cumin	Seed	*Cuminium cyminum*	Iran	2.3	—	2.6[f]	—	
Dill	Leaf	*Anethum graveolens*	S. Europe	0.3–1.5	—	—	—	
Dill	Seed	*Anethum graveolens*	Russia	2.3–3.5	3.6	—	—	
Eucalyptus	Leaf	*Eucalyptus globulus*	Russia	1–1.5	1.8	—	—	
Fennel	Seed	*Foeniculum dulce*	Europe	2.5–3.5	5.8	—	15	(ethanol)
Fenugreek	Seed	*Trigonella foenum graecum*	India	<0.01	2	—	8	(ethanol)
Galangal	Root	*Alpinia galanga*	India	0.5–1.0	1.4	—	—	
Galbanum	Resin	*Ferula galbaniflua*	Iran	12–22	—	20[e]	—	
Geranium	Leaf	*Pelaragonium odoratisimun*	Mediterranean	0.15–0.2	0.2	—	—	
Ginger	Root	*Zingiber officinalis*	Australia	1–2	2	—	5	(acetone)
Ginger	Root	*Zingiber officinalis*	China	1–2	2	—	5	(acetone)
Ginger	Root	*Zingiber officinalis*	West Indies	1.5–3.0	2.5	—	6	(acetone)
Ginger	Root	*Zingiber officinalis*	Jamaica	1.5–3.0	2.5	—	6.5	(acetone)
Ginger	Root	*Zingiber officinalis*	Nigeria	1.5–3.0	3	4.6[a]	7	(acetone)
Hazel	Nut	*Nuces. Coryli avellanae*	Europe	—	2	—	—	
Hop	Fruit	*Humulus lupulus*	England	0.3–0.5	12	16[d]	20	(ethanol)
Hypericum	Leaf	*Hypericum* sp.	Russia	—	2.2	—	—	
Hyssop	Leaf	*Hyssopus officinalis*	S. Europe	0.1–0.3	—	—	—	
Jasmine	Flower	*Jasminum grandiflorum*	India	1.2–1.5	1.4	—	3	(hexane)
Juniper	Berry	*Juniperus communis*	Yugoslavia	0.7–1.6	2.7	7.2[b]	—	
Labdanum	Resin	*Cistus ladaniferus*	Spain	1–2	—	—	—	
Laurel	Leaf	*Laurus nobilis*	Russia	0.6	2.5	—	—	
Lavandin	Leaf	*Lavandula grosso*	France	1–2	—	3.5[d]	—	

continues

Table 2–2 continued

Botanical	*Part*	*Source*	*Origin*	*Steam dist*	*L. CO_2*[†]	*S. CO_2*	*Solvent (specified)*
Lemon	Peel	*Citrus limonum*	Europe	0.5–1	0.8	0.9[a]	1 (expression)
Lemongrass	Leaf	*Cymbopogon citratus*	China	0.3–0.4	0.8	—	—
Lilac	Flower	*Syringa vulgaris*	Europe	not possible	—	0.0024[a]	—
Lily of valley	Flower	*Convallana majalis*	Europe	not possible	—	0.25[a]	—
Lovage	Root	*Levisticum officinale*	European	0.1–0.2	0.9	—	8 (ethanol)
Mace	Aril	*Myristica fragrans*	W. Indies	4–15	13	—	40 (expression)
Marjoram	Leaf	*Majorlana hortensis*	N. Europe	0.2–2.0	—	1.7[b]	—
Massoia	Bark	*Crytocaria massoia*					
Mushroom	Dried	*Boletus edulis*	Italy	—	—	2.5[e]	—
Myrrh	Gum	*Commiphora mol mol*	Somalia	4–6	—		
Nutmeg	Seed	*Myristica fragrans*	W. Indies	7–16	13	—	45 (expression)
Olibanum	Gum	*Boswellia* sp.	Somalia	4–6	—		
Oregano	Leaf	*Origanum vulgare*	European	3–4	—	5[b]	—
Orris	Root	*Iris pallida*	Italy	0.2–0.3	0.7	—	—
Paprika	Fruit	*Capsicum annum*	Spain				
Parsley	Seed	*Petroselinum crispum*	India	2.0–3.5	3.6	9.8[c]	20 (ethanol)
Peach	Leaf	*Prunus persica*					
Peanut	Legume	*Arachis* sp.	U.S.A.				20 (expression)
Pepper	Fruit	*Piper nigrum*	India	1.0–2.6	6.7	—	18 (acetone)
Pepper	Fruit	*Piper nigrum*	Malaysia	1.0–2.0	3.5	10	10 (acetone)
Peppermint	Leaf	*Mentha piperita*	U.S.A.				
Peru balsam	Resin	*Myroxylon pereirae*	S. America	3.3–4.5	4.5	5.3[e]	50 (ethanol)
Pimento	Berry	*Pimenta officinalis*	Jamaica	3.3–4.5	4.5	5.3[b]	6 (ethanol)
Poppy	Seed	*Papaver somniferum*	India	—	2.5	—	50 (ethanol)
Rose hip	Fruit	*Rosa* sp.					
Rose pepper	Fruit	*Piper rosa*					

continues

Rosemary	Leaf	*Rosemarinus officinalis*	Europe	0.5–1.5	1.9	7.5[b]	5	(methanol)
Sage	Leaf	*Salvia officinalis*	Europe	0.5–1.1	—	4.3[b]	8	(ethanol)
Sandalwood E.I.	Wood	*Santalum album*	India	3–6	4.8	—	—	
Sandalwood	Wood	*Santalum spicatum*	Australia	2–4	—	—	—	
Savory	Leaf	*Satureja hortensis*	Europe	1–2	0.8	—	—	
Saw palmetto		*Serenoa repens*						
Sea buckthorn		*Hippophae rhamoides*						
Spearmint	Leaf	*Mentha spicata*	U.S.A.					
Tea	Leaf	*Thea sinensis*	Africa	<0.01	0.2	—	35	(ethanol/water)
Thyme	Leaf	*Thymus vulgaris*	Europe	1–2	2.1	—	10	(ethanol)
Tobacco	Leaf	*Nicotinia* sp.	U.S.A.					
Turmeric	Root	*Curcuma longa*	India	5–6	3.4	—	—	
Vanilla	Fruit	*Vanilla fragrans*	Madagascar	<0.01	4.5	—	25–45	(ethanol/water)
Vetiver	Root	*Vetiveria zizanoides*	W. Indies	0.5–1	1.0 (1.0)	—	—	
Wormwood	Root	*Artemesia absinthum*	Europe	0.5–1	1.0	—	—	
Yarrow	Leaves	*Achillea millefolium*	Russia	—	1.1	—	—	

[†]CO_2 yields are own results except: [a]supercritical CO_2 reported yield (Moyler, *Proceedings*, 1989); [b]supercritical CO_2 + entrainer yield (Calame and Steiner, 1987); [c]subcritical CO_2 (Naik, Lentz, et al., 1989; Naik, Maheshwari, et al., 1989); [d]D. Gerard, supercritical total; [e]D. Gerard, supercritical fractionated. [f]Goodarznia et al., 1998.

2.2.6 *Preparation of Plant Material*

Prior to extraction, many preparation techniques are used to optimize the flavor of plant material.

2.2.6.1 *Drying*

This removes the bulk of water, enabling a larger batch of material to be processed. Drying also minimizes mold spoilage, e.g., fresh culinary green ginger rhizomes are considerably more susceptible to molds than dried root.

2.2.6.2 *Grinding*

This reduces particle size to increase surface area and allows better penetration of solvent. Care should be exercised to ensure that finely divided material is not left unprotected so that evaporation and oxidation take place. Reclaimed gaseous carbon dioxide is most suitable for protection or "blanketing" of ground plant materials.

2.2.6.3 *Chopping*

Some plants are better chopped than ground before extraction. This is the case when tannins or other astringent components are present that are not required in an extract intended for use in, for example, dairy products.

2.2.6.4 *Roasting*

Examples of pre-roasting before flavor extraction are coffee and kola nuts. It should be noted that decaffeinated coffee is roasted after the removal of caffeine from the green beans with supercritical CO_2. This ensures that the full roast flavor is present in the coffee even though the caffeine has been removed.

2.2.6.5 *Fermentation*

Often fermentation of plant material takes place naturally during drying, e.g., grass to hay, plums to prunes, etc. and is an essential process for some foods to exist at all, e.g., black tea from green tea leaves.

2.2.6.6 *Curing*

Forms include salting of fish (Scandinavian gravlax salmon), hot and cold smoking of meats, fish, cheese and preservation of meats. Some curing is only a controlled fermentation, e.g., vanilla beans.

2.2.7 *Vanilla Bean Curing (Classical Method)* (Gebruder Wollenhaupt, *Real Vanilla*)

After the harvest the vanilla beans have no aroma. This develops only during the course of the curing, which should begin immediately upon picking in order to prevent the beans from bursting (splitting). The preparation consists of three processing steps.

2.2.7.1 *Scalding*

Batches of 25–30 kg beans are set into plaited wire baskets and dipped into hot water at 60° C for 2 min. This not only interrupts the maturing process in the cellular tissue but also distributes the sap and enzymes, which are important for the formation of aroma.

2.2.7.2 *Fermentation process*

The drained, still hot beans are laid in cases lined with wool blankets, in which they perspire for 24 h and lose part of their excess moisture. With the fermentation process, the forming of aroma begins and the beans now turn a chocolate brown color.

2.2.7.3 *Drying and conservation*

On the following day, the beans are dried on trays in the sun from 10 am until 3 pm. Then the trays are stacked in storerooms until the next morning in order to maintain a consistent temperature for the efficient fermentation by enzymes for as long as possible. This procedure is repeated, depending on the location and weather, for up to 8 days. Then the beans are dried on racks in a bright and airy storeroom for up to 3 months, before they are packed in wooden trunks to season and develop their aroma for another 2–3 months. The main task of the preparer is the constant checking of the beans to determine the right time for these operations and to notice possible mold formation at a very early stage, removing any moldy beans quickly.

As the classical curing method is strongly dependent on the weather conditions in Madagascar, an industrial method has been developed which is occasionally used. Figure 2–1 describes a building for the preparation of 180 tons of green vanilla, equivalent to 40 tons of prepared vanilla, for use during the 2-month harvesting season.

(a) Arrival of green vanilla.
(b) Storage of green vanilla. The two pre-storage rooms each take 3 tons of beans piled 1 m high for a period of 48 h (to avoid uncontrolled fermentation).
(c) Sorting. With the help of a conveyer belt (5 m long), ten workers sort the beans into the following grades: 1, burst (split beans); 2, beans shorter than 12 cm; 3, unripened beans; 4, beans of 2nd choice; 5, beans of 1st choice.
(d) Basketing. After sorting, the beans are put into cylindrical wire baskets each holding up to 50 kg.
(e) Scalding. The wire baskets are dipped into 63–65° C hot water, heated by a wood fire.

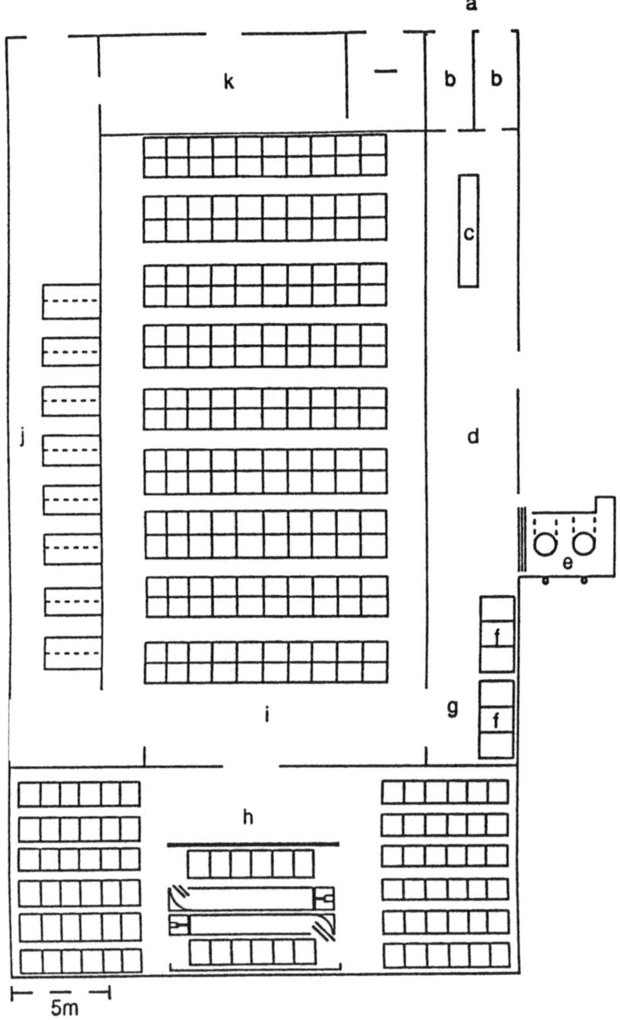

Figure 2–1 Plan of vanilla preparation building (Gebruder Wollenhaupt, *Real Vanilla*).

(f) Perspiring cases. After draining, the baskets are emptied into cases lined with wool blankets. The beans must be carefully and evenly piled in order to maintain a uniform temperature of about 50° C within the hermetically closed cases. This procedure kills the growth substances in the bean.

(g) Evaporating. After 48 h the beans are spread out on plaited shelves that are spaced vertically at 6 cm distances in a mobile rack so that air may circulate between the shelves. Every rack contains 22 shelves each of 11–12 kg vanilla beans.

(h) Drying tunnels (electrically heated to 65° C). Both drying tunnels take six mobile racks. Every 30 min a new rack is driven into the tunnel, so that the passage of each rack takes 3 h. For the period of 8 days every rack is passed each day through the drying tunnel.
(i) Drying room. The beans are now oily and ductile and are further dried depending on their moisture content. They are stored for 2 months on shelves of mobile racks, in a shady and well ventilated drying room.
(j) Storage room.
(k) Dispatch area.

2.3 SOLVENTS

The choice of solvents available for food use is becoming more and more restricted. Recent recommendations of the EEC Solvents Directive have suggested that only the following solvents be used and they give advice on levels of residual solvents which are "tolerable" in finished foods (B.E.M.A., *Guidelines)* (Tables 2–3 and 2–4).

Table 2–3 Solvents and allowable residues in foodstuffs

Solvent	*dated 1/97 (ppm)*	*Residues (ppm)*
Ethanol		Technically unavoidable quantities
Butane, propane		Technically unavoidable quantities
Acetone		Technically unavoidable quantities
Butyl and ethyl acetates		Technically unavoidable quantities
Nitrous oxide		Technically unavoidable quantities
Carbon dioxide		Technically unavoidable quantities
Dichloromethane	0.02	0.1 ppm (1 ppm confectionery/pastry, 5 ppm decaffeinated tea, 10 ppm decaffeinated coffee)
Hexane	2	1 ppm (5 ppm oil/cocoa, 10 ppm defat flour, 30 ppm defat soya products)
Methanol, propan-l-ol, propan-2-ol	1	
Methyl acetate		1 ppm (20 ppm in decaffeinated tea and coffee)
Ethyl methyl ketone	1	1 ppm (5 ppm fat or oil, 20 ppm decaffeinated tea and coffee)
Cyclohexane, butanols	1	1 ppm
Diethyl ether	2	2 ppm
Phytone (arcton 134)*		

*Any trace residues if combusted in products like cigarettes would form HF which is corrosive. The makers state that these levels are well below current safety standards.

Table 2–4 Solvent comparison (Moyler, *Chem. Ind.*, 1988)

	Viscosity (cP)				
	0° C	*20° C*	*Latent heat evap (cal/g)*	*Boiling point (°C)*	*Polarity (ϵ)*
Carbon dioxide	0.10	0.07	42.4	−56.6	0
Acetone	0.40	0.33	125.3	56.2	0.47
Benzene[a]	0.91	0.65	94.3	80.1	0.32
Ethanol	1.77	1.20	204.3	78.3	0.68
Ethyl acetate	0.58	0.46	94	77.1	0.38
Hexane	0.40	0.33	82	68.7	0
Methanol	0.82	0.60	262.8	64.8	0.73
Dichloromethane	—	0.43	78.7	40.8	0.32
Pentane	0.29	0.24	84	36.2	0
Propan-2-ol (IPA)	—	2.43	167	82.3	0.63
Toluene[a]	0.77	0.59	86	110.6	0.29
Water	1.80	1.00	540	100.0	>0.73
Propylene glycol	—	56.00	170	187.4	>0.73
Glycerol	12,110.0	1490.0	239	290.0	>0.73

[a]Not food grade solvents.

2.3.1 *Polarity*

A useful guide to the profile of components extractable by a solvent is its polarity. Armed with the knowledge of the constituents of a plant, it is possible to predict which components will be extracted under a given set of extraction conditions, e.g., if leaf waxes would or would not be extracted (yes with non-polar solvents, no with polar solvents).

2.3.2 *Boiling Point*

The temperature at which a solvent will boil under a given pressure or vacuum system will influence the profile of an extract. For instance, the change from methanol to ethanol to isopropanol to butanol at atmospheric pressure will alter the profile of molecular weight extracted. The higher the boiling point of reflux conditions, the higher the molecular weight profile that will be extracted.

2.3.3 *Viscosity*

The ability that a solvent has to penetrate plant material will depend on its mobility, hence viscosity. This property enables efficient extraction because of lower residence time of contact between solvent and plant material. The lower the viscosity of the solvent, the more efficient the commercial extraction.

2.3.4 *Latent Heat of Evaporation*

Costs of extraction are often limited to energy costs, and solvents of low latent heat of evaporation have the advantage of being removed using less energy than solvents of higher latent heats.

2.3.5 *Temperature/Pressure*

Besides being linked to boiling points, the temperature of reflux of a solvent in a given system is also dependent on pressure. As a guide, the temperature of a distillation will drop by 15° C every time the pressure is halved, so it is possible to accomplish some distillations at room temperature if the vacuum is high enough. This property is utilized for removal of residual solvent in the "falling film evaporator." This technique can be maximized as it is in "molecular distillation" or short path distillation (Moyler, *Molecular Distilled Oils,* 1988).

2.4 TINCTURES

2.4.1 *Water Infusions*

The use of water to extract or "infuse" plant material such as roasted coffee and fermented tea leaves has been known since antiquity. Infusions of cocoa beans and vanilla beans were treasured by the Aztec Indians of Mexico and indeed were served to the conquistadors of Spain as a mark of respect. They introduced the beverage to Europe and the rest of the world.

The technique for the preparation of infusions is simple, using either hot or cold water that remains in the finished extract. The plant material is ground to a broad spectrum particle size to facilitate the flow of water through the bed of material. This is important because if the particles are all too fine, the bed will compact and not allow the passage of solvent. In the specific case of infusions, there is often some swelling of the material when it comes into contact with water which slows the flow of solvent. This softening is known as "maceration." The material is allowed to stand in a "percolator" (Figure 2–2) for approximately 1 week. The liquid "menstruum" is drawn off from a tap which is located at a position low on the side of the percolator. Fresh solvent is added to the percolator to give successive "coverings." All the menstruums of varying strengths are then combined and standardized for flavor and non-volatile residue. This is the completed infusion. Some examples are listed in Table 2–5.

Figure 2–2 Solvent extraction plant.

2.4.2 *Alcoholic Tinctures*

By a process analogous to infusions but using ethanol, propan-2-ol, propylene glycol or glycerol as a solvent, a tincture is prepared. Unlike with an oleoresin, the solvent is left as part of the product which is then incorporated directly into a food, beverage or pharmaceutical product.

An adaptation of the "maceration" technique is sometimes used where the chopped plant material is placed in a gauze bag and the solvent is pumped through it. This "circulatory maceration" is usually more rapid than static percolation and offers process advantages with some materials, e.g., vanilla beans and other expensive raw materials.

Many such tinctures of herbs and spices are used in the soft drinks and alcoholic beverage industries (Moyler, *Distilled Beverage Flavour,* 1988) to impart mouthfeel and taste, as well as the flavor that could have been achieved with a

Table 2–5 Water infusions

Examples	*Source*	*Applications*
Cola nuts	*Cola acuminata*	Soft drinks, cola drinks
Chrysanthemum flowers	*Chrysanthemum* sp.	Chrysanthemum tea drink
Coffee	*Coffea robusta, Arabica*	Soft drinks, candy, liqueurs
Tea	*Thea sinensis*	Soft drinks
Cocoa	*Theobroma cacao*	Soft drinks, liqueurs
Licorice	*Glycyrrhiza glabra*	Candy, pharmaceutical products
Carob bean	*Ceratonia siliqua*	Chocolate flavorings, baked goods, tobacco flavorings, nut and rum flavorings
Rhubarb	*Rheum officinale*	Soft drinks, pharmaceutical products
Elderberries	*Sambucus nigra*	Red fruit flavorings

solution of the essential oil. This list is extensive, from the exotic southernwood, dittany of Crete and chrysanthemum flowers to the more mundane ginger, kola and quassia bark products (Table 2–6).

2.5 OLEORESINS

Some oleoresins occur naturally as gums or exudates directly from trees and plants. Examples are benzoin (*Styrax benzoin*), myrrh (*Commiphora molmol*), copaiba balsam (*Copaefera* sp.), olibanum (*Boswellia frereana*) and other balsams. They are usually only partially solvent soluble and are usually extracted before use, when they are known as "prepared oleoresins."

Table 2–6 Alcoholic tinctures

Examples	*Source*	*Applications*
Benzoin resin	*Styrax benzoin*	Tobacco flavors, chocolate and vanilla flavors
Quassia bark	*Picrasma excelsa*	Bittering agent for soft drinks and alcoholic beverages
Buchu leaves	*Barosma betulina*	Blackcurrant flavors
Strawberry leaves	*Fragaria vesca*	"Green" strawberry flavors
Blackcurrant leaves	*Ribes nigrum*	Blackcurrant flavors
Elderflowers	*Sambucus nigra*	Beverages
Banana fruit	*Musa* sp.	Tropical fruit flavors
Ginger root	*Zingiber officinale*	Beverages
Orris root	*Iris pallida*	Tobacco flavors, soft drinks

2.5.1 *Solvents*

Those commonly employed have already been listed, i.e., ethanol, water (and mixtures), isopropanol, methanol, ethyl acetate, acetone. The preparation of oleoresins closely follows the methodology for tinctures and infusions.

Oleoresins, resins and concretes have their extraction solvent removed with only minimal residues remaining. Attempts to remove these last traces of solvents usually result in the loss of some of the "top-notes" from the oleoresin. This is not important if the primary use of the oleoresin is for pungency, e.g., capsicum (*Capsicum annuum*). For most other applications, such as pepper and ginger oleoresins, it is preferable to tolerate solvent residues rather than lose top-notes from the oil fraction that make a valuable contribution to flavor in a food application. Concretes differ from oleoresins only in that they are more crystalline in physical form and are made by the same technique. An exception is orris concrete (*Iris pallida* and *Iris germanica*) which is steam distilled but contains myristic acid (tetradecanoic acid) and sets to a soft crystalline mass at ambient temperature, but is technically an essential oil.

2.5.2 *Solubility*

Depending on the boiling point and polarity of the solvent, a balance between the molecular weight and polarity of the component profile of the plant material will be extracted with time. For example, with a hydrocarbon solvent, the low molecular weight (MW), low polarity components extract first, and presuming that sufficient solvent is available so that saturation does not occur, medium MW, low polarity components and low MW, polar components then extract. The successively high MW, low polarity and medium MW, polar components and so forth are extracted until saturation is reached. In practice, sufficient solvent is circulated from the start of the extraction to remove all desirable components and this fractionation effect is not normally observed.

2.5.3 *Commercial Solvent Extraction System*

A commercial solvent extraction system is shown in Figure 2–2. The oleoresins listed in Table 2–7 are typical examples of the many produced commercially, but there is considerable overlap between the drug, pharmaceutical, food and beverage industries.

2.5.3.1 *Standardized oleoresins* (Figure 2–3)

Some commercial oleoresins are not what they seem. The high level of lipids in some plants means that a total extract would be very weak in flavor. Such extracts are standardized by addition of steam distilled oil to reinforce the "flavor" as well as providing the "taste" of the spice. Examples of these "standardized oleo-

Table 2–7 Commercially produced oleoresins

Examples	*Source*	*Applications*
Pepper oleoresin	*Piper nigrum*	Snacks, curries, sauces, pickles
Pimento oleoresin	*Pimenta officinalis*	Pickles, chutneys, sauces, spice blends
Tonka concrete	*Dipteryx odorata*	Tobacco flavors (restricted)
Vanilla oleoresin	*Vanilla fragrans*	Liqueurs, natural baked goods, sweet sauces
Ginger oleoresin	*Zingiber officinale*	Baked goods
Fenugreek oleoresin	*Trigenella* f.g.	Spice blends, curries
Lovage oleoresin	*Levisticum officinale*	Soups, savory sauces
Nutmeg oleoresin	*Myristica fragrans*	Soups, savory and sweet sauces
Turmeric oleoresin	*Curcuma longa*	Spice blends, natural color
Oregano oleoresin	*Origanum vulgare*	Pizza toppings, herb sauces
Rosemary oleoresin	*Rosemarinus officinalis*	Natural antioxidant, meat sauces

resins" are coriander, bay, nutmeg, mace, pepper. They are also available in the form of emulsions, spray dried encapsulated or simply "spread" or "plated" onto salt, rusk or maltodextrin. It should be remembered that because of the blending with steam distilled oils, these "standardized spices," although functional, are not a true representation of the plant material. Several manufacturers offer ranges of such standardized spices: a compilation is given in Table 2–8.

2.6 ABSOLUTES

These are alcohol soluble extracts of plant materials which represent the "heart" of the flavoring. Some examples are listed in Table 2–9.

2.6.1 *Solvents*

Absolutes can be prepared by one of the following:

(a) Extraction of the plant material with ethanol and evaporating the solvent.
(b) Extracting a crude whole oleoresin with a suitable solvent, evaporating the solvent and re-extracting the absolute from the oleoresin with alcohol. This has the process advantage of utilizing readily available resins of the base and so minimizing the use of alcohol.
(c) Some naturals can be extracted with CO_2 to give extracts with all of the attributes of an absolute but processed without organic solvents (see Section 2.7).

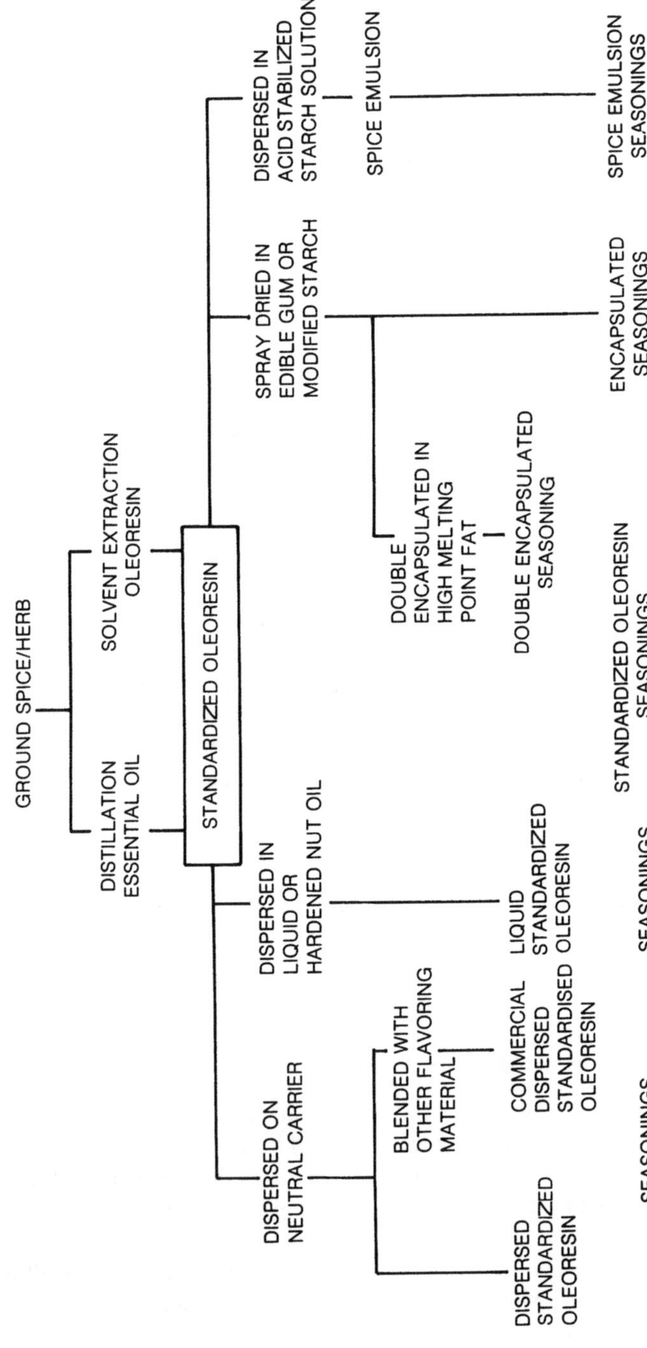

Figure 2–3 Flow chart of a typical standardized oleoresin range of spices.

Table 2–8 Commercial standardized oleoresins

Botanical	*Oleoresin*	*Emulsion*	*on Salt*	*on Dextrose*	*on Rusk*	*Encapsulated*
Aloes	H					
Angelica	H					
Aniseed	H	U				B
Basils	BHK	BUK	BH	H	H	B
Bay	BHEK	BUK	BH	BH	BH	B
Borage	H					
Bouquet garni			BH	H	H	
Capsicum	BEK	K	H	H	H	
Caraway	BH	BU	BH	H	H	B
Cardamom	BHEK	BUK	BH	H	H	B
Cassia	BH	B	BH	H	H	
Cayenne	BH	BU	B	B	B	B
Celerys	BHEK	BUK	BH	BH	BH	B
Chamomile	H					
Chervil	BH	U	B			
Chinese five spice	H					
Cinnamon	BHK	BUK	BH	BH	H	B
Clove	BHEK	BUK	BH	BH	BH	B
Coffee	H					
Corianders	BHEK	BUK	BH	BH	BH	B
Cumin	BHEK	BUK	BH	BH	H	B
Curry	HE	U	H	H	H	
Dills	BH	B	BH	H	H	B
Fennel	BHK	BUK	B			B
Fenugreek	BHE	B	BH	H	H	B
Galangal	H					
Garlics	BH	BUK	B	B		B
Gentian	H					
Gingers	BHEK	BUK	BH	BH	BH	B
Herb blends	H					
Juniper berry		U	B			
Laurel	H					
Leek			B			B
Lovage	BH`		B			
Mace	BHEK	BUK	BH	BH	BH	B
Malt	H					
Marjoram	BHEK	BUK	BH	BH	BH	B
Massoia	H					
Mint		U	BH	H	H	
Mixed spice	B			B		
Mushrooms	B	U	B			B
Nutmeg	BHEK	BUK	BH	BH	BH	B
Onion	BHE	BUK	B			B
Orange	H					
Oregano	BHEK	BUK	BH	H	H	B
Origanum	B					

continues

Table 2–8 continued

Botanical	*Oleoresin*	*Emulsion*	*on Salt*	*on Dextrose*	*on Rusk*	*Encapsulated*
Paprika	BHEK	BK	BH	H	H	B
Parsleys	BHE	BU	BH	H	H	B
Peppers	BHEK	BUK	BH	BH	BH	B
Pimento	BHEK	BUK	BH	H	BH	B
Rosemary	BHEK	BUK	BH	H	H	B
Sages	BHEK	BUK	BH	H	H	B
Savory	BH	B	B			B
Shallot		U				
Smoke		U				
Tarragon	BHK	BUK	BH	H	H	B
Tea	H					
Thyme	BHEK	BUK	BH	BH	BH	B
Tolu	H					
Turmeric	BHK	BK	BH	H	H	B
Vanilla	H					
Valerian	H					

Legend: B = B.B.A;. H = L. Hitchin; U = Universal; E = East Anglian; K = Kalsec.

Table 2–9 Alcohol soluble extracts

Examples	*Source*	*Applications*
Rose absolute	*Rosa centifolia* (de Mai) *Rosa damascena*	Confectionery, fantasy, soft drinks
Jasmine absolute	*Jasminum grandifolia*	Top-note fruit flavors
Tobacco absolute	*Nicotiana* sp.	Tobacco flavors
Ginger absolute	*Zingiber officinalis*	Beverages, liqueurs
Spearmint absolute	*Mentha spicata*	Oral hygiene
Boronia absolute	*Boronia megastigma*	Fruit flavors
Deer tongue absolute	*Trilisa odoratissima*	Tobacco flavors
Vanilla absolute	*Vanilla fragrans*	Dairy, liqueurs
Orange blossom, absolute from waters	*Citrus aurantium*	Orange flavors modifier
Labdanum absolute	*Cistus ladaniferus*	Soft fruit flavors
Cassis absolute	*Ribes nigrum*	Natural blackcurrant flavors
Mimosa absolute	*Acacia decurrens*	Fruit flavors
Genet absolute	*Spartium junceum*	Fruit flavors

2.7 EXTRACTION WITH CARBON DIOXIDE AS A SOLVENT

2.7.1 *Introduction*

Fundamentally all dry botanicals with an oil or resin content can be extracted with carbon dioxide. This pressured solvent behaves during extraction in a similar way to any of the other solvents previously discussed. As a solvent it has some significant advantages compared to alternatives:

It is odorless and tasteless.
It is food safe and non-combustible.
It easily penetrates (because of its low viscosity).
It is easily removed with no solvent residue (because of its low latent heat of evaporation).
It can be used selectively (by varying the temperature and pressure of extraction).
It is inexpensive and readily available.

A comparison with traditional forms of extraction shows that carbon dioxide is versatile.

Traditional extraction

CO_2 extraction

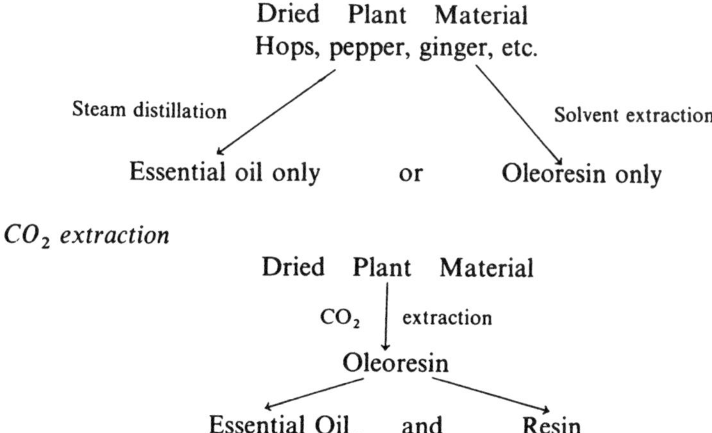

Commercially CO_2 can be used in two distinct modes of extraction which are dependent on its operation above or below the critical point in the phase diagram for CO_2 (Figure 2–4).

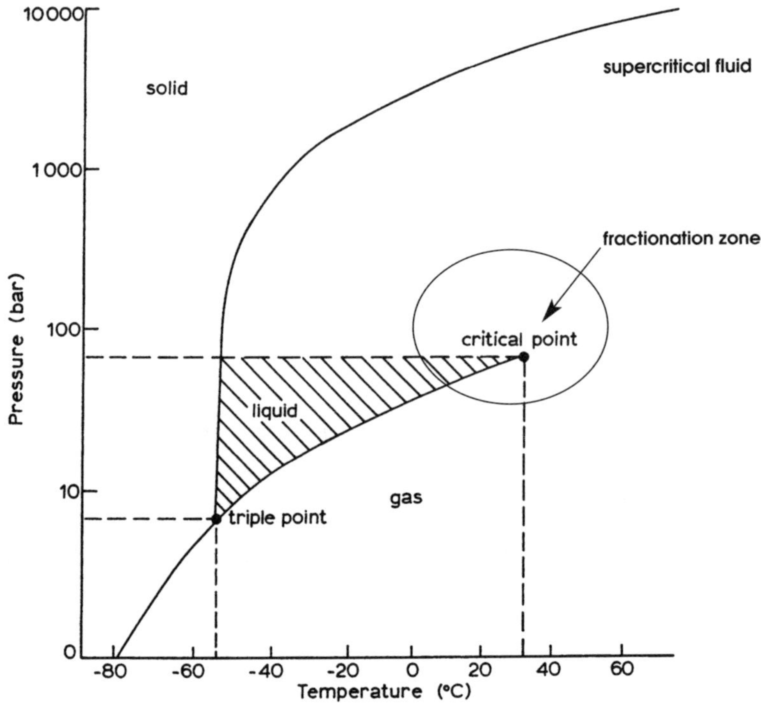

Figure 2–4 Phase diagram for CO_2 (Moyler, 1987).

2.7.2 *Subcritical CO_2*

Using extraction conditions of 50–80 bar pressure and 0 to +10° C, it is commercially viable to extract essential oils as an alternative to steam distillation. The energy savings of CO_2 explained in the section on solvents offset some of the capital expenditure of the extraction equipment.

2.7.2.1 *Solubility of components*

See Table 2–10.

2.7.2.2 *Extraction circuit (Gardner, 1982)*

See Figures 2–5 and 2–6.

2.7.2.3 *Entraining solvents*

"Entrainers" can be injected into the flow CO_2 to increase its polarity for certain extractions, e.g., juniper berry oil (Moyler, 1984). Ethanol is the only suitable entrainer for food grade products, although others have been reported in the

Table 2–10 Solubility of botanical components in liquid CO_2. Liquid CO_2 extracts all of the useful aromatic components from botanical materials (Moyler, 1987)

Very soluble	*Sparingly soluble*	*Almost insoluble*
Low MW aliphatic hydrocarbons, carbonyls, esters, ethers (e.g., cineole), alcohols, monoterpenes, sesquiterpenes	Higher MW aliphatic hydrocarbons, esters, etc.; substituted terpenes and sesquiterpenes; carboxylic acids and polar N and SH compounds; saturated lipids up to C12	Sugars, protein, polyphenols, waxes; inorganic salts; high MW compounds, e.g., chlorophyll, carotenoids, unsaturated and higher than C12 lipids
MW up to 250	MW up to 400	MW above 400

technical literature. The use of hexane and similar organic solvents obviates a major advantage that CO_2 offers for foods, that of no solvent residues.

2.7.2.4 *Properties of extracts*

These include: no solvent residues; no off notes; more top-notes; more back notes; better solubility; concentration of aromatics.

2.7.2.5 *Applications*

See Table 2–11.

2.7.3 *Supercritical CO_2*

Conditions are 200–300 bar pressure and +50 to 80° C for extraction. *In situ* fractionation is possible at 80–100 bar and at +10 to 50° C.

2.7.3.1 *Solubility of components*

As a generalization, supercritical CO_2 can extract all the soluble components from plant material in a similar way to organic solvents, giving oleoresins. These oleoresins are free of organic solvent residues and can be further fractionated (see Section 2.7.3.3) to give oils. The profile of any "supercritical CO_2 extract" should be studied carefully to ensure the desired characteristics are present and have not been fractionated out.

2.7.3.2 *Costs*

These extracts, although desirably giving an oleoresin free of solvent residue, are at a considerable cost disadvantage to conventional solvent extracts. Daily processing costs are similar but the capital cost of a commercial scale, high pressure, supercritical plant is millions of dollars, whereas solvent extraction

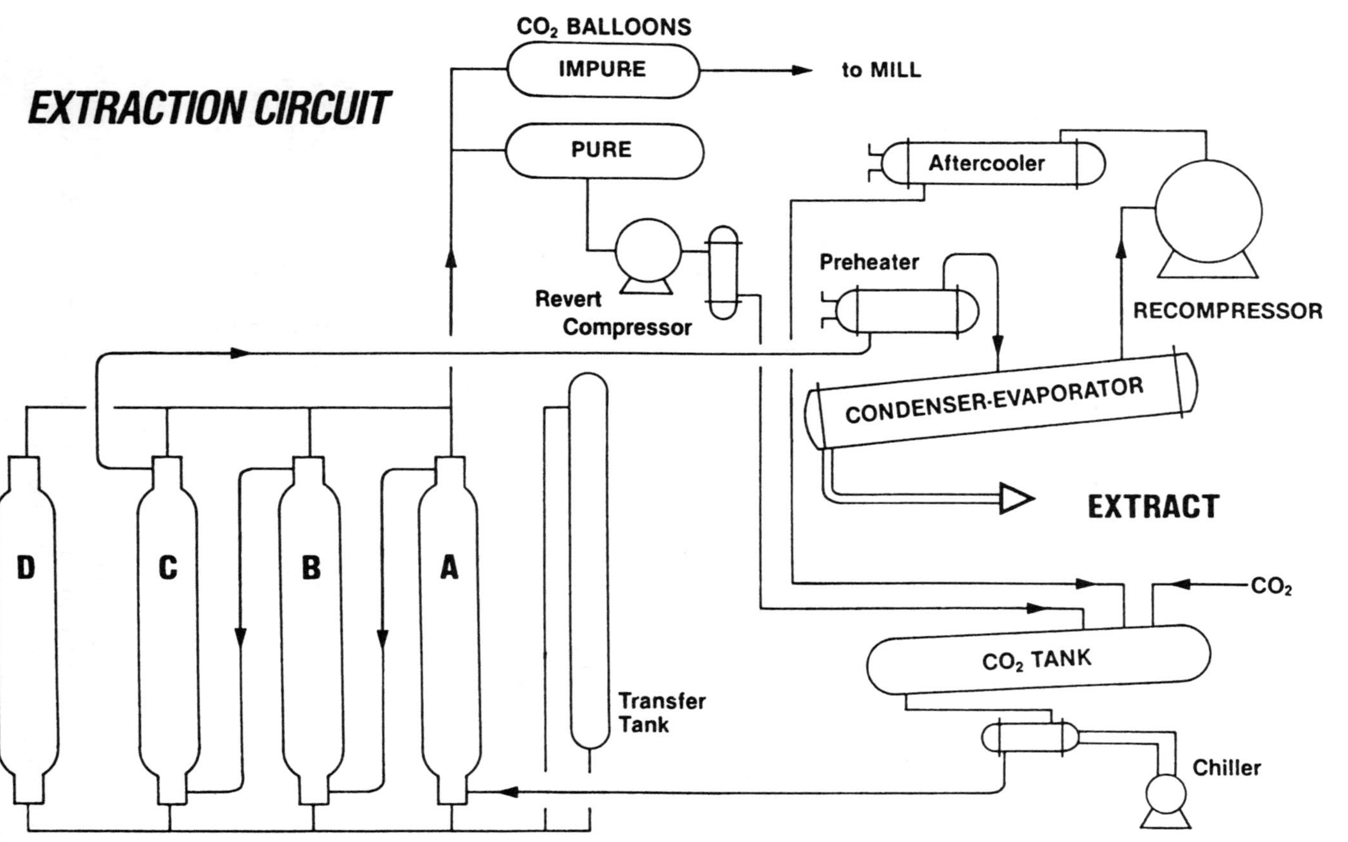

Figure 2–5 Extraction circuit CO_2.

Figure 2–6 CO_2 extraction plant.

equipment costs just thousands of dollars. Solvent oleoresins are more cost effective than supercritical extracts.

2.7.3.3 *Fractionations*

Supercritical CO_2 extraction does have one significant advantage over subcritical CO_2 or solvent extraction. This is the ability to fractionate the oleoresin by altering the pressure of the system and "tapping" out the desired fraction. This concept was reported by Brogle (1982) and applied in several experimental publications including a paper by Sankar (1989) on pepper oil from supercritical CO_2 oleoresin.

Table 2–11 Applications of Co_2 extracts

Examples	Source	Applications
Ginger oil	*Zingiber officinalis*	Oral hygiene, beverages, sauces
Pimento berry oil	*Pimenta officinalis*	Savory sauces, oral hygiene
Clove bud oil	*Syzygium aromaticum*	Meats, pickles, oral hygiene
Nutmeg oil	*Myristica fragrans*	Soups, sauces, vegetable juices
Juniper berry oil	*Juniperus officinalis*	Alcoholic beverages, gin
Celery seed oil	*Apium graveolens*	Soups, vegetable juice (tomato)
Vanilla absolute	*Vanilla fragrans*	Liqueurs, pure dairy, tobacco
Cardamom oil	*Elletaria cardamomum*	Meats, pickles, spice blends
Aniseed oil	*Illicium Verum*	Liqueurs, oral hygiene
Coriander oil	*Coriandrum sativum*	Curry, chocolate, fruit flavors

2.7.3.4 *Applications*

Some applications and examples of supercritical CO_2 extracts are listed in Table 2–12.

Table 2–12 Supercritical CO_2 extracts

Examples	Source	Applications
Pepper oleoresin and oil (Sankar, 1989)	*Piper nigrum*	Spices, meat, salad dressing
Clove bud oleoresin (Gopalakrishnan et al., 1988)	*Syzygium aromaticum*	Oral hygiene, meats
Ginger (Chen et al., 1986; Chen & Ho, 1988)	*Zingiber officinale*	Spices, sweet products
Cinnamon bark (Calame & Steiner, 1987)	*Cinnamomum zeylanicum*	Baked goods, sweet products
Lilac flower (Calame & Steiner, 1987)	*Syringa vulgaris*	
Cumin (Goodarznia et al., 1998)	*Cuminium cyminum*	Mexican and Indian cuisines
Marjoram	*Majorlana hortensis*	Soups, savory sauces
Savory	*Satureja hortensis*	Soups, savory sauces
Rosemary	*Rosemary officinalis*	Anti-oxidant, soups
Sage	*Salvia officinalis*	Meat, sauces, soups
Thyme	*Thymus vulgaris*	Meat, pharmaceutical products

2.7.3.5 *Commercial CO_2 extracts*

Commercial CO_2 extracts (liquid, supercritical and fractionated supercritical) are listed in Table 2–13, which shows how the range available has expanded since

Table 2–13 Commercial CO_2 extracts suppliers

		Universal[a]	*CAL*[b]	*Flavex*[c]	*SKW*[d]	*Daniel*[e]
Ambrette seed	*Hibiscus abelmoschus*	✓		✓		
Angelica seed	*Angelica archangelica*	✓				
Anise seed	*Anisum pimpanellum*					
Anise star	*Illicium verum*	✓	✓	✓		✓
Apple	*Malus pumila*				✓	
Arnica root	*Arnica montana*			✓		
Basil	*Ocimum basilicum*			✓		
Bergamot	*Citrus bergamia*			✓		
Black currant buds	*Ribes nigrum*		✓			
Butter	*Lactis diacetylactis*			✓		
Calamus root	*Acorus calamus*			✓		
Calendula flower	*Calendula off.*			✓		
Caraway	*Carum carvi*	✓		✓		✓
Cardamom	*Elletaria cardamomum*	✓		✓		✓
Carrot root	*Radix claucus carota*			✓		✓
Carrot seed	*Radix claucus carota*		✓			
Celery seed	*Apium graveolens*	✓	✓			
Chamomile	*Matricaria chamomilla*			✓		
Cinnamon bark	*Cinnamomum zeylanicum*	✓	✓	✓		✓
Clove bud	*Syzygium aromaticum*	✓	✓	✓		✓
Cocoa	*Theobroma cacao*	✓			✓	
Coffees	*Coffea arabica*	✓	✓		✓	
Coriander seed	*Coriandrum sativum*	✓		✓		
Cubeb	*Piper cubeba*	✓				
Elemi	*Canarium* sp.			✓		
Fennel	*Foeniculum vulgare*			✓		
Fenugreek	*Trigonella f.g.*	✓				
Galbanum	*Ferula galbaniflua*		✓	✓		
Gingers	*Zingiber off.*	✓	✓	✓	✓	✓
Hazelnut	*Nuces coryli avellanae*			✓	✓	✓
Hop cones	*Humulus lupulus*	✓	✓	✓	✓	✓
Hypericum flower	*Hypericum perforatum*			✓		
Juniper berry	*Juniperus communis*	✓	✓	✓		✓
Lovage root	*Levisticum off.*			✓		
Mace	*Myristica fragrans*	✓		✓		✓
Marjoram	*Majorlana hortensis*			✓		
Massoia	*Crytocaria massoia*			✓		
Mushroom	*Boletus edulus*			✓		

continues

Table 2–13 continued

		Universal[a]	*CAL*[b]	*Flavex*[c]	*SKW*[d]	*Daniel*[e]
Myrrh	*Commiphora mol mol*			✓		
Nutmeg	*Myristica fragrans*	✓		✓		✓
Oakwood	*Quereus alba*				✓	
Olibanum	*Boswellia carteri*			✓		
Orris	*Iris pallida*			✓	✓	
Paprikas	*Capsicum annum*		✓	✓		
Peach leaves	*Prunus persica*			✓		
Peanut	*Arachis* sp.			✓	✓	✓
Pear	*Pyrus communis*				✓	
Peppermint	*Mentha piperita*	✓		✓		
Pepper oil	*Piper nigrum*	✓		✓		
Pepper extract	*Piper nigrum*	✓		✓		
Pimento berry	*Pimenta off.*	✓		✓		
Rose hip	*Rosa* sp.			✓		
Rose pepper	*Piper rosa*		✓	✓		
Rosemary	*Rosmarinus off.*		✓	✓	✓	
Sage	*Salvia off.*		✓	✓		
Saw palmetto	*Serenoa repens*			✓		
Sea buckthorn	*Hippophae rhamoides*			✓		
Spearmint	*Mentha spicata*	✓				
Strawberry juice	*Fragaria vesca*	✓				
Tea	*Thea sinensis*	✓			✓	
Thyme	*Thymus vulgaris*			✓		
Tobacco	*Nicotinia* sp.			✓		
Vanillas	*Vanilla fragrans*	✓	✓	✓		✓
Ylang ylang	*Cananga*			✓		

[a]Universal Flavors Ltd., Bilton Road, Bletchley, Bucks MK1 1HP UK.

[b]CAL, 86 Rue de Paris, Orsay, Grasse 91400 France.

[c]Flavex GmbH, Nordstrasse 7, D 66780. Rehlingen, Germany (UK agents, Guinness Chemicals, 33 London St. Reading).

[d]SKW GmbH, Postfach 1262, D 83303, Trostberg.

[e]H.E. Daniel Ltd. Belasis Avenue, Billingham, Teesside UK.

their launch in 1982. Further details are given in the following references: Moyler et al., 1992; Moyler, 1993; Kandiah & Spiro, 1990; Spencer et al., 1991; Barton et al., 1992; Reverchon et al., 1993, 1994; Sovova et al., 1994, 1995; Stastova et al., 1996; Oszagyan et al., 1996; Reverchon, 1996; and Roy et al., 1996. World-wide production of CO_2 commercial extracts is now available for 64 botanicals.

2.8 SUMMARY

When selecting a natural product extract for inclusion into a flavor, the choice of extraction technique, based on a consideration of the end-product, will often prevent problems at a later stage of product development. The selection of solvent is often the key to obtaining an extract with the desired properties of flavor, taste and solubility required. The most natural, true tasting extracts of plant materials are obtained by the techniques that utilize low temperature and avoid degradative heat processes and reactive solvents.

REFERENCES

Analytical Methods Committee, *The Analyst* (London) 113 (1987) 1125.

Analytical Methods Committee of Chemical Society, Essential Oils Subcommittee, *The Analyst* (London) 109 (1984) 1343.

P. Barton, R. Hughes, M. Hussein, *Ibid.* 5 (1992) 157.

B.E.M.A. British Essence Manufacture Assoc., *Guidelines on Allowable Solvent Residues in Foods and Natural Flavouring Definition*, BEMA, 6 Catherine Street, London, WE2B 5JJ.

H. Brogle, *Chem. Ind.* 12 (1982) 385.

C.A.L., *Technical Data Sheet*—Celery Seed Oil, C.A.L. Grasse, France (1986).

J.P. Calame and R. Steiner, in *Theory and Practice in Supercritical Fluid Technology*, eds. M. Hirata and T. Ishikawa, Tokyo Metropolitan Univ. (1987) 277–318.

C.C. Chen and C.T. Ho, *J. Agric. Food Chem.* 36 (1988) 322.

C.C. Chen, M.C. Kuo and C.T. Ho, *J. Agric. Food Chem.* 34 (1986) 477.

D.S.J. Gardner, *Chem. Ind.* (London) 12 (1982) 402.

Gebruder Wollenhaupt, *Real Vanilla*, Vanille Import-Export GmBH, 2057 Reinbek, Hamburg W.G. (undated).

I. Goodarznia et al., CO_2 Cumin oil, *Flavor and Fragrance Journal* 7/733 (1998).

N. Gopalakrishnan, P.P.V. Shanti and C.D. Narayanan, *J. Food Sci. Agric.* 39 (1988) 3.

H.B. Heath, in *Source Book of Flavors*, AVI (1982).

M. Kandiah, M. Spiro, *Int. J. Food Sci. Technol.* 25 (1990) 328.

B.M. Lawrence, *Perfumer and Flavorist* 10 (1985) 1.

P.A.P. Liddle and P. de Smedt, *Parfum Cosmet Arome* (1981) 42.

D. McHale, W.A. Laurie and J.B. Sheridan, *Flavour and Fragrance J.* 4 (1989) 9.

D.A. Moyler, in *Developments in Food Science*, Vol. 24 ed. G. Charalambous, Elsevier, Amsterdam (1989) pp. 263–80.

D.A. Moyler and H.B. Heath, Flavors and fragrances world perspective, in *Developments in Food Science*, Vol. 18, eds. B.M. Lawrence, B.D. Mookherjee and B.J. Willis, Elsevier, Amsterdam (1988) pp. 41–64.

D.A. Moyler, in *Proceedings ICEOFF*, New Delhi, India (1989) Abstract 84.

D.A. Moyler, *Chem. Ind.* 18 (1988) 660.

D.A. Moyler, in *Distilled Beverage Flavour*, eds. J.R. Piggott and A. Paterson, Ellis Horwood (1988).

D.A. Moyler, in *Theory and Practice in Supercritical Fluid Technology*, eds. M. Hirata and T. Ishikawa, Tokyo Metropolitan Univ. (1987) 319–341.

D.A. Moyler, in *Extraction of Natural Products using Near-Critical Solvents*, eds. M.B. King and T.R. Bott, Blackie, Glasgow (1993) pp. 140–183.

D.A. Moyler, *Perfumer and Flavorist* 9 (1984) 109.

D.A. Moyler, *Molecular Distilled Oils,* Daniel, UK (1998).

D.A. Moyler, R.M. Browning and M.A. Stephens, "10 Years of CO_2 extracts-review" in *Proceedings ICEOFF*, Vienna/Austria (1992) pp. 51–100.

S.N. Naik, H. Lentz and R.C. Maheshwari, *Fluid Phase Equilibria* 49 (1989) 115.

S.N. Naik, R.C. Maheshwari and A.K. Gupta, in *Proceedings ICEOFF* New Delhi, India (1989) Abstract 202.

M. Oszagyan et al., *Flavour & Fragrance Journal II* (1996) 157.

J.A. Pickett, J. Coates and F.R. Sharpe, *Chem. Ind.* (1975) 7.

E. Reverchen et al., *Ind. Eng. Chem. Res.* 32 (1993) 2721.

E. Reverchen et al., *J. Supercrit. Fluids* 7 (1994) 185.

E. Reverchen, *A. I. Ch. E. J.* ***42*** (1996) 1765.

A.H. Rose, ed., *Alcoholic Beverages, Economic Microbiology*, Vol. 1 Academic Press (1977).

B.C. Roy, M. Goto, T. Hirose, *Ind. Eng. Chem. Res.* 35 (1996) 607.

RIFM 'Safrole' Food Cosmetics Toxicology 20, (1982) 825–6.

K.U. Sankar, *J. Food Sci. Agric.* 48 (1989) 48.

N. Shaath and P. Griffin, Frontier of flavor, in *Developments in Food Science*, Vol. 17, ed. G. Charalambous, Elsevier, Amsterdam (1988) pp. 89–108.

H. Sovova, R. Komers, J. Kucera, J. Jez, *Chem. Eng. Sci.* 49 (1994) 2499.

H. Sovova, J. Jez, M. Bartlova, J. Stastova, *J. Supercrit. Fluids* 8 (1995) 295.

J.S. Spencer et al. *J. Supercrit. Fluids* 4 (1991) 40.

J. Stastova, J. Jez, M. Bartlova, H. Sovova, *Chem. Eng. Sci.* 51 (1996) 4347.

Chapter 3

Fruit Juices

Philip R. Ashurst and Barry Taylor

3.1 INTRODUCTION

There has been a long association between fruit juices and flavorings. Traditionally, fruit flavorings were some of the earliest types available and because of their relative simplicity, they have often been used to enhance or substitute for fruit juices in beverages.

Fruit juices and their components also play a very important part in many flavorings, with concentrated juices frequently used as a base to which other components may be added. With the growth in interest and demand for natural flavors, fruit juice components are an essential source of these ingredients, although they are rarely, if ever, combined in the same proportions as in the original fruit juice.

The biological function of fruits is to be attractive to animals to ensure distribution of the seed via animal feces or, in the case of larger fruits, to provide a bed of rotting humus in which the seed may develop. In contrast to many other vegetable products such as cereals, the starting point for juice production is the tender fleshy fruit which is prone to more or less rapid decay. This instability is increased once the fruit is broken to initiate a process and in consequence, all man's early attempts to utilize fruit juices ended in fermented products such as wine or cider. Early in the nineteenth century, Appert (1725–1841) showed that fruit juices could be stabilized by heat treatment after bottling and in 1860 the discoveries of Pasteur provided a scientific background for this observation.

In both Europe and the United States, the commercial production of pasteurized fruit juices began late in the nineteenth century, but it was not until the second quarter of the twentieth century that technical and commercial development of fruit juices really began to be significant. With availability of fruit juices came increasing consumption and their incorporation into other products such as soft drinks. The world-wide availability of fruit juices is now taken for granted and the manufacturing industry is large, complex and well organized.

There has always been a close link between flavorings and fruit juices, with synthetic materials used to extend and enhance juices. Juices themselves, particularly concentrates and volatile fractions, are being increasingly used as components of flavorings. This trend has become more noticeable as the demand for natural flavorings has increased.

3.2 FRUIT PROCESSING

3.2.1 *General Considerations*

It is necessary to review typical manufacturing processes for various types of fruit in some detail in order to become familiar with the different types of fruit raw materials that may be available. A typical fruit processing operation is shown in Figure 3–1.

Botanically, fruits are formed from floral parts of plants with or without associated structures. The tissues bearing juice are either thin walled parenchyma or special sacs such as the modified placental hairs of citrus. Fruit is often classified on the basis of the hardness of its pericarp (the wall of the ovary from which the fruit is derived). A typical classification is shown below (see also Appendix II):

Dry fruits	Dehiscent (bursting fruits), e.g., legumes Indehiscent (non-bursting fruits), e.g., nuts
Fleshy fruits	Berry, e.g., blackcurrant, citrus Drupe (stone fruits), e.g., peach Aggregate, e.g., raspberry
False fleshy fruits	e.g., banana

Most juices are derived from fleshy fruits and these may be conveniently subdivided for processing into three categories.

1. fruits which are pulped and their juices removed by pressing, e.g., apple, berry fruits
2. fruits requiring the use of specialized extraction equipment, e.g., citrus fruits, pineapples
3. fruits requiring heat treatment before processing, e.g., tomatoes, stone fruits

In most large fruit processing operations the plant is usually dedicated to one type of fruit.

Citrus fruits are unusual because the outer skin or flavedo is rich in essential oils and other tissues such as the albedo or carpellary membranes contain substances that give rise to bitter flavors. The processing of citrus typically involves separation of these various components as an important principle and for certain products such as comminuted bases, the various components are recombined in different proportions (see Section 3.3.2).

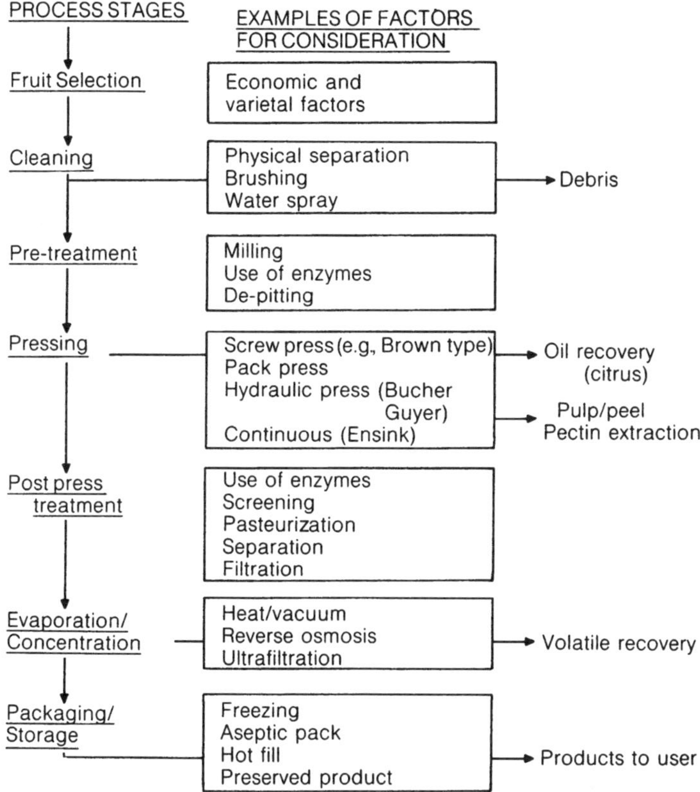

Figure 3–1 Block diagram of stages in a typical fruit juice processing operation.

The flesh of stone fruits is separated from the stones or pits, not only to facilitate ease of handling, but also because the stones are further processed to obtain both fixed oils and glycosides. Fixed oils, such as those from peach, have application in the cosmetics industry and glycosides may be used as a source of other natural flavoring ingredients such as benzaldehyde.

Another important aspect of the processing of fruit is the presence of pectins. These substances contribute to the viscosity of fruit juices and assist in the suspension of colloidal material and tissue fragments that make up its cloud.

When most fruits are pressed, pectolytic enzymes are released and these will, if not destroyed, clarify and or cause gelling of the juice. Rapid initial processing of freshly pressed juice is therefore an important factor in determining whether cloudy juice is obtained (in which case the enzymes must be destroyed by pasteurization to at least 95° C) or clear juice, in which case enzymes are allowed to act and may be enhanced by the addition of synthetic enzymes in further quantity.

3.2.2 *Soft Fruit Processing*

Many different types of press have been developed for processing soft fruits, especially apples, on batch and continuous operation. The conventional pack press (Figure 3–2) sandwiches layers of pulped fruit enclosed in press cloths between slotted wooden racks. The layers of racks containing fruit are then subjected to mechanical or hydraulic pressure by an ascending ram working against the fixed frame.

3.2.2.1 *Pre-processing*

Pectin in fresh apples is mostly insoluble and this enables fruit to be pressed immediately after cleaning and milling. One of the characteristics of stored fruit is that the pectin tends to become more soluble (Smock & Neubert, 1950), leading to difficulties in pressing. Most apples are therefore processed in the United Kingdom in the period between September and December when the fruit is fresh.

For processing of soft berry fruits, there are similarities in the pressing but significant differences in the pre-treatment. Soft berry fruits cannot, for example, be

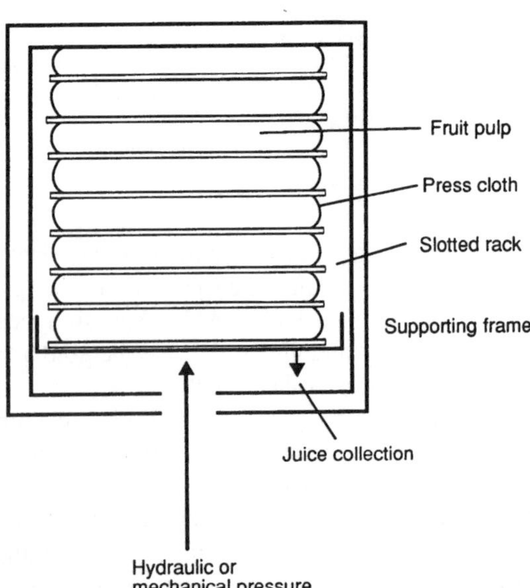

Figure 3–2 Diagram of a typical "pack press" used for processing apples and soft fruits.

subjected to washing procedures because of potential mechanical damage, and must therefore be subjected to higher standards of harvesting.

In some berry fruits, notably blackcurrants, pectinesterases are present that demethylate pectin to pectic acid. This in turn may react with calcium present to form calcium pectate which forms a firm gel, making the fruit more difficult to press and reducing juice yields significantly.

By warming the milled fruit to about 50° C and adding commercially available pectolytic enzymes that contain polygalacturonases, pectic acid can be broken down into galacturonic acid or its low molecular weight polymers before calcium pectate can be formed. By carrying out this process at around 45° C the reaction can normally be completed in 2–4 h to give a pulp from which juice may be freely expressed in good yield. The reactions involved are summarized in Figure 3–3.

In addition to the presence of pectins, many soft fruit juices contain polyphenolic substances. These components may be involved in the formation of hazes and precipitates at almost any stages of the life of the products. They are also implicated in many non-enzymic browning reactions.

Polyphenols may, however, be removed by interaction with protein (e.g., gelatin). Hydrogen bond formation occurs between the phenolic hydroxyl group of the tannins and the carbonyl group of the protein peptide bond to give an insoluble complex. This reaction is affected by factors such as pH, concentration of polyphenols (Van Buren & Robinson, 1969) and the type of protein used (Wucherpfennig et al., 1972). After formation the precipitate may be removed by filtration on filter aid or centrifugation.

3.2.2.2 *Juice processing after pressing*

Reference has already been made to the need for rapid pasteurization of expelled juice. This process performs two functions; where cloudy juice is required it destroys pectolyic enzymes which would otherwise clarify the juice, and the naturally occurring yeasts and molds that are a characteristic of all fruits and their untreated juices are largely killed off.

Various alternatives are available for further processing of expelled juice but, typically, soft fruit juices are flash heated to boiling point and some 10–15% of the volume vaporized. Vapor and hot juice are separated in a cyclone with juice then passing to an evaporator where it is reduced to between 12–17% of its original volume (i.e., concentrated between six and eight times).

Separated vapor is normally passed to a fractionating column with reflux where the low boiling constituents giving the juice most of its aroma and flavor are separated from water which forms the bulk of the vapor. The volume of this fraction would be typically around 1% of the original juice volume and this fraction would be referred to as "100 fold volatiles." This mixture of substances is a very important source of components for natural flavors (see Section 3.6.2).

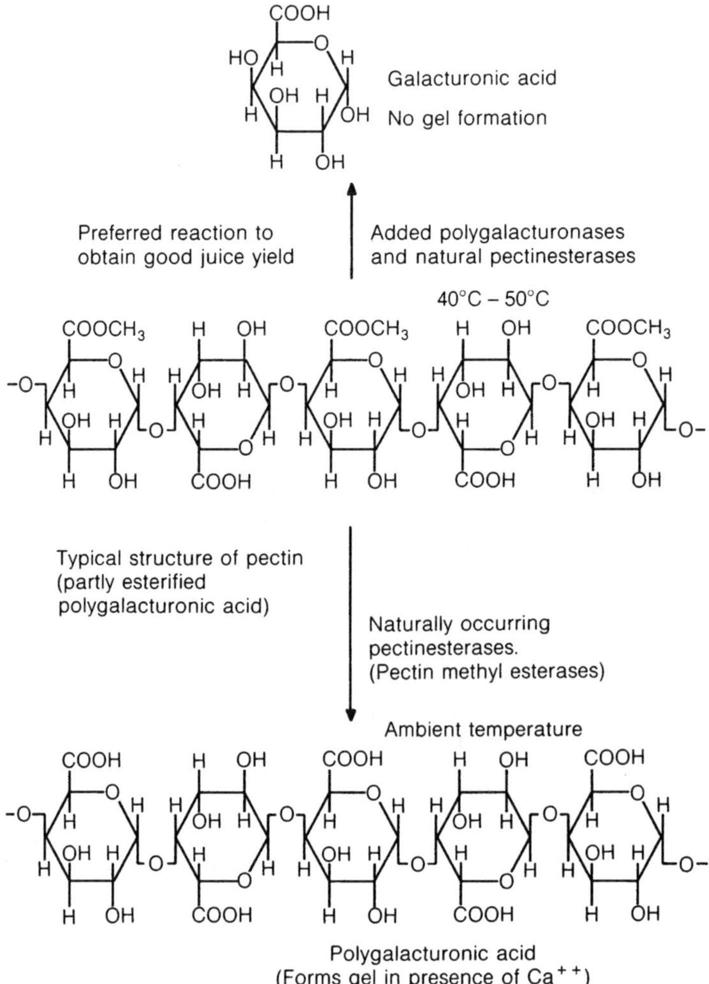

Figure 3–3 Reactions involved in enzymic degradation of pectin.

As an alternative to evaporation and removal of volatile substances, the expelled juice may be filled into containers for the domestic market.

Some processors link de-aeration to the process of pasteurization before concentration takes place. Dissolved oxygen has a degradative effect on juices and it has been assumed that this will have a rapid spoilage effect when the juice is heated. The de-aeration process, when used, involves flowing the juice over a large surface area in a vacuum chamber where pressure is reduced to at least 100 mbar.

De-aeration is undoubtedly successful in removing air entrained in the product and this has the effect of reducing foaming in an evaporator. The link between dissolved oxygen and rapid spoilage during heating has not, however, been successfully proved (Kefford et al., 1959).

Many processors do not find de-aeration to be of much practical value because of the lack of observed degradation and especially because of the loss of volatile substances that often results.

3.3 SPECIALIZED FRUIT PROCESSING

3.3.1 *Citrus*

Botanically, citrus varieties are forms of berry fruits in which the hairs inside the ovary walls form juice sacs. Epicarp is the familiar highly colored, oil bearing outer layer. Both juice and oils are now valuable commodities in all citrus varieties and the recovery of both materials is important to the economics of processing.

Many different processes are used world-wide for citrus types with a two-stage operation being widely employed. In a typical process, the fruit passes over an abrasive surface or roller where the sacs in the epicarp are pierced and oil washed away by water spray. The resulting oil-in-water emulsion is screened to remove vegetable debris and oil is separated by centrifugation and then dried and packed.

The rasped fruit then move onto an extractor where the juice is removed leaving albedo (pith) and peel (flavedo). Various juice extractors have been used with fruit being compressed in a roll mill or screw press, or the juice bearing material reamered out. Expressed juice is subjected to screening (sometimes referred to as finishing) before being further processed.

Other juice processing machines have been devised to minimize the handling of fruit and to incorporate both oil and juice extraction into a single operation. Of these the FMC “in line juice extractor” is probably the best known. Its operation is based upon pairs of stainless steel cups with radiating fingers that intermesh when the upper and lower halves of the cup are brought together. Both cups are fitted with centrally mounted circular cutters; that in the bottom half is hollow and connected to a perforated tube beneath. A single fruit is fed automatically into the opened lower cup; the upper cup then descends forcing the fruit into the perforated tube from which the juice escapes. Residual albedo is collected separately.

Simultaneously, the peel is abraded and flaked by intermeshing fingers. These peel flakes contain the bulk of essential oils and, when compressed in a screw press, yield an emulsion from which the essential oil may be collected by centrifugation.

Other pressing equipment has been developed for lemons which yield a more valuable oil and for which the demand for juice, although significant, is not as great as for orange.

Limes are normally processed in a slightly different manner. Washed fruit is compressed in a screw press to yield a pulpy juice that typically also contains the oil emulsion. Larger pieces of pulp are screened out and, in the classical process, juice and oil emulsion are fed to large tanks where a natural separation process occurs. Naturally occurring enzymes, which may be enhanced by the addition of commercially available synthetic pectolytic enzymes, clarify the juice while the oil bearing emulsion and pulp settle to the top of the tank. At the same time other debris settles below the clarified layer.

This process is normally completed in 10–30 days although in adjacent processing factories, there may be variations in time and juice clarity.

Clarified juice is then typically filtered and concentrated while the oil bearing emulsion is steam distilled. The process often requires a period of some hours of heating before actual distillation starts and it is in this way that the oil is brought to the specification required by the consumer. Stream distillation of lime oils usually brings about a number of changes to the components present in the undistilled oil.

Many variations of the classical process for limes have been developed with emphasis on the production of juice of low color by ultrafiltration and on the production of cold pressed oil by direct centrifugation of the oil bearing emulsion. Cold pressed lime oil is, however, of different character to distilled lime and is often used in different applications.

The fate of citrus juices both during and after processing will vary with their subsequent use. Whereas lime juice is normally available only as a clarified juice, lemon and orange juices are available both clear and cloudy. Because of the very high natural acidity (up to 8% as citric acid) of lime and lemon juices, clarification can take place using unpasteurized juices which will not normally ferment. Orange juice must be treated differently. Clarification is carried out by addition of pectolytic enzymes after pasteurization to destroy microbiological activity.

As with soft fruit juices, the bulk of citrus juices are subjected to concentration to facilitate shipping and, during this process, volatile components are usually collected separately from the juice concentrate. Both oil and water phase volatile fractions are collected from processes and these are widely used in flavorings.

3.3.2 *Comminuted Citrus Bases*

There was until 1995 a special category of soft drinks in U.K. food legislation (*The Soft Drinks Regulations*, 1964) for which comminuted bases are an integral part. Some of the early processes which were developed in the first half of this century were patented and have been reviewed (Charley, 1963). In an original process, whole oranges were shredded into sugar syrup which extracted both juice and oil emulsion. By screening the syrup, peel, seed and coarse rag were removed and the syrup pasteurized to yield the finished product.

More recent developments, which are almost always confidential, work on the principle of taking different parts of the fruit (except seeds) in varying proportions, blending them and milling the resulting mixture through a carborundum stone mill which "comminutes" or disintegrates the solid components, releasing pectin and emulsifying the juice and oil components. The mixture must be rapidly pasteurized at 95° C to destroy enzyme activity, stabilize cloud and prevent microbiological spoilage.

Comminuted bases have a number of advantages for the soft drinks industry and, to a lesser extent, elsewhere in the food industry. The principal advantages are summarized as:

1. a more intense flavor that is often preferred to the corresponding concentrated juice
2. excellent cloud characteristics because of incorporated oil, pectin and other fruit materials
3. good raw material stability

The main disadvantage of comminuted bases is in products exposed to oxygen, when the oil incorporated in the comminute base will often oxidize with corresponding characteristic flavor deterioration.

In more recent years, the boundary between cloudy juices and comminutes has become much less distinct. By incorporation of a small percentage of peel components or pulp extractives into juice, it is possible to make significant cost reductions and although such materials may be satisfactory or even preferable for some manufacturing uses, they have found their way into pure fruit juice (particularly orange) at times of raw material shortage and corresponding high costs. This has led a number of European countries to set up sophisticated analytical testing to confirm juice quality.

3.3.3 *Pineapple Juice*

Pineapples are somewhat unusual because the process to manufacture juice is mainly a byproduct of the pineapple canning industry. Although pineapple juice does have a characteristic flavor, the overall perception of this product relies heavily on the texture associated with the presence of tissue.

Pineapple is also typically low in pectin content and, in consequence, enzymatic degradation of this pectin can result in rapid clarification. It is then essential in pineapple processing to ensure rapid deactivation of enzymes by heat treatment.

The major processing areas of the world are Hawaii and the Philippines although the relative ease with which pineapples may be grown and harvested means they are grown very widely in many developing nations.

In pineapple processing, the fruit must be peeled, cored and sliced. The cores, trimmings, small and outsize fruit are typically milled into a puree which is pasteurized in a tubular unit before being fed to screw presses which extract the juice. This juice will, depending upon the final quality required, then be screened or centrifuged before being passed to the evaporator for concentration or canning as single strength juice. Some products may be subjected to homogenization to further stabilize the cloud components.

During evaporation to produce the concentrated juice, volatile components are normally collected for use with either the reconstituted juice or as flavor ingredients.

3.3.4 *Processes Requiring Heat*

Probably the best example of fruit products requiring heat treatment before processing into juice is the tomato, although this process is normally applied to stone fruits as well. Tomato, like pineapple juice, is dependent to a large extent upon suspended solids to provide the characteristic flavor sensation, and rapid enzyme deactivation by heat treatment is essential to avoid clarification.

In the process, cleaned fruit is usually coarsely chopped and heated in a tubular heat exchanger following which it is screened to remove peel, seeds and other debris. Secondary screening through paddle finishers with a mesh size of about 0.5 mm will normally give the typically pulpy juice.

Other processing may be used such as stone-milling or homogenization to improve smoothness or cloud and a limited evaporation of water in scraped surface heat exchanges may be used to produce concentrated paste.

Stone fruits are often processed in a similar manner although juice production may, as with pineapple, be associated with waste and rejected fruit from a canning operation. The stones (or pits) from the fruit are subjected to drying and may be used in a further process as previously described.

Fruit processes involving heating may change the character of the flavor associated with the fruit and the flavorist must establish whether a "fresh" or "processed" character is required when using or matching these products.

3.4 PRODUCTS AND PACKAGING

A wide range of fruit products is available for the flavorist although their range of use is often affected by their degree of concentration and the packaging in which they are available.

As indicated previously, almost all juice products must be pasteurized to control enzyme activity and the levels of microflora that are responsible for spoilage. Heat treatment is often incorporated as part of the packaging process although this is usually a second process to that carried out immediately after the juice is expelled.

Because of their low pH, most fruit juices and their concentrates will not support the growth of pathogenic micro-organisms and, in consequence, the pasteurization of products at between 80 and 85° C for around 30 s is normally sufficient to remove spoilage organisms. In more recent packaging developments using aseptic containers, ultra high temperature (UHT) conditions (e.g., 2–3 s at 120° C) may be used to sterilize the juice before it is packed in aseptic conditions to ensure asepticity. Low acidity juice, especially tomato and some tropical fruits with higher pH values, require more stringent conditions to give sterile products.

Most fruit products, whether concentrated or not (except possibly the volatiles referred to earlier), are available commercially in a number of alternative packaging types.

3.4.1 *Frozen Juices*

Probably the largest commercial volume of trade in concentrated fruit juices is in frozen form. Orange juice concentrate accounts for the largest volume of frozen juice concentrates.

The process of freezing is not practically applicable to other than very concentrated (normally at least six times) juice because of the ice crystal/block size. With highly concentrated juices, the product generally remains as a firm slush at around −20 to −25° C. With a single strength juice the whole package would form into a single unmanageable block of frozen juice and cloud would probably be reduced. Typical packaging for frozen juice employs 200 liter steel drums with removable heads, the juice being held in two large polythene bag liners inside the drum.

Freezing of juices has many advantages as it maintains flavor and color for long periods while keeping enzymic and microbial activity at a minimal level. The principal disadvantages of frozen juices are the practical problems of shipping and storage at around −25° C (although the industry is now well able to accommodate these) and the need to allow adequate time (normally around 48 h) for thawing to ambient temperature to take place before the juice is used.

A further disadvantage is that some processors may, to enhance flavor, concentrate a product to a degree greater than that required and then dilute or "cut back" the concentrate to say a six times value with fresh unpasteurized juice. This will give the product an excellent flavor and appearance but may re-introduce active pectolytic enzymes and also microbiological contamination. These are held in check as long as the product remains frozen but, upon thawing, gelatin or fermentation may subsequently occur. If the frozen juice is to be diluted and packed for direct consumption it will have to be resubjected to a thermal process and this potential problem is dissipated. If, however, the juice is to be used as a concentrated base for a flavoring, the problems indicated may arise.

3.4.2 *Aseptic Packaging*

Packaging developments over the past twenty years or so have led to aseptically packed juice in containers of sizes up to about 1000 liters becoming a technically and commercially alternative form of packing. The most familiar form of aseptic packaging for fruit juices is the 250 ml and 1 liter pack for domestic consumption and the principle can readily be applied to different pack sizes and to different degrees of concentration.

Pasteurization of juice, either as a concentrate or at single strength, is best carried out under UHT conditions (e.g., 2–3 s at 120° C) although satisfactory products can be achieved by using more conventional conditions such as 85° C for 30–45 s. This juice is carried to the aseptic filler, the key part of the operation, and filled into a flexible (and often multilayer) bag that is supported by some external rigid container (e.g., openhead steel drum, strong board case, etc.). The bag will have probably been sterilized by γ-irradiation and it is opened, filled and resealed in an aseptic atmosphere which may be steam, sterile air or a chemical spray such as hydrogen peroxide/peracetic acid.

Juice packed in this way will normally be microbiologically stable for an infinite time (unless the inner sterile container is opened or punctured for any reason). Commercially, this form of packaging is attractive because of ease and lower cost of storage and shipping. There is also an effective guarantee that the product will be aseptic at reconstitution and this ensures that the microbial load is minimal when further processing is started.

The principal disadvantage of aseptic packaging arises because juices packed in this way are usually shipped and stored at ambient temperatures, and browning frequently occurs. This may be a significant disadvantage for juice that is to be used for consumption as such; it may be less of a disadvantage when the juice concentrate is to be used as a flavoring component. Browning of color is sometimes accompanied by unacceptable flavor changes.

3.4.3 *Self-Preserved Juice*

Many clarified soft fruit juices may be concentrated to around 68–70° Brix (see specifications) at which level of solids they are effectively self-preserving. Such juices may be, in some circumstances, vulnerable to spoilage by molds; the growth of molds usually occurs in closed containers subjected to heating and cooling. Water vapor from the concentrate condenses inside the container and drains back to the surface of the concentrate allowing the development of conditions to support mold growth. Self-preserved juices, which are particularly important for use in flavorings, are best stored in cool, 5–10° C, even temperatures in full containers. Occasional problems occur with osmotolerant yeasts which cause slow fermentation even at high solids levels. These problems are usually self-evident.

3.4.4 *Preserved Juice*

A frequent method of storing fruit juice concentrates is by the use of chemical preservatives, although the *U.K. Fruit Juices and Nectars Regulations* do not allow significant levels in juices for consumption as such. Preserved juices are probably most widely used in the preparation of bases for use in the flavoring and formulation of soft drinks. The presence of preservatives which may interact with other flavoring components may restrict other uses to which these products may be put.

Probably the most widely used preservative in concentrated juices is sulfur dioxide at levels of 1500–2000 mg/kg. In juice concentrates with high levels of natural sugars the amount of "free" sulfur dioxide diminishes with storage. Sulfur dioxide reacts with many flavoring ingredients such as aldehydes, and any juice concentrate required for use as a flavoring base should not contain this ingredient.

3.4.5 *Hot Pack Products*

This process relies on raising the temperature of a juice to at least 70° C, usually by means of a plate heat exchanger, filling the hot product into its final container, closing the container and inverting it or otherwise heating the closure to pasteurizing temperature. The pack is maintained at a required temperature to achieve pasteurization and then cooled.

Relatively small amounts of juices are available in a sterile "hot pack" form. This type of packaging is typically used for single strength (unconcentrated) juices where a high quality product is required. Most forms of packaging other than aseptic packs are inappropriate for unconcentrated juices and aseptic packs are rapidly displacing hot packs (usually cans).

3.5 PRODUCT SPECIFICATION(S)

As indicated above, most juices and their components that are used in flavorings are concentrated in some measure because unconcentrated juices are unlikely to contribute much flavor except in a significant proportion (e.g., 5–10%) in the finished product.

At the time of writing, official methods of fruit juice analysis are under discussion within Europe and some have been adopted as national standards. Methods published by the International Federation of Fruit Juice Producers (*Methods of Analysis*) and the Association of Official Agricultural Chemists (*Official Methods of Analysis*) are, however, widely used.

3.5.1 *Soluble Solids Content*

The solids content is a measure of the amount or degree of concentration and thus an important factor especially when fruit or juice content is to be claimed in a finished product. The most common analytical measure applied to juices of whatever concentration is their soluble solids content. This is determined classically by filtering a known quantity of the juice to remove suspended solids and then evaporating the resulting solution at 105° C. The solids content is then determined by weight. This classical method is cumbersome and not regularly used; the most widely used routine method of estimating solids content is by the use of the refractometer.

The results obtained from the refractometric observations are based on Brix values for sucrose solutions. For juices with relatively low acidity, a reasonable relationship exists between actual soluble solids content and observed Brix value. The relationship is improved by the use of acidity correction tables.

For pure sucrose solutions, the relationship between the Brix value (%w/w) and the solids content in grams per liter (%w/v), which is generally of greater practical significance in experimental work, is a function of density. Thus

$$(\%\text{ solids content w/w}) \times \text{density} = \%\text{ solids content w/v (g/ml } 20^\circ\text{ C)}$$

These relative values are set out in Table 3–1. The relationship between observed Brix value and soluble solids content is much less satisfactory for highly acidic juices such as lime and lemon although the use of the refractometer for control purposes in any one factory remains valid. In establishing the degree of concentration of a fruit juice, it is thus not possible to relate strictly the solids content (degrees Brix) to this value. A 60° Brix orange juice is normally referred to as a 6:1 concentrate and this may be true for some juices. If, however, the typical observed Brix value of single strength juice (as obtained from the fruit) is 11.0°, then concentration is as follows:

	Brix (observed) %w/w	*Solids content (g/liter) %w/v*
Single strength juice	11.0	114.5
Concentrated juice	60.0	771.72
Degree of concentration		$\frac{771.72}{114.5} = 6.74{:}1$

These degrees of concentration are more significant when the fruit juice content to be added to a product is of special significance, e.g., to meet a legislative standard for juice content. For the more practical considerations of a juice concentrate that will give a good flavor base without undue risk of fermentation, 60° Brix would normally be considered a minimum level (except for acidic juices such as lemon or lime).

Use of the hydrometer as a rapid method for estimating the soluble solids content for fruit juices should also be mentioned.

Table 3–1 Table of constants for sugar solutions at 20°B. B = °Brix (%w/w), GPL = grams per liter, SG = specific gravity

B	*GPL*	*SG*	*B*	*GPL*	*SG*
1	10.0	1.000	36	416.3	1.156
2	20.1	1.005	37	429.7	1.161
3	30.3	1.010	38	443.2	1.166
4	40.6	1.015	39	456.8	1.171
5	50.9	1.018	40	470.6	1.177
6	61.3	1.021	41	484.5	1.182
7	71.8	1.026	42	498.5	1.186
8	82.4	1.030	43	512.6	1.192
9	93.1	1.034	44	526.8	1.197
10	103.8	1.038	45	541.1	1.202
11	114.7	1.043	46	555.6	1.208
12	125.6	1.047	47	570.2	1.213
13	136.6	1.051	48	585.0	1.219
14	147.7	1.055	49	599.8	1.224
15	158.9	1.059	50	614.8	1.230
16	170.2	1.064	51	629.9	1.235
17	181.5	1.068	52	645.1	1.241
18	193.0	1.072	53	660.5	1.246
19	204.5	1.076	54	676.0	1.252
20	216.2	1.081	55	691.6	1.257
21	227.9	1.085	56	707.4	1.263
22	239.8	1.090	57	723.3	1.269
23	251.7	1.094	58	739.4	1.275
24	263.8	1.099	59	755.6	1.281
25	275.9	1.104	60	771.9	1.287
26	288.1	1.108	61	788.3	1.292
27	300.5	1.113	62	804.9	1.298
28	312.9	1.118	63	821.7	1.304
29	325.5	1.122	64	838.6	1.310
30	338.1	1.127	65	855.6	1.316
31	350.9	1.132	66	872.8	1.322
32	363.7	1.137	67	890.1	1.329
33	376.7	1.142	68	907.6	1.335
34	389.8	1.146	69	925.2	1.341
35	403.0	1.151	70	943.0	1.347

3.5.2 *Titratable Acidity*

The measurement of titratable acidity in a fruit juice is a very important indicator to the flavorist of the “sharpness” of the juice. It is normally measured by direct titration with a standardized aqueous alkali solution (e.g., 0.1*N* sodium hydroxide solution) using either a pH meter or phenolphthalein indicator to establish the end point.

For citrus and many other juices, acidity is calculated as citric acid (care must be taken to specify the result either in terms of monohydrate or anhydrous form) and results are conveniently expressed as grams per liter (gpl). Many juices contain acids other than citric (e.g., malic, oxalic, etc.) and for products such as apple juice concentrates, results will normally be expressed in terms of malic acid (gpl).

For highly acidic juices, determination of acidity is the most important criterion in establishing the degree of concentration of the juice in a parallel way to that described above.

When consideration is being given to the use of a fruit juice concentrate as base for a flavor, acidity is important in establishing whether any (rapid) changes are likely to occur to the added flavor ingredients, natural or otherwise. Many reactions are catalysed or otherwise affected by acids and it is important to avoid changes in flavor character. For example, when high levels of some essential oils are compounded into acidic juice bases, there is often a rapid oxidative degradation of terpene constituents.

A further consideration to the use of acidic juice concentrates in compounded flavorings is the application for the flavoring. If, for example, a flavoring is to be added to dairy products, consideration must be given to whether the level of the acidity will affect proteins present. In some sugar confectionery applications, the presence of acids will cause some partial inversion of sucrose that may have a significant and, in some situations, undesirable effect on the finished product.

3.5.3 *Brix/Acid Ratio*

The Brix/acid ratio is a concept that has been widely used in the juice packing industry. It is an arithmetical proportion of soluble solids to (citric) acid content. It is used to indicate whether a juice is sweet, sharp, or "median" in character and is also a useful indicator of the degree of ripeness of the fruit from which the juice originates.

Juice packers often specify the "ratio" of the product required to ensure that there is minimal batch to batch variation. The ratio is calculated as follows:

$$\frac{\text{Observed Brix value}}{\text{Acid content}} = \text{Ratio}$$

It should be noted that the acid content used in calculating ratios must be as %w/w and not %w/v. In this way, the result is not changed with increasing Brix.

3.5.4 *Other Specifications*

Depending upon the flavor applications to which a juice is to be put, various other specifications will be more or less important. The main factors to be considered, with some comments on their significance, are as follows.

3.5.4.1 *Routine quality factors*

pH value: This may be important in determining the application to which the juice is to be put (see Section 3.5.2).

Color: Juice concentrate color is especially important for some flavor applications, especially red fruit flavors where the natural color of the juice may be used as an (undeclared) coloring material. Deterioration of color (browning) will also then become an important consideration.

Pulp (screened or suspended): The presence of pulp is very important in some fruit drinks as consumers often associate it with overall quality. In the manufacture of flavorings, pulp is normally unacceptable as it will usually cause difficulties in manufacture. It will therefore be important to specify the concentrate accordingly.

Oil content: In fruit juices other than citrus, no oil is normally present and, in this respect, citrus juices are unique. In many citrus juices (and especially comminutes) there will often be residual or added oil and it will be most important to know how much oil is present in a compounded flavor.

Glycosides: In some fruit juices, especially citrus, glycosides may be present (e.g., naringin in grapefruit and hesperidin in orange). Many glycosides are bitter and may affect the character of the juice concentrate. Glycosides too may interact with other flavor components. Juices are not normally supplied with specific glycoside levels but methods are available for estimating them.

Ascorbic acid: Many juices have naturally occurring ascorbic acid and, in some processing plants, ascorbic acid may be added as an anti-oxidant. In juice concentrate to be used as flavoring bases it will often be important to have a level of ascorbic acid to minimize anti-oxidant effects; 200–400 mg/kg would be a typical addition.

Preservatives: Most juice concentrates for flavoring manufacture will be required free from preservatives and this will normally not present any difficulties to the supplier apart from ensuring juice processing takes place before any microbiological spoilage occurs. Some juices (e.g., lemon and lime) are often supplied with high levels of sulfur dioxide to both prevent spoilage and minimize browning. As already indicated, sulfur dioxide will react with many added oil and flavoring components and it will normally be essential to ensure juices for use in compounding are free from this preservative. Other preservatives such as benzoic and sorbic acids are less likely to interact with flavor components but are best avoided in juice for flavoring use as they both have a distinctive flavor characteristic.

3.5.4.2 *Stability criteria*

Viscosity: The viscosity of a juice may not be an important factor in most flavor applications although if other components are to be emulsified into the juice, viscosity may increase still further to a level which is unacceptable. Increasing viscosity may, however, be an early indicator of gelling (see below).

Separation: This factor only applies to cloudy juices but may be important as no separation will normally be acceptable in the finished flavoring. Separation of the juice or concentrate may also be indicative of deteriorating or aging juice.

Turbidity: Turbidity measurements, usually carried out after centrifugation of a cloudy juice, will indicate the amount of cloud present. Cloud levels in a juice will normally be more important for applications and uses other than flavoring manufacture.

Gelling and enzyme activity: The level of sugar in many juice concentrates is often ideal for gel formation to occur. The occurrence of gels is normally associated with pectolytic enzyme activity at some stage during the juice manufacture. Methoxyl groups are removed from pectin by the activity of the enzyme pectinesterase and, in this state, gel formation may occur on storage, particularly in the presence of calcium ions (see Section 3.2.1).

Gelling is undesirable for most applications and tests are available to establish the tendency of the concentrate to form a gel. Formation of a gel indicates usually either delay in processing pressed juice or inadequate thermal processing. The effect may not become apparent until the juice has been stored for some time and may also indicate a low level of residual enzyme activity.

3.5.4.3 *Microbiological evaluation*

Although the addition of many flavor ingredients to juice concentrates will have the effect of preventing further spoilage occurring and will thus render the juice based flavoring stable, it is essential to ensure that the juice is of good microbiological condition before manufacture is undertaken. Tests will normally be established for viable counts as 20 and 37°C and especially for yeast and mold counts. If yeasts are present, the presence of osmotolerant yeasts can cause problems in concentrates. The presence of some acid tolerant organisms may also cause problems (e.g., development of diacetyl and acetylmethyl carbinol).

Of particular significance in fruit juice concentrates is the level of contamination by mold spores. Not only can they cause difficulties in concentrates before the product is used (e.g., flavor deterioration), but even in a compounded flavoring containing juice concentrate, the mold spores may, in some circumstances, survive to contaminate the final foodstuff in which the flavoring is used.

3.5.5 *Juice Adulteration*

The adulteration of fruit juices may be of greater concern to the packer intent on selling the juice as such than to the flavorist. In some circumstances it may be important to the flavoring compounder to know whether the juice being used has been adulterated as the possibility of unwanted reactions between possible adulterants and flavoring ingredients may have to be considered. Alternatively, the use of a lower cost juice that could not be used for consumption, as such, may be acceptable for a flavoring application.

The following analytical checks are some of those widely used to indicate possible juice adulteration in citrus juices and may have wider applications in other juices. In detection of adulteration it is vital to establish a wide database as the key parameters of a juice may vary naturally from country to country and from season to season within a country. In order to establish the probability of adulteration, mathematical methods are used to compare the parameters.

Unless adulteration is of a gross nature, it may be extremely difficult to detect objectively without access to a method such as isotope analysis. Some of the parameters important in establishing adulteration are as follows:

- spectrophotometric examination
- phosphorus
- citric acid/isocitric acid ratio
- polyphenolic substances
- sugar analysis
- reducing sugars
- amino acids (total, by formol titration)
- spectrum of individual amino acids
- protein content
- ash

Individual values of the above parameters may not have particular significance. Comparative values of all the parameters, particularly against historical data are important and the experienced analyst can usually readily identify an adulterated juice.

To the flavoring compounder, perhaps the most likely compounds of interest are the reducing sugars and amino acids which could enter into Maillard reactions. Aroma and flavor of the juice remain the ultimate consideration.

Typical composition values of fresh orange juice are shown in Table 3–2.

Table 3–2 Typical comparative data on the composition of fresh orange juice

Parameter	*Units*	*Europe [11]*	*United States [10]*
Relative density	20°/20°C	Min 1.045	Mean 1.046
Brix corrected	°Brix	Min 11.18	Min 10.5
Volatile acid (as acetic)	g/liter	Max 0.4	—
Ethanol	g/liter	Max 3.0	—
Lactic acid	g/liter	Max 0.5	—
Sulfurous acid	mg/liter	Max 10	—
Titratable acidity	mEq/liter	106–160	—
Citric acid	g/liter	7.6–11.5	6.3–19.7
D-Isocitric acid	mg/liter	65–130	—
Citric/isocitric	Ratio	Max 130	—
L-Malic acid	g/liter	11–2.9	1.05–1.77
L-Ascorbic acid	mg/liter	Min 200	270–870
Ash	g/liter	2.9–4.8	2.8–7.3
Potassium	mg/liter	1400–2300	1200–2770
Sodium	mg/liter	Max 30	2.0–25
Magnesium	mg/liter	70–150	102–178
Calcium	mg/liter	60–120	65–305
Nitrate	mg/liter	Max 10	—
Total phosphorus	mg/liter	350–600 (as PO_4)	84–315
Sulphate	mg/liter	Max 150	—
Formol index per 100 ml	ml/1 N NaOH	15–26	12–30
Hesperidin	mg/liter	Max 1000	835–1230
Water soluble pectin	mg/liter	Max 500	—
Glucose	g/liter	20–28	—
Fructose	g/liter	22–30	—
Glucose fructose	Ratio	Max 1	—
Sucrose	g/liter	Max 45	30–65

3.5.6 *Specifications for Essence/Volatiles/Citrus Oils*

Because of the particular significance of juice volatiles and citrus oils in the manufacture of flavorings, there are a number of specifications and analytical parameters that should be considered in the use of these raw materials. As in all flavor ingredients, the ultimate specification is the subjective evaluation of aroma and flavor by experienced tasters. Some may also consider the gas chromatography profile to be of particular significance.

3.5.6.1 *Aldehyde content*

This characteristic is of particular importance in lemon and orange oils where it is normally calculated as citral and decanal, respectively, as a measure of the aromatic character of the oils. As with many such measurements, the sample to sample variation is probably more significant than an isolated specific value.

3.5.6.2 *Refractive Index*

This measurement is widely used in the specification of flavor ingredients and essential oils and is often a good indicator of the change in composition. For accurate measurement, the refractometer should be used at a constant 20°C and monochromatic light from a sodium source should be used as the source of illumination.

3.5.6.3 *Optical rotation*

This physical property of substances containing asymmetrically substituted carbon atoms is widely used although more in the evaluation of essential oils than flavoring ingredients. A beam of polarized light usually from a sodium lamp source is passed through a tube containing the substance under examination. Rotation of the beam to the right (dextrorotation) indicated by a + value or to the left (levorotation) and indicated by a − sign is then used as a characteristic of the sample. The quantitative value is in degrees of rotation but reported for a tube containing 100 mm of the liquid under examination. If shorter or longer tubes are used the value is corrected back to 100 mm.

3.5.6.4 *Residue on evaporation*

This characteristic is not widely used in the evaluation of fruit essential oils and volatile fractions, but may be useful in detecting the additions of adulterants.

A weighed quantity of oil is heated on a steam bath for 5 h followed by 2 h in an oven at 105°C. During the heating it is possible to detect, by smelling, unusual volatile substances and any residue will often indicate whether adulterants are present.

3.5.6.5 *Organoleptic quality*

This is possibly the most important if subjective evaluation of a fruit oil or essence because it is this characteristic that determines the use to which the material may be put. In the case of fruit juice volatiles, a 1.0-ml sample is added to 100 ml of a solution containing 125 g/1 of sucrose and 2 g/1 of citric acid. The final solution is tasted against a standard and evaluated both qualitatively and quantitatively.

In the case of citrus oils, 1.0 ml of a solution of 2% of the oil in either ethanol or isopropanol is added to 100 ml of the above sucrose/citric acid solution and again evaluated by taste.

3.5.6.6 *Peroxide value*

This value is of significance for citrus oils generally as an indicator of the amount of deterioration undergone by the oil. It is often an indication of its age.

3.5.6.7 *Ester content of volatile fractions*

The ester content of the volatile fractions of fruit juices is determined by converting the esters to hydroxamic acids which are then reacted with the appropriate reagent and determined colorimetrically.

The above characteristics may not individually be of great significance but, as with analytical parameters for juices themselves, may be used to build up a clear picture of the juice volatile or citrus oil. Against a databank of historical information they will, in experienced hands, give surprisingly clear information about origin, variety, adulteration, etc.

The above measurements are not an exhaustive list and the flavorist may well wish to consider other examinations such as the ultraviolet absorbance of citrus oils and the profile by gas liquid chromatography in order to broaden the picture still further.

3.6 VOLATILE COMPONENTS OF FRUIT JUICES

3.6.1 *Production*

So far this chapter has dealt with the production of different kinds of juices and some of their characteristics. A vital aspect of the selection and processing of fruit juices for use as flavorings is the isolation and use of the volatile components. In the case of citrus where oil components are also present the volatile components are usually subdivided into oil phase volatiles and water phase volatiles. In most other fruits the volatile substances are obtained only from the concentration of the juice and are thus known as water-phase volatiles.

These volatile fractions of juices have long been recognized as containing a major part of the distinctive aroma and flavor of a juice but it was not until 1944 that a successful process was developed in the United States for their recovery from the concentration process. The equipment then known as an "essence recovery unit" is now commonplace in many of the juice processing plants around the world. A typical aroma recovery unit is shown in diagrammatic form in Figure 3–4.

As indicated above, a typical aroma or essence obtained from a fruit juice contains a large number of individual substances and as techniques for their detection and recognition become more sophisticated, the number increases. The number of substances present may vary from 50 or less to several hundred with typical levels ranging from 1 ppb to 50 ppm. Control of the variation of these components is a major problem in the use of fruit juice volatiles (see Section 3.7).

Many individual components have a low solubility in aqueous systems and, when concentration occurs to the typical "100" fold volatiles, physical separation may sometimes occur. However, solubility of components in the concentrated volatile fraction is usually enhanced by the presence of ethanol, which, after wa-

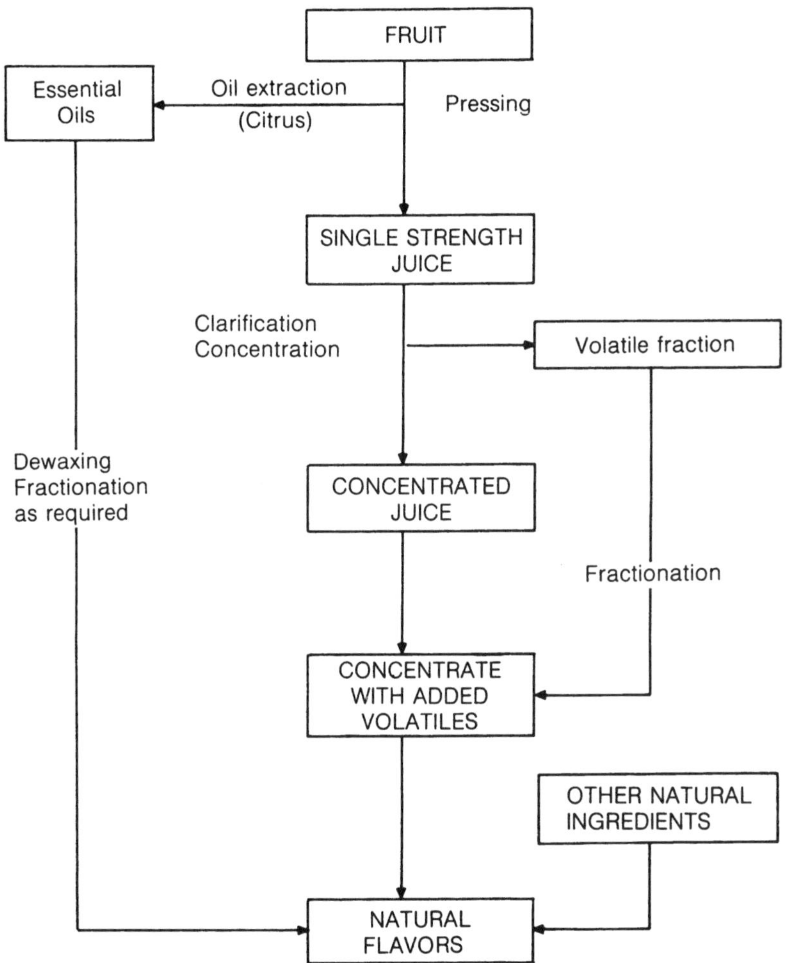

Figure 3–4 Block diagram of fruit processing operation to produce natural flavorings.

ter, is frequently the major component of juice volatiles. Ethanol levels can range from as low as 1% v/v to 15% or even 30% v/v in extreme cases. The amount of ethanol is often a useful indicator to the quality of the juice prior to concentration as it normally arises from fermentation by yeasts of the natural sugars in the whole juice.

Most volatile fractions would be expected to contain at least 1% of ethanol and values up to 5 or 6% would be considered both normal and acceptable. Above 6%, many consider that the amount of fermentation that has occurred will have materially affected the characteristics of the aroma although the net result may still be acceptable to the flavorist. From a commercial standpoint, ethanol content in volatile fractions can have a dramatic effect on costs in countries (like the United Kingdom) where levels of excise duty are high. There can often be a significant commercial benefit in reblending the volatile fraction with concentrated juice before importation because this can usually be arranged to reduce ethanol levels to below the threshold of excise duty. After ethanol, all the aroma substances together rarely exceed 0.5% of the concentrate.

Stability of the various volatile fractions from different fruits may vary considerably although this can be significantly influenced by the choice of plant design. In an appropriately designed unit, thermal stability of volatiles is usually excellent although even in the most sophisticated plant, substances such as ethyl acetate will usually undergo at least partial hydrolysis and this can result in changes in the aroma if the hydrolyzed esters have a significant effect on the flavor.

An important aspect of the collection of juice volatiles is whether or not their components have any tendency to form azeotropic mixtures with water.

Class I juices contain flavor constituents that exhibit strong azeotropic properties (e.g., Concord grape, pineapple). Class II contains important flavor constituents exhibiting weak azeotropic properties (e.g., strawberry). Class III contains no important flavor ingredients that exhibit azeotropic characteristics (e.g., apple, pear).

For the design of equipment to collect volatile fractions which exhibit any azeotropic properties it is necessary to have information about the azeotropic equilibria between the volatile substances and water, as well as with juice. It also follows that a system designed to collect volatiles from a specific juice may not necessarily perform well with other juices. This may be important in determining the quality of volatiles for flavoring uses.

When juices are processed through a volatile recovery system it is, under some circumstances, possible to cause chemical changes to the non-volatile residues (the concentrated juice) that cause undesirable changes in color (browning) and the production of unwanted cooked taste characteristics.

In early processes, recovery of volatile components involved heating the juice with between 5 and 50% of the volume being flash-evaporated at atmospheric pressure. Vapor removed in this way was then fed to a distillation column where rectification took place to yield an "essence" of about 1/150th of the volume of the feed juice. Current techniques involve flash distillation of a juice under vacuum to remove volatile substances before concentration of the non-volatile ingredients to the required level (also under vacuum).

The volatile substances (or essences) are not only used for the manufacture of natural flavorings but also for re-addition to concentrated juice such that on reconstitution, more of the aroma and taste of the original juice is recreated. This effect is also achieved by over-concentrating a juice and adding single strength juice to the required level of concentration. Such a product when re-constituted also shows flavor characteristics that are greatly enhanced compared with vacuum concentrated products.

3.6.1.1 *Fruit volatiles by "front end stripping"*

Although the majority of fruit juice processing plants utilize removal of the volatile fraction from the expressed, clarified juice at the initial stage of the concentration (evaporation) process (see Figure 3–4), it is evident that the quality of the isolated aroma volatiles will reflect, as referred above, not only the type, but the condition of the fruit at the point of processing and also the way in which the processing has been carried out. From the moment the fruit has been pulped, during enzymation and the subsequent pressing stage, there will also be changes occurring due to the disruption that has taken place and the consequent release of indigenous enzymes.

As already discussed, the presence of ethanol relates to breakdown of natural sugars present in the juice. It is unlikely that ethanol can be completely eliminated from the recovered volatiles, but providing the fruit is carefully selected for soundness, a more authentic product can be obtained by stripping the volatiles directly from the pulped fruit prior to enzymation and pressing. This approach will minimize the formation of other "breakdown" products, including methanol which also arises during the "traditional" process from the enzymatic destruction of pectin in preparation for pressing of the fruit pulp.

The removal of aroma volatiles by pulp stripping, or front end stripping is, as one might expect, subject to a great deal of specialization in terms of plant requirements. The objective of front end stripping is to remove volatiles with the minimum of damage to flavor components and to the remaining fruit itself; it is therefore important to pay special attention to the input of heat throughout the stripping process to avoid localized effects and burning of the product.

Front end stripping is best geared to a continuous pulp feed. Tubular or scraped surface heat exchangers are used for pre-heating the pulp which is fed at constant velocity into a separator (Figure 3–5) from which the volatile fraction is collected via a chilled condenser system.

Design of the plant is critical; rate of heat input is carefully controlled to enable a steady state and optimum level of volatile strip to be reached for a particular fruit system. Separator designs vary from the simple cylindrical chamber with side entry and steady input rate of heated pulp to high speed tangental entry units to create a thin film effect and therefore increased efficiency into the removal of volatiles, or to the more sophisticated counter current column systems providing multi-stage effects, of which the Australian spinning cone column is a prime example (Figure 3–6).

The spinning cone column (SCC) utilizes the thin film effect of pulp/liquid flowing downward over a spinning cone assembly against an upward gas or vapor

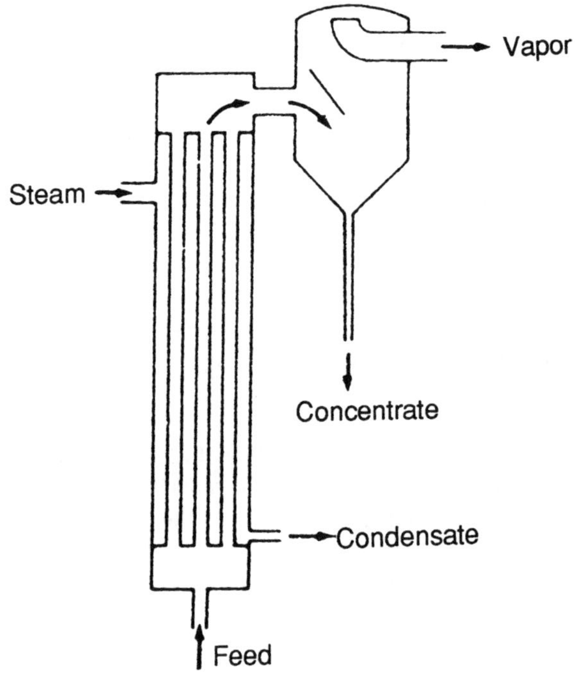

Figure 3–5 Separator assembly.

flow (usually steam). The volatile fraction is removed at the top of the column via a condenser system. Residence time for fruit pulp in the column is quite short, and can be varied, as can also the gas/vapor phase. Plant parameters can be adjusted to requirements and generally the system can run at a lower pulp entry temperature than the more conventional designs. The efficiency of separation is impressive; the degree of evaporation required for 90% removal of the total fruit aroma as calculated from the relative effective volatilities with respect to water is shown in Table 3–3. It is evident that the ease of removal of the volatile fraction is fruit specific. Conventional procedures already described rely upon further rectification in order to concentrate the single stage volatiles fraction, particularly with soft fruits such as strawberry, raspberry and so on, and, although this can be achieved, the overall efficiency of the process will suffer due to a certain level of degradation of some volatile components. The column technique avoids this problem and is capable of producing authentic and entirely characteristic natural flavorings.

Fruit juice and fruit volatile thus possess an accessible source of natural flavoring chemicals and as separation techniques improve it is highly likely that,

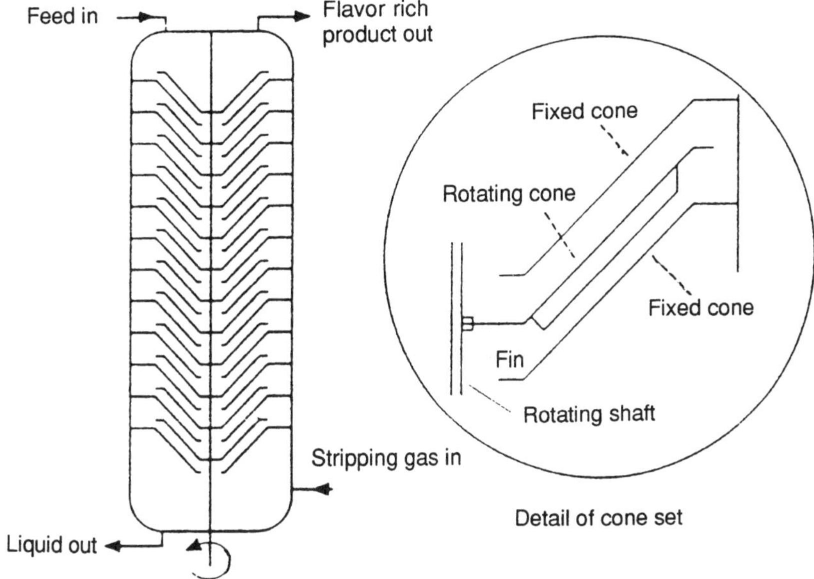

Figure 3–6 Australian spinning cone column.

whereas at present the volatile fractions comprise a mixture of these ingredients, in the not too distant future it will be possible to isolate individual ingredients on a commercial scale for a wider use within the flavor industry.

3.6.2 *Composition of Fruit Juice Volatile Fractions*

The broad composition of the volatile fractions of fruit juices was referred to in Section 3.6.1. The qualitative composition of the actual volatile substances is now covered in greater detail here. As previously indicated, a large number of substances are usually present in the volatile fraction obtained during concentration of fruit juices. The substances found are principally esters, alcohols, car-

Table 3–3 Degree of evaporation required for 90% removal of total fruit aroma

Fruit type	*Single stage (%)*	*SCC-multi-stage (%)*
Apple	10	0.3–1.0
Orange	20	1.0–2.0
Grape	42	2.0–3.0
Apricot	55	3.0–4.0
Strawberry	82	5.0–6.0

bonyl compounds and free (volatile) acids. Most of these substances are of relatively simple structure and a list, not exhaustive, of the commoner components found in many juice volatiles appears in Table 3–4.

The pattern of substances found in the volatile fraction of a juice is characteristic of a number of factors.

3.6.2.1 *Components that are characteristic of the genus*

Clear differences occur in the pattern of volatile substances that differentiate, for example, apple juice volatiles from those of pear or blackcurrant juices.

3.6.2.2 *Components characteristic of the species*

Marked differences occur in the pattern of volatile substances that may be obtained from different varieties of fruits.

3.6.2.3 *Components characteristic of processing methods and/or technology*

Quite large differences may occur if the same variety of fruits are processed in different ways in the same factory or in different factories using various technological alternatives.

Table 3–4 Typical components of juice volatiles

Alcohols	*Acids*	*Carbonyl compounds*
Methanol	Formic acid	Formaldehyde
Ethanol	Acetic acid	Acetaldehyde
n-Propanol	Propionic acid	Propanal
*Iso*propanol	Butyric acid	*n*-Butanal
n-Butanol	Valeric acid	*Iso*-butanal
2-Methylpropan-1-ol	Caproic acid	*Iso*-valeraldehyde
n-Amyl alcohol	Caprylic acid	Hexanal
2-Methyl butan-1-ol		Hex-1-en-2-al
n-Hexanol		Furfural
		Acetone
		Methyl ethyl ketone
		Methyl propyl ketone
		Methyl phenyl ketone
Esters:		
Amyl formate	Methyl acetate	Ethyl acetate
n-Butyl acetate	*n*-Amyl acetate	*Iso*-amyl-acetate
n-Hexyl acetate	Ethyl propionate	Ethyl butyrate
n-Butyl butyrate	*n*-Amyl butyrate	Methyl *iso*valerate
Ethyl-*n*-valerate	Ethyl caproate	*n*-Butyl caproate
Amyl caproate	Amyl caprylate	

It has already been indicated that if a juice is allowed to stand for some time between pressing and pasteurization, naturally occurring yeasts that are present on the fruit will begin to ferment the juice. This not only gives rise to the development of ethanol but also to a range of substances that are characteristic more of the fermentation products than of the juice itself.

Where relatively large amounts of ethanol develop, further infection, by, for example, acetobacter, can, in the right conditions, lead to the production of significant amounts of volatile acids, especially acetic acid. The presence of these two substances in some quantity can lead to further amounts of esters being produced or the transesterification of esters already present.

Further changes may occur within a specific concentration plant.

With all the possible areas for variation of the volatile components in a fruit juice, the flavorist may feel that it is almost impossible to obtain a consistent quality of volatile fraction for use in a formulation. In practice, however, individual processing plants are able to exercise a high degree of consistency.

Control of quality of volatile fractions of juices is almost always achieved by a combination of methods. Gas chromatography (GC) is often employed to give a "fingerprint" of the active flavor ingredients present although the final selection and use of a consignment of fruit juice volatiles is by subjective evaluation of aroma and flavor. Aroma, as with other aromatic substances, is typically evaluated by use of the "smelling strip," an absorbent strip of firm paper which is dipped into the mixture of volatile materials and then smelled at regular intervals. Flavor is typically evaluated by experienced taster(s) using a solution of 12.5% sucrose, 0.25% citric acid anhydrous and up to 0.5% of the mixture of volatile substances.

3.7 THE USE OF FRUIT JUICES IN FLAVORINGS

3.7.1 *Fruit Juice Compounds*

This chapter is not devoted to the application of ingredients but the use of juice concentrates in the manufacturing of fruit juice compounds for use in the preparation of juice containing beverages, an extremely important sector of their use. In the manufacture of a beverage, typical constituents in descending order of quantity would be water, sugar, fruit juice, carbon dioxide, citric acid, flavoring, ascorbic acid, coloring and, in some cases, preservative.

The function of a compound is to pre-mix all the required components, other than sugar, water and carbon dioxide, in a concentrated form. This enables the bottler to purchase a compound to which he or she adds sugar, water and carbon dioxide in order to obtain the finished beverage.

A 20-fold compound for a 10% juice drink would, for example, mean that by the addition of appropriate amounts of the missing ingredients, 1 liter of compound would contain sufficient of the key components to make 20 liters of finished beverage. A typical composition of the compound would be:

Orange juice concentrate 60° Brix	300 liters
Citric acid anhydrous	50 kilos
Natural cloudifier	20 liters
Natural orange flavoring	20 liters
Coloring	q.s.
Ascorbic acid	q.s.
Preservative	q.s.
Water to	1000 liters

Thus the compound would contain 300 liters of juice at 60° Brix (0.772 g/liter of solids), i.e., 231 g/liter of orange juice solids.

Assuming single strength juice to contain 115 g/liter of solids, 1 liter of compound would contain $(231 \times 100)/115 = 200\%$ juice solids. This diluted by a factor of 20 would give 10% juice solids in the finished drink.

The market span for beverage compounds in the United Kingdom is probably greater than in almost any other country in the world. This arises from the complexity of former U.K. Soft Drink Regulations which allowed for both concentrated soft drink and ready to drink versions.

A particular category in both strengths is for products containing comminuted citrus bases, the so-called whole fruit drinks. A variety of appropriate compounds are available in different strengths with all the ingredients necessary except water, sweetener and, where appropriate, carbon dioxide. Comminuted citrus bases (see Section 3.3.2) contribute fruit to this product but also contain dispersed essential oils to enhance the flavor as well as peel and albedo components to enhance cloud and stabilize the overall beverage.

3.7.2 *Flavorings*

Fruit juice components are widely used in flavorings as such. They are the mainstay of natural flavorings of many types. The compositions of individual formulations are confidential, but typical recipes would be based on a high proportion of a fruit juice concentrate to which would be added appropriate juice volatile fractions as well as individual flavor components. For natural flavorings, these individual ingredients would be separated normally by physical processes such as fractionation.

A surprising degree of control may be exercised by the use of these techniques although, as indicated, consistency of supply from selected processing plants is an essential component of consistent flavor. A list of the typical components of fruit juice volatiles appears in Table 3–4.

3.7.2.1 *The use of fruit juices as flavoring components*

Single strength fruit juices are not widely used as ingredients of flavorings although concentrates and fractions obtained during processing are very important in terms of the quantity used.

Fruit juices and, to some extent, purees and pulps are, however, often used in various finished food and beverage products as part of an overall flavor system. Whole fruit is an essential part of the manufacture of jams and conserves; it is impossible to distinguish the effect, in products of that nature, of the fruit itself and the flavor it brings to the product. Fruit juices are nevertheless added to products for a number of reasons. From a marketing standpoint it is often valuable to be able to claim that a product contains a certain percentage of fruit juice.

A related reason is the enhanced nutritional status that fruit juice content brings to a product. Many authorities are critical of some products, e.g., flavored soft drinks which are described as having empty calories. The addition of fruit or fruit juice largely dispels such criticisms. The product with, say, a 10% juice content is seen as having a significant nutritional benefit when compared with simply flavored products.

The addition of fruit juices to food and drink products whether beverages, ice cream and other dairy products, or other foods, is almost always by use of concentrates appropriately reconstituted. Single strength juices are rarely used in products because of the very high cost of transporting them to the point of use and the volumetric problems often posed by most aspects of their use.

3.8 SUMMARY

Fruit juices are normally considered in their own right or at least as direct ingredients of food or beverages and much of this chapter is relevant to them in any of these applications as well as ingredients of flavorings. There is much published on fruit and fruit processing although even this is sparse when their economic importance is taken into account. Much fruit juice technology remains in the hands of engineering companies and those associated with growing and processing fruit and is often considered commercially sensitive. The determined researcher can normally have access to most of the relevant data in a given area of juice technology but it remains surprisingly difficult to find a collected concise source of relevant basic data on the subject.

This chapter has therefore, examined some of the more important methods of manufacture of fruit juices, their composition and some methods of analysis. It has outlined the main forms of packaging used and parameters for specifications. The objective has been to provide the flavorist, applications technologist and others involved with flavorings with the basic information to enable a more systematic approach to be made to the procurement, evaluation and use of fruit juices.

The information is slanted toward juices for use in flavorings but is equally relevant in most cases to other juice uses.

REFERENCES

V.L.S. Charley, *Food Technol.* 18 (1963) 33.

J.F. Kefford, H.A. McKenzie and P.C.O. Thompson, *J. Sci. Food Agric.* 10 (1959) 51.

Methods of Analysis, International Federation of Fruit Juice Producers, Zug, Switzerland, various dates.

Official Methods of Analysis, 14th edn. A.O.A.C., Arlington, VA.

R.M. Smock and A.M. Neubert, in *Apples and Apple Products,* Interscience Publications, London (1950).

The Soft Drinks Regulations, H.M.S.O., London, Statutory Instrument (1964) No. 760.

The *U. K. Fruit Juices and Nectars Regulations,* H.M.S.O., London, Statutory Instrument.

J.P. Van Buren and W.B. Robinson, *J. Agric. Food Sci.* 17 (1969) 772.

K. Wucherpfennig, P. Possmann and E. Kettern, *Flussiges Obst.* 39 (1972) 388.

FURTHER READING

J.B. Redd, C.M. Hendrix and D.L. Hendrix, in *Quality Control for Processing Plants,* Intercit Inc. (1988).

R.S.K. *Values, The Complete Manual,* Association of German Fruit Juice Industry, Bonn, Flussiges Obst. (1988).

Chapter 4

Synthetic Ingredients of Food Flavorings

H. Kuentzel and D. Bahri

4.1 GENERAL ASPECTS

4.1.1 *Introduction, Definitions and Documentation*

The need for synthetically prepared flavor compounds arises from the fact that during the storage of foodstuffs a certain loss of flavor is inevitable. These losses can be compensated for by adding synthetically produced flavor compounds. Besides this, synthetic flavor compounds have the great advantage of being available in the required quantity and quality, irrespective of crop variation and season.

A constant quality permits standardization of the flavorings. Synthetically produced flavor compounds make it possible to vary the proportions of single components and thereby create new flavor notes.

In a description of this vast field, some definitions are essential. All substances added to a foodstuff are called ingredients, i.e., flavors, colors, emulsifiers, salt, etc. We distinguish between natural and synthetic ingredients according to their origin. In this chapter we concentrate only on flavor-enhancing or -modifying ingredients, and more specifically on the synthetic substances. Equally useful to circumscribe this subject are the definitions of IOFI (International Organization of the Flavour Industry) for natural, nature identical (synthetic) and process flavors. The anticipated new guidelines of the EEC will also be of great importance for future flavor definitions (see Section 4.3).

To illustrate the relationship between sensory perception such as flavor, taste and mouthfeel on the one hand and parameters related to the chemical structure such as volatility and molecular weight on the other, some examples of ingredients are listed in Table 4–1. In this chapter we concentrate mainly on volatile flavor components, thus omitting the polar and non-volatile ingredients which are responsible for taste and mouthfeel.

Table 4–1 Classification of food ingredients

Volatility	*Organoleptic function*	*Molecular weight*	*Examples*
High	Flavor, taste	< 150	Esters, ketones, simple heterocycles
Low	Taste, flavor	< 250	Carbonic acids, amides, extended C-skeletons
None	Taste	$50 \simeq 10{,}000$	Sugar, amino acids, SMG, salt, nucleic acids, bitter agents, natural sweeteners
	Mouthfeel	> 5000	Starch, peptides
	Unknown	$> 10{,}000$	Biopolymers, melanoids

There are today several thousand flavor compounds known and described in a multitude of books, papers, patents, etc., and an efficient way of documentation is therefore mandatory. To fulfill the different needs of users, a number of specialized documentation systems were created. A most valuable tool is *Volatile Compounds in Food*, a compilation of about 5500 compounds found in 255 food products. This list edited by the TNO, Division for Nutrition and Food Research (Zeist, The Netherlands), is annually updated by a supplementary Volume (van Straten & Maarse, 1988).

Another regularly updated work is the so-called GRAS-List, a world-wide reference list of materials used in compounding flavors and fragrances with sources of supply and data of the legislative status of each compound (Chemical Source Association, 1989).

A number of classical books listing flavor and fragrance compounds, their use and description should also be mentioned here. This list, by no means exhaustive, contains such well-known names as Arctander (1969), Fenaroli (1975), Gildemeister and Hoffmann (1968) and Guenther (1952). In more recent years several authors have covered the whole field of ingredients and composition in a comprehensive manner (Birch & Lindley, 1986; Ziegler, 1982; Heath, 1978, 1981; Heath & Reineccius, 1986).

Another source of information on synthetic flavor components is the catalogues of the flavor compound producers (a list of which can be found in Chemical Source Association, 1989). In addition to the information in these sales catalogues a wealth of know-how and research results can be retrieved from the appropriate patents of these companies. The patent literature can be found in *Chemical Abstracts*, a documentation system which will be further explained below.

A review work covering a large range of relevant results is edited by CRC Press. In the periodical entitled *Critical Reviews in Food Science and Nutrition*, each year about a dozen articles summarize a topic of the indicated field (Clydesdale, 1989). For more than 40 years, Bedoukian (1989) has written a concise annual review article about perfumery and flavor materials.

A very extensive information system (including hard copies, micro films or electronically stored data) is offered by Chemical Abstract Services. It is possible to retrieve, in an efficient way, specific data concerning synthetic, analytical or physical results. A search can be initiated by chemical formulae, key words, name of author(s), date of publication, etc. The possibility of combining any of these questions further enhances the value of this data system (Chemical Abstracts Service, 1989).

Additionally, there exists a broad variety of journals covering scientific and applied aspects as well as marketing and business questions of foodstuffs.

4.1.2 *Flavor Generation*

To gain a better insight into food flavorings it is important to understand the generation of flavors in foodstuffs. Four methods of formation can be distinguished:

enzymatic
non-enzymatic
fermentative
autoxidative } formation of flavor compounds

In Section 4.2.3, some examples of each type are covered in more detail but a more general overview has to suffice here. With the enzymatic formation it is important in the first instance to know from which metabolic cycle the flavor compounds in question originate. In this context the most important metabolic cycles are those of fatty acids, amino acids, carbohydrates and terpenoids (the latter being secondary metabolites). Some general examples will further explain the situation.

The biosynthesis of fatty acids starts with acetic acid in an activated form. Chain elongation is effected by several enzymatic steps within the fatty acid-synthetase complex, adding a C_2-unit in each synthetic cycle, thereby producing even-numbered carbon chains. Hydrolysis and decarboxylation of the intermediate β-keto-ester generates the corresponding methyl ketone with one carbon less (Figure 4–1a). These ketones may be the starting material for the corresponding secondary alcohols, which in turn may be esterified. Fatty acids are generally accepted as the origin of these flavor-active compounds (Schultz et al., 1967; Tressl et al., 1970).

A simple and generally occurring transformation of amino acids to aldehydes with one carbon less is the so-called Strecker degradation. This reaction needs a carbonyl compound as counterpart for the transamination. α-Diketones are especially suitable for this sequence and give rise to α-aminoketones, which in turn are the origin of further reaction products (Figure 4–1b). With radioactive labeling, the relationship between the aldehyde formed and its parent amino acid can be proven as exemplified in Section 4.2.3.5. The so-called Strecker aldehydes are very important flavor compounds, either as volatile constituents themselves or as reactive intermediates for further transformation.

A very large and well investigated field within enzymatic flavor formation is that of the biosynthesis of terpenoids. The basic building blocks for the hundreds of known terpenoids are the "isoprene units" shown in Figure 4–2. This C_5-unit is generated from mevalonic acid, which in turn is formed from three acetic acid units with every single step catalyzed by a specific enzyme. The activated form of the C_2-starting material is a thioester of acetic acid with Coenzyme A (not shown in the figure). This is similar to the starting point of the biosynthesis of fatty acids. A pyrophosphate group (abbreviated as OPP) is the activated "leaving group" for the OH group used by nature. Decarboxylation of mevalonic acid drives the reaction to the unsaturated C_5-unit, which readily undergoes further reactions. Fusion of two or three such units in a head-to-tail manner gives rise to a carbon chain with a characteristic array of the extra methyl groups. Their pattern can be found in a plethora of monoterpenes (C_{10}) and sesquiterpenes (C_{15}), acyclic or cyclized, oxygenated or reduced, and is known to generations of chemists as the "isoprene rule" (Wallach, 1914; Ruzicka, 1953).

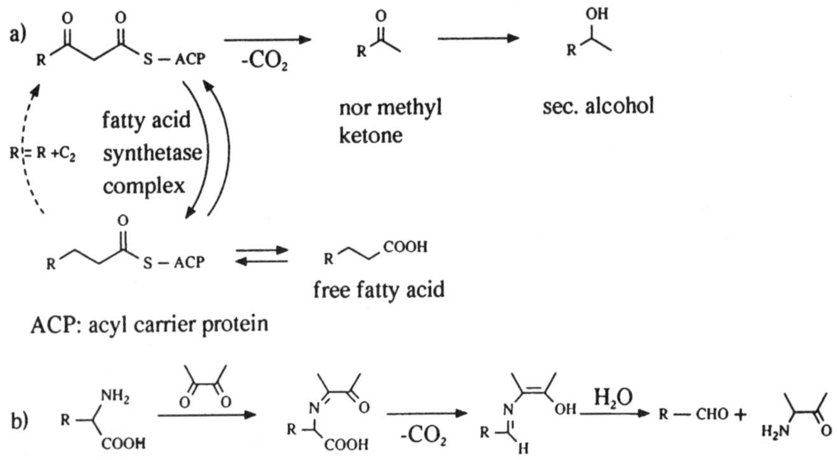

Figure 4–1 Generation of flavor compounds by different metabolic pathways: (a) from fatty acids (Schultz et al., 1967; Tressl et al., 1970); (b) from amino acids (Strecker degradation) (Belitz & Grosch, 1987).

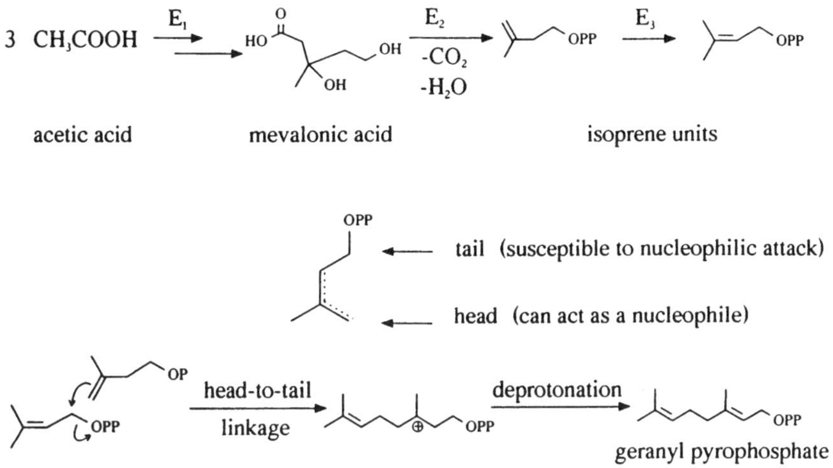

Figure 4–2 The first steps in the biogenesis of terpenes. E, enzyme or enzyme system; OPP, pyrophosphate.

Besides these "regular terpenes" a wealth of irregular compounds are formed, either by a Wagner-Meerwein type rearrangement or by fusion of the C_5-units in another way to the normal head-to-tail linkage. In Sections 4.2.3.3, 4.2.3.4 and 4.2.3.5 and in Figures 4–14, 4–15 and 4–20 some common examples of the biosynthesis of terpenoids are outlined. It is far beyond the scope of this article to give an overview of terpene biochemistry and reference to some general and review articles is sufficient (Solomons, 1988; Schreier, 1984; Manitto, 1981).

The generation of flavor compounds by a Maillard reaction (or a caramelization) is described as non-enzymatic formation. The Maillard reaction is characterized as the thermal reaction of a reducing sugar with an amino acid. Thereby an overwhelming number of substances are formed. Their range covers the simplest degradation products such as H_2O, NH_3, H_2S, together with typical flavor compounds such as furans and pyrazines as well as the highly complex brown pigments known as melanoidins.

Characteristic of this type of reaction is the rich variety of substances formed (a variety related to chemical classes as well as to homologous series) and the very low concentration of the single compounds (usually in the range of parts per million).

In the formation of substances by thermal processes during a non-enzymatic reaction, the only reactions to occur are those generally considered in a chemical sense as simple (e.g., eliminations, additions, isomerizations). The above introductions may be helpful to explain the formation of many of these flavor compounds, but are not adequate as a guide for their practical preparation. The main reason is the low specificity of such thermal reactions. Chemical syntheses are normally expected to proceed in a way that yields are several orders of magnitude higher than in the undirected manner of these thermal processes.

The exploration of the most abundant browning reaction goes back to 1912, when Maillard investigated this process. Since then many researchers have worked on the reaction and nowadays the first steps are fairly well known, but consecutive steps await further elucidation (Eriksson, 1981; Waller & Feather, 1983; Ledl et al., 1986). In Figure 4–3 a general scheme and a specific example of Mills and Hodge (1976) are given. The amine function of proline reacts with the carbonyl function of glucose and rearranges to the Amadori compound 1-desoxy-l-prolino fructose. In this model reaction system, a number of cyclic enolones could be identified, all containing the C_6-skeleton of the starting glucose.

It is easy to imagine that with a more complex mixture of starting materials (several sugars and amino acids) or under a higher reaction temperature, a very large range of products can be formed, partly by further degradation to smaller fragments, partly by incorporation of heteroatoms like nitrogen and sulfur.

Less well-known but nevertheless an important part of Maillard products are the "meta-stable" flavor compounds. Their restricted lifetime reflects the familiar time-dependent increase and decrease of flavor sensations during roasting, baking and so on. The caramelization reaction is the transformation of sugars or carbohydrates by strong heat treatment. Volatile components are important for the flavor whereas non-volatile substances form the brown color. The structure and

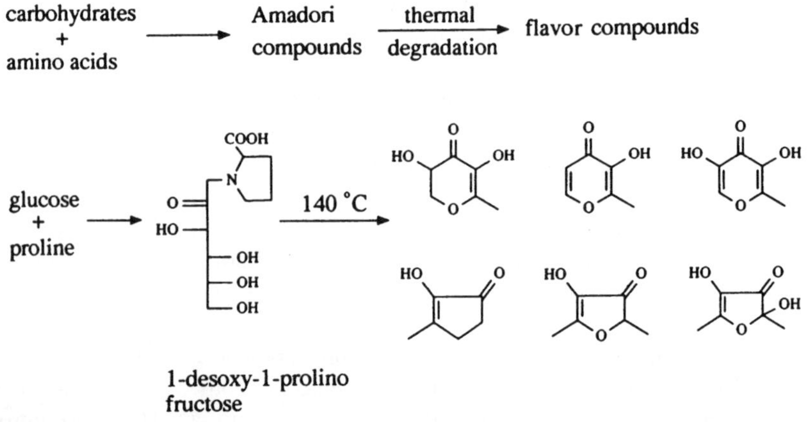

Figure 4–3 The Maillard reaction: general scheme and an example (Mills & Hodge, 1976).

formation of the products responsible for the desired effects of flavor and color are well documented (Houminer, 1973; Belitz & Grosch, 1987).

Fermentative formation covers classical processes such as the ripening of cheese, fermentation of wine and modern biotechnology. The old processes are based on ancient experience of mankind and probably also on trial and error. The modern processes of biotechnology, however, allow the production of high value products in a highly specific way. Biotechnology is a very fast growing science with an increasing influence on future food preparation and flavor generation (Moo-Young, 1985; Rehm & Reed, 1989).

Autoxidative formation is mostly unwanted because it normally gives rise to off-flavors. These are often indicative of deteriorated food and act as a sign of warning against consumption. Chemically, radical reactions with oxygen take place, which produce short chain carboxylic acids, aldehydes and ketones, responsible for rancid notes. These reactions occur especially with polyunsaturated fatty acids as substrates (Porter et al., 1981; Bemelmans & ten Noever de Brauw, 1975).

4.1.3 *Flavor Analysis*

Analytical chemistry is the basic science to elucidate new flavor compounds. The identification of a new compound is the successful end of an isolation procedure which is schematically shown in Figure 4–4. It is up to the skill of the analyst to choose the right conditions to concentrate, separate and isolate a new substance without its destruction. It can be a fascinating adventure to trace a new compound at the start of an analysis from its aroma until finally the spectroscopic data of an isolated sample allow elucidation of its chemical structure.

The relevance of a new flavor compound can be the novelty of its chemical structure or the importance of its odor note ("impact chemical"). For more detailed descriptions of separation procedures the reader is referred to several monographs (Maarse & Belz, 1985; Schreier, *Techniques of Analysis*, 1984,

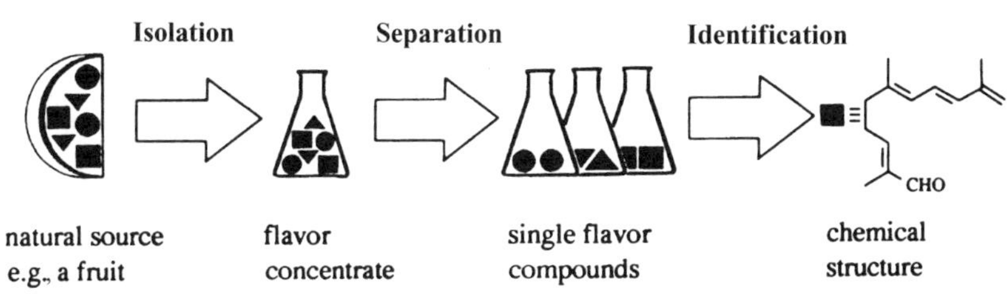

Figure 4–4 The principal steps in flavor analysis.

Analysis of Volatiles, 1984). Further references on instrumental methods and flavor analysis research in food science are also listed (Macleod, 1973; Adda, 1985; Berger et al., 1985).

A particularly efficient method for flavor research is the "headspace technique" combined with GLC-MS coupling. In this way it is possible to collect enough volatile materials from a handful of berries or flowers to determine the constituents. It was especially the combination of mass spectrometry (MS) with gas liquid chromatography (GLC) which brought about an enormous increase in sensitivity of detection in the 1970s.

A "first revolution" in the analysis of volatile compounds had been brought about in the 1950s by the development of GLC. Using this tool it was possible to separate and isolate tiny amounts of substances. It complemented existing procedures because, in this technique, separation is improved by using very small amounts. (For separation by distillation, the converse is true.) The technical improvements have reached an impressive level during the last 40 years. An apparatus of early days equipped with short, thick columns (e.g., 3 m $\times$ 5 mm) and a thermal conductivity detector had a practical detection limit in the range of milligrams to micrograms (1×10^{-3} to 1×10^{-6} g). Its modern successor equipped with a capillary column (e.g., 40 m $\times$ 0.3 mm) and a flame ionization detector (or a nitrogen or sulfur sensitive detector) detects traces in the range of picograms (1×10^{-12} g).

Another development of the 1980s is high performance liquid chromatography (HPLC) which makes possible separation at normal temperatures of thermally labile compounds in a way that is independent of volatility and polarity. Each method has its own complementary value within the repertory of analytical procedures.

The modern development of equipment and interpretation of results would not have been possible without electronically based data handling and storage. For example, the measured mass spectra of a substance can be compared within a very short time with thousands of reference spectra making possible a crude analysis (~90%) within a few days. Flavor analysis is carried out in the research facilities of universities (departments of Food Science, Nutrition, Agriculture and so on) or in those of the food and flavor industry. Research topics at universities are usually of a more academic and fundamental nature and the results are published in scientific journals, whereas industrial research is often more applied and the results are patented or sometimes temporarily withheld to maintain the "know-how." Continuous analytical research activities result in a steady increase of knowledge on constituents in all kinds of foodstuffs. The number of volatile compounds in food as published by TNO (van Straten & Maarse, 1988) amounted to 800 in 1967, tripled to 2400 by 1974, and reached 5500 in 1988 (see Figure 4–55). Thanks to new analytical techniques and instrumental methods there is no end to this development, but efforts to identify new and valuable impact chemicals are becoming greater and greater.

4.1.4 *Flavor Manufacture*

There are three principal methods to obtain flavor materials: to gain the substance from natural sources by physical methods such as extraction, steam distillation and so on; to produce it by a chemical synthesis; or to obtain the material as a reaction flavor. A number of criteria determine which method is applicable in a given situation.

For synthetic production of a compound the following conditions should be fulfilled:

(a) The chemical structure of the flavor compound responsible for the desired note must be known. Structural elucidation of the flavor compounds will be the target of flavor analysis, and in many instances this work will have been carried out and published (van Straten & Maarse, 1988). But in cases of very low concentration of a flavor compound, due to its low threshold value or in cases of a delicate balance of several flavor compounds, the indentification of the real impact chemicals can be very expensive or extremely difficult. Yet this is still the best way to elucidate structures of valuable flavor notes, because, on one hand, the pure "trial and error" approach is even more laborious and, on the other hand, there is still no generally valid correlation of structure and flavor. Hence the design of a molecule with tailor-made olfactory properties is not yet in sight. There are few examples of compounds obeying such a rule (cf. the molecular theory of sweet taste proposed by Kier [1972]) and they are rather the exception in structure-activity relationship in human olfaction (Amoore, 1982).

(b) The purchasing cost of the flavor material must not exceed a certain limit. More specifically, it is the utilization cost (i.e., the price per kilogram combined with the used dosage) that is important. In the free market every odor or taste has a price range depending on the available flavor materials. A new substance must be placed favorably in this price structure in order to be considered for introduction.

(c) The legislative situation must be clear. National regulations must be considered. In most countries it is easier to introduce a synthetic flavor compound, if it has been identified in one or more common foods, by the argument that intake of a foodstuff is a form of long-term toxicological observation.

(d) Last but not least, the consumer of the flavored end-product must accept and buy it.

If the above considerations are all in favor of synthetic production, then the synthetic work in its strict sense may start. Synthetic chemistry is a science of its own, and only some outlines are given here. Theoretical knowledge, practical expertise and endurance are essential for a successful synthesis. Normally three phases are carried out: from the laboratory (gram scale preparation) to the pilot plant (kilogram scale) and to the production plant (up to ton scale). Each stage has its own problems and difficulties, but let us assume that the desired flavor compound is finally produced and is now at the disposal of the flavorist.

It is estimated that about 80% of all flavorings sold contain at least one synthetic compound, and this indicates the importance of synthetic flavor ingredients.

4.1.5 *Composition and Formulation*

The development of powerful techniques like GC, coupled with GC-MS and HPLC in the last two decades, has led to the identification of hundreds of individual flavor components in food and beverages. The work of flavorists is gradually moving away from the traditional approach of composition by imagination and subjective interpretation of flavor ingredients to the current use of formulation, and in the future flavor design. The blending together of flavor ingredients to create a satisfying flavor still remains a highly sophisticated process based on a long practical experience coupled with a very deep knowledge of the subject. Flavorists and flavor chemists have to collaborate very closely to achieve a fully satisfying flavor, capable of meeting all the required processing conditions.

To meet specific applications, tailor-made flavors may seem to some to be today only a matter of raw material purchasing; actually much work remains to be done to reach this aim. Substances identified in food have to be evaluated, either singly or in a mixture according to their threshold level, toxicologically tested and correctly synthesized in an economical way. More trace impact substances occurring in natural products have yet to be characterized and identified.

Until this work is done, the flavorist will continue to solve his or her problems by using global ingredients as bases in combination with single ingredients, although the computer aided flavor creation (CAFCR) is in view. In the future more and more synthetic flavor ingredients will play a very important role (Figure 4–5).

4.2 SYNTHETIC FLAVOR INGREDIENTS

4.2.1 *Classification*

The many reasons for use of synthetic flavor ingredients are basically the same as those for the use of flavors themselves (enhancing, replacing, economical price, varying, rounding up, masking, etc.).

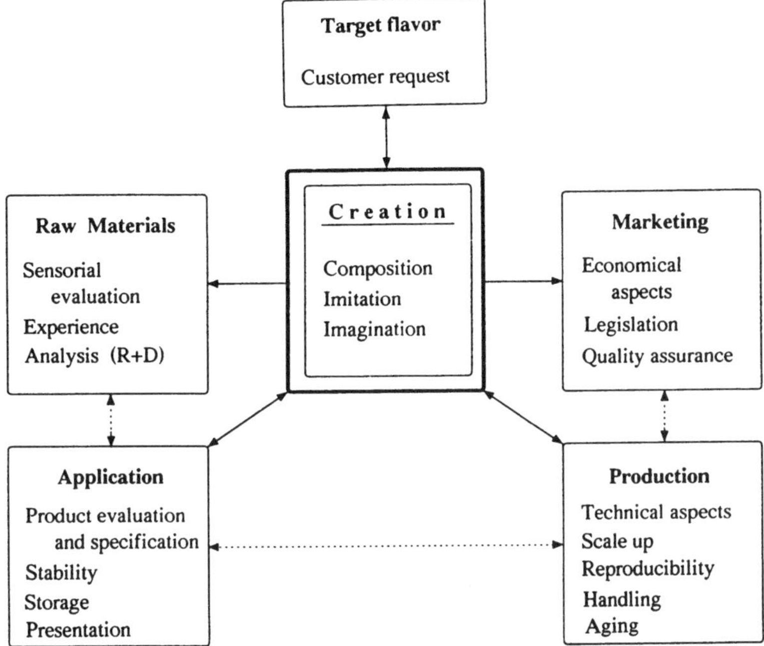

Figure 4–5 The flavor creation.

Synthetic flavor ingredients, i.e., nature identical products, cover the whole range of organic substances but no obvious relationship has been established between structure and flavor properties. Some components of similar structure have broadly similar odors but there are many exceptions. The experience of many flavorists "in the field" is rather discouraging, because the same molecule may be differently perceived at different concentrations.

One of the many possible ways to describe flavor substances or ingredients used in the process of flavor composition is to classify them in groups with parent flavor characteristics as shown in Figure 4–6 (flavor wheel). The same component which belongs to a typical group may be used in quite different flavors. This flavor wheel is not always representative because every flavorist has his own subjective perception of flavor notes, but experience shows that compositions achieved in this way always give similar overall results.

About 2000 single synthetic ingredients are known to be used in flavor compositions, most of them having been identified in natural materials. A small number have not yet been found in nature but are recognized as safe and therefore permitted by most countries (e.g., ethyl vanillin, ethyl maltol). Another 3000 substances

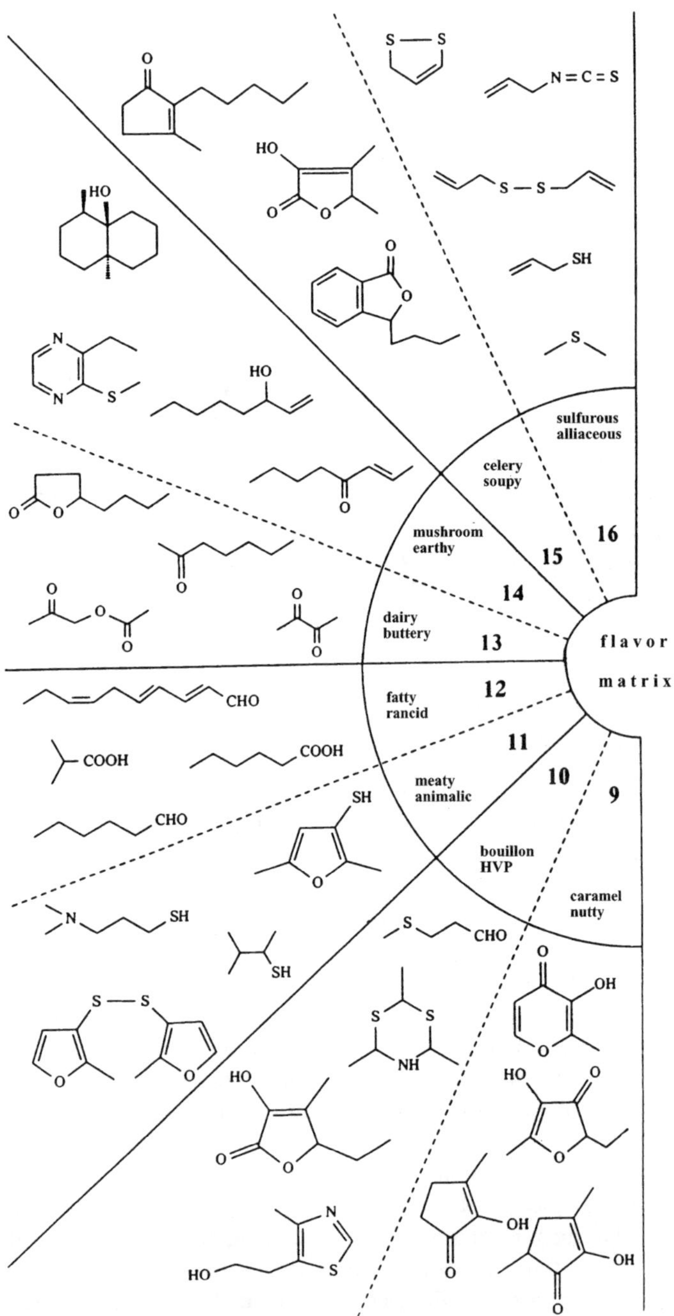

Figure 4–6 The flavor wheel.

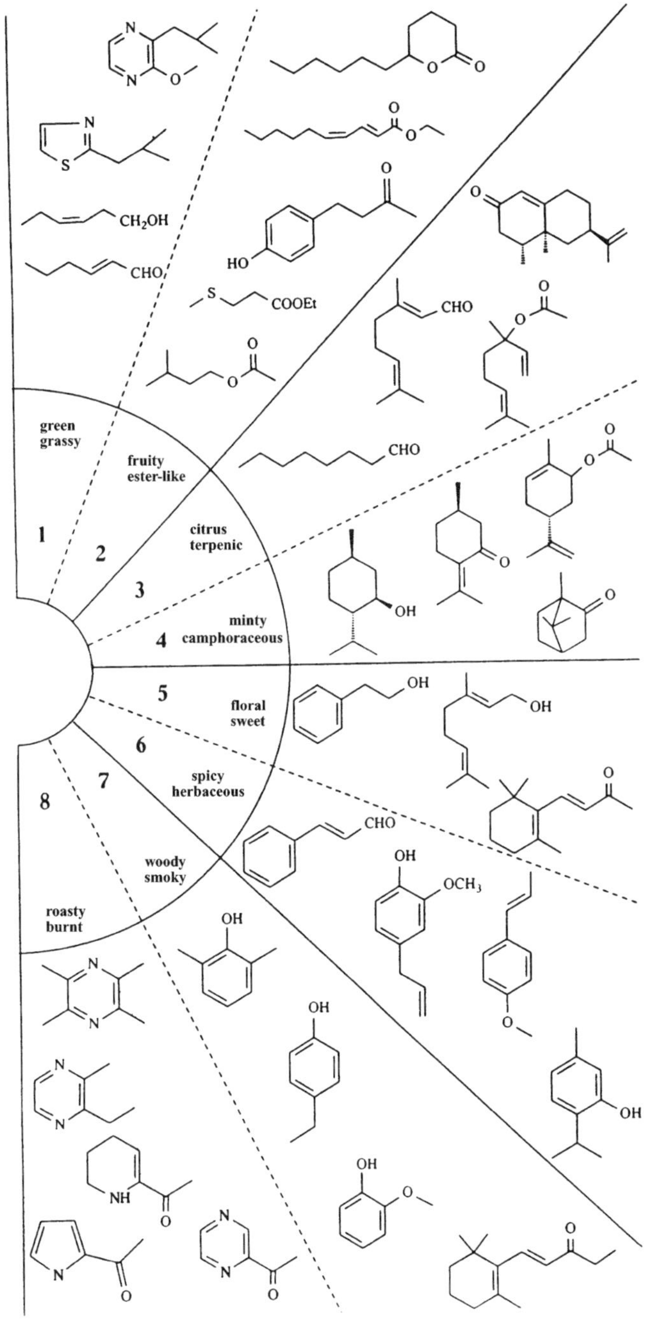

1 green grassy

2 fruity ester-like

3 citrus terpenic

4 minty camphoraceous

5 floral sweet

6 spicy herbaceous

7 woody smoky

8 roasty burnt

arising from new raw materials such as process or biotechnological flavors are used as building blocks. As mentioned above, flavor notes are not necessarily bound to specific structures of chemical classes. For flavorists, classification according to sensory properties is more important and we discuss the different synthetic flavor ingredients according to their properties as shown in Figure 4–6.

4.2.2 *The Flavor Wheel*

The flavor wheel is a pictorial illustration of some basic flavor relations. At the center is the flavor matrix which represents the body of the flavor to be created. It contains all the ingredients needed to support, dilute, enhance and protect the single flavor components, which cannot be applied in a pure state.

Depending on the application, different carriers (solid, liquid, polar, apolar) and enhancers (sweet, savory, salty) as well as intermediate rounding up compounds (extracts, building blocks) are widely utilized. Around the flavor matrix, segments are arranged according to the individual flavor notes and illustrated by some typical examples. The sequence of the segments is chosen in such a way that the flavor notes are adjacent to kindred characteristics, comparable to the array of the spectra colors in a rainbow. In principle a global flavor perception can be partitioned into a number of "pure" flavor notes, and rearranged again by composing these notes in the right proportions in an appropriate flavor matrix. Theoretically a correct flavor profile description is simultaneously a recipe for the reconstruction of this flavor. The flavor wheel will give some insights into the most important flavor notes with their possible reactions. This may be useful for flavor chemists but is not intended to replace the experience and creativity of flavorists.

The sections treating the flavor notes from 4.2.3.1 to 4.2.3.16 give a short general description of each flavor, some information about occurrence and origin of typical flavor compounds and examples of their use. Paragraphs and schemes on biogenesis and synthesis of specific examples conclude each flavor segment.

4.2.3 *The Different Flavor Notes*

4.2.3.1 *The green grassy flavor note*

The green flavor note is described as the odor of freshly cut grass or ground leaves and green plant materials. Typical substances representing this class are short chain unsaturated aldehydes and alcohols such as *trans*-2-hexenal and *cis*-3-hexenol. Esters and heterocycles like alkyl substituted thiazoles and alkoxy pyrazines with very low flavor threshold levels also belong to this group (Figure 4–7).

Figure 4–8 shows the well-known biogenetic pathway for the formation of C_6-unsaturated aldehydes and alcohols (Tressl et al., 1981). *Cis*-3-Hexenal, *trans*-2-hexenal, *cis*-3-hexenol and *trans*-2-hexenol are very important impact substances occurring in many fruits and vegetables (apple, tomato, grapes, etc.). The enzy-

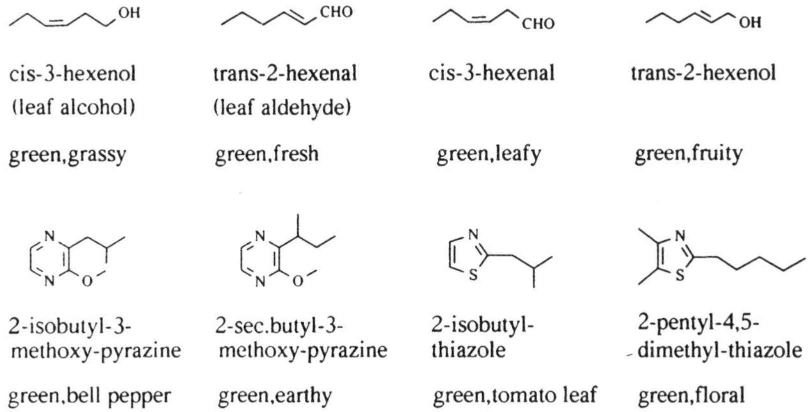

Figure 4–7 Examples of green grassy flavor notes.

matic formation of thiazoles and pyrazines is suggested as originating from amino acids, but their biogenetic pathways have not yet been clarified.

Many other compounds like esters, acids and terpenoids could be placed in this group as "rounding up" green notes (e.g., hexyl 2-methyl-butyrate, α-pinene). Only *trans*-2-hexenal and *cis*-3-hexenol are commonly used in flavor compounding because of their stability and commercial availability. 2-Isobutyl thiazole is

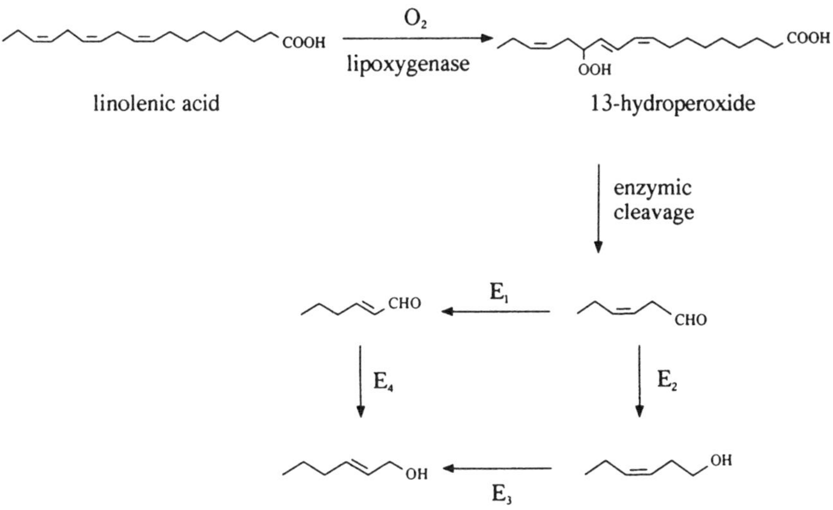

COOH
O_2
lipoxygenase
COOH
OOH
linolenic acid
13-hydroperoxide
enzymic
cleavage
CHO
E_1
CHO
E_4
E_2
OH
OH
E_3

Figure 4–8 Enzymatic formation of C_6-compounds from linolenic acid (Tressl et al., 1981). E_1, E_2, . . ., different enzyme systems.

used only in specific notes (e.g., tomatoes). Figure 4–9 shows synthetic routes for the preparation of *cis*-3-hexenol and 4,5-dimethyl-2-pentylthiazole, respectively. The synthesis of *cis*-3-hexenol (Figure 4–9a) is straightforward, the important step being the selective hydrogenation of the triple bond to the *cis* isomer (Kerr & Suckling, 1988). Several different preparations are known today. The first manufacturing method introduced in the early 1960s started a "green period" in the flavor and fragrance industry. The annual world consumption of leaf alcohol is estimated to be 40 tons (Bedoukian, 1986).

The thiazole synthesis follows a procedure of Dubs and Stüssi (1976), beginning with the appropriate starting materials as shown in Figure 4–9b. The same procedure can be used to prepare other alkyl substituted thiazoles, usually in good yields. Many of these alkyl thiazoles display green flavor notes in numerous "shades."

4.2.3.2 *The fruity ester-like flavor note*

The fruity ester-like note is characterized as the sweet odors occurring generally in ripe fruits such as banana, pear, melon, etc. Typical ingredients representing this group are esters and lactones, but ketones, ethers and acetals are also involved (Figure 4–10). Tropical fruits are known to contain sulfur compounds responsible for the typical "exotic" notes (e.g., the methyl and ethyl ester of 3-methylthiopropionic acid).

Isoamylacetate is responsible for the sweet character in almost all fruity notes. 2,4-Decadienic acid ester is the impact substance found in Bartlett pears (Jennings & Tressl, 1974). 3-Methylthiopropionic acid esters are characteristic for pineapple. Those components are also used with many other flavor notes to give special characters. Additional compounds with non-specific fruity characters but generally described as sweet and fresh (like lactones, aromatic aldehydes, and terpenoids) can also be used to round up fruity notes.

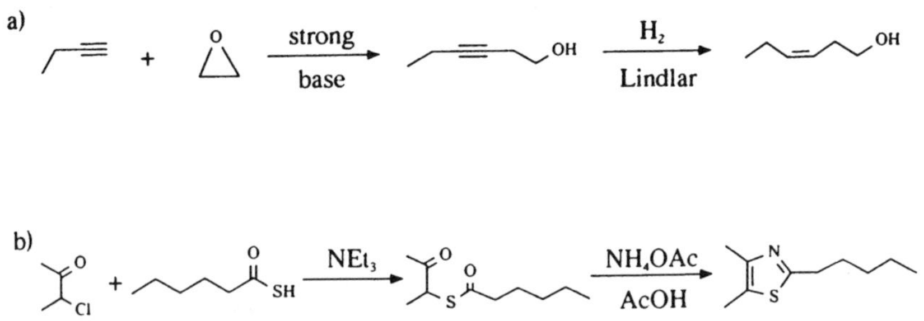

---

a) + strong base → OH; H_2 Lindlar → OH

b) O Cl + O SH; NEt_3 → O S O; NH_4OAc AcOH → N S

Figure 4–9 Synthesis of (a) *cis*-3-hexenol (Bedoukian, 1986) and (b) 4,5-dimethyl-2-pentythiazole (Dubs & Stüssi, 1976).

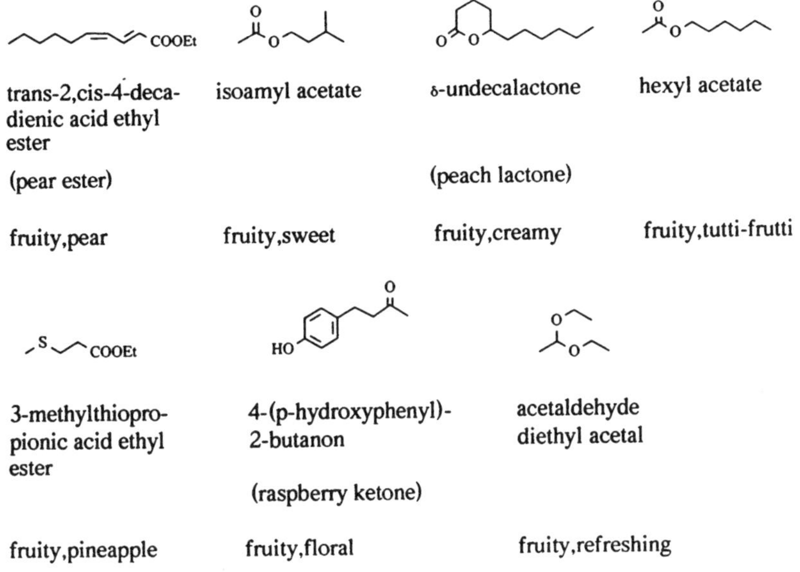

Figure 4–10 Examples of fruity ester-like flavor notes.

The formation of 3-methylthiopropionic esters can be assumed to start from methionine by Strecker degradation, followed by oxidation and esterification. Figure 4–11 shows the formation of the impact substance *trans*-2, *cis*-4-decadienic acid ethyl ester found in pear, apple and grape. Jennings and Tressl (1974) proposed an interesting scheme to explain their analytical findings during post-

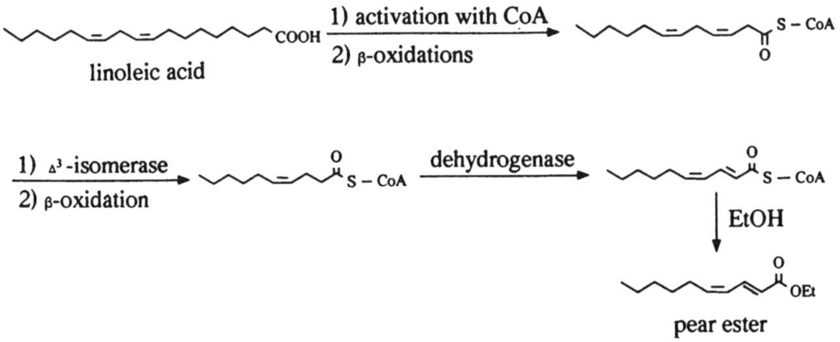

---

linoleic acid COOH → 1) activation with CoA 2) β-oxidations → S – CoA

1) Δ³-isomerase 2) β-oxidation → S – CoA → dehydrogenase → S – CoA

EtOH ↓

OEt

pear ester

Figure 4–11 Formation of ethyl *trans*-2,*cis*-4 decadienoate (pear ester) (Jennings & Tressl, 1974).

harvest ripening of Bartlett pears. Linoleic acid is degraded by several β-oxidations, each degrading cycle shortening the chain by two carbons. The degradation can be explained using the same enzymes of the fatty acid metabolism as in the formation, each enzyme-catalyzed reaction can occur in both directions.

Figure 4–12a demonstrates, by example with δ-undecalactone, a simple and efficient synthesis of 6-membered lactones with a variable chain in the 5-position (Ijima & Takahashi, 1973). In Figure 4–12b, an extension of a Claisen rearrangement is applied to prepare the "pear ester." The reaction conditions and the kind of base in the isomerisation step are crucial for a good yield of the wanted 2-*trans*, 4-*cis* double bond system (Oberhänsli, 1976). The other possible isomers are less desired, due to more fatty and green flavor notes.

4.2.3.3 *The citrus terpenic flavor note*

The citrus-like flavor note is characterized as the typical note occurring in citrus fruits and plants (citrus, lemon, orange, grapefruit), but also certain terpenoids can occur in this group. One of the most important components with a strong impact odor is citral (a mixture of geranial and neral), which can be either isolated from natural raw materials (lemon grass) or synthesized (Figure 4–13). Another terpenic

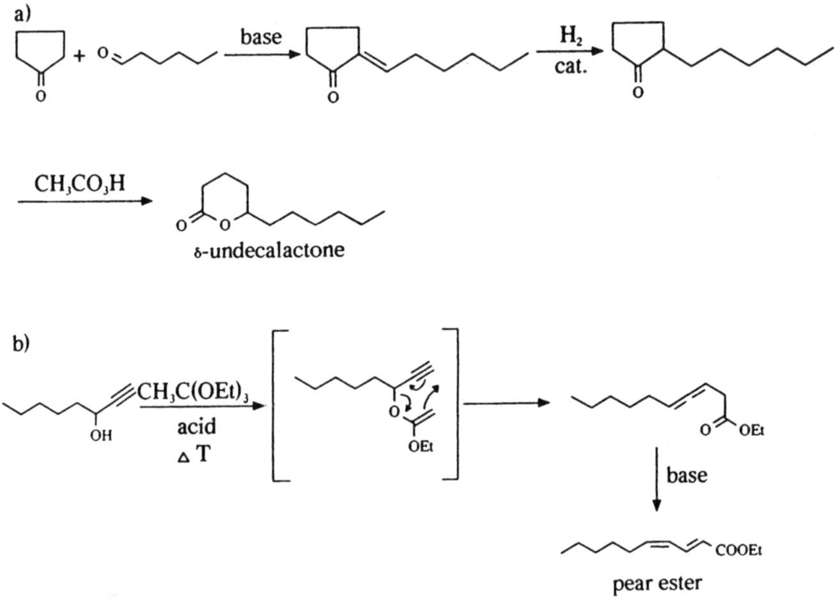

a)

base

H_2 cat.

CH_3CO_3H

δ-undecalactone

b)

$CH_3C(OEt)_3$ acid ΔT

OH

OEt

base

COOEt

pear ester

Figure 4–12 Synthesis of (a) δ-undecalactone (Ijima & Takahashi, 1973) and (b) ethyl *trans*-2,*cis*-4-decadienoate (pear ester) (Oberhänsli, 1976).

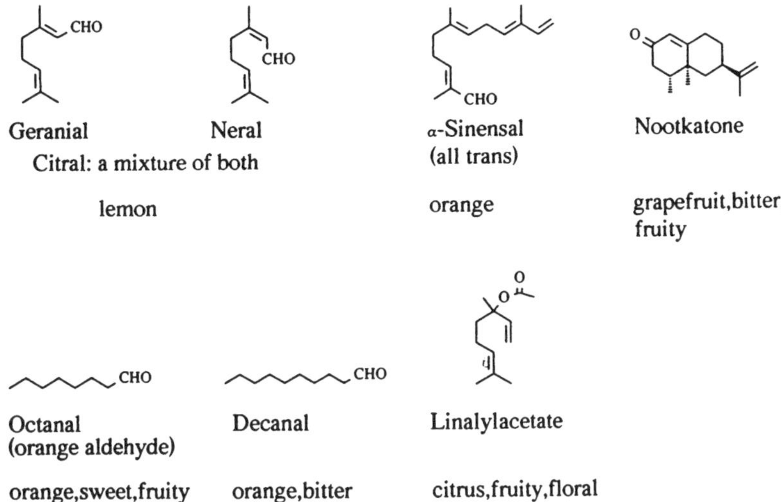

Figure 4–13 Example of citrus terpenic flavor notes.

component with a rather more bitter character is nootkatone, which is an impact substance found in grapefruit. Simple aliphatic aldehydes of medium chain length (octanal, decanal), but also sinensal, an unsaturated C_{15}-aldehyde, and some esters of monoterpenic alcohols (linalyl acetate) are used to round up this note.

In Figure 4–14, the biogenetic pathways of three representatives of terpenoid structure and their common precursors, i.e., mevalonic acid and "isoprene units" are shown. The formation of geranial can be easily explained by an oxidation of geranyl pyrophosphate, which in turn is the regular head-to-tail linkage product of two isoprene units as outlined in Figure 4–2. α-Sinensal is a regular sesquiterpene formed from farnesyl pyrophosphate (a further head-to-tail elongation of geranyl pyrophosphate), though the oxidation took place at the "head" of the molecule. Whereas geranial and α-sinensal are examples of regular acyclic terpenes, nootkatone must have been formed by two-cyclization steps and a [1,2]-methyl shift, which ultimately ends up in an irregular isoprene pattern.

Besides natural citral which is frequently preferred for its harmony, standardized synthetic citral is generally used when large amounts and low costs are needed. Citral is an important synthetic ingredient with a total consumption of about 100 tons in 1985 for flavors, and an annual production of several thousand tons in the United States (Bedoukian, 1986). Figure 4–15a shows a technical synthesis of citral (Nissen et al., 1981) and one of the published syntheses of all-*trans-α*-sinensal (b) is shown in Figure 4–15b (Helmlinger & Naegeli, 1976; Buechi & Wuest, 1974). The citral synthesis is very elegant and short, because

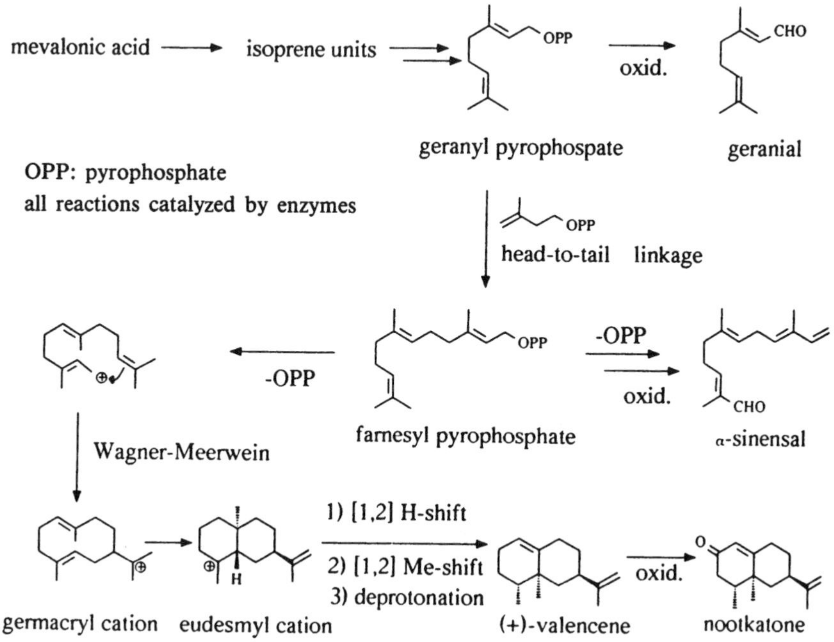

Figure 4–14 Biogenesis of geranial, α-sinensal and nootkatone (Solomons, 1988). OPP, pyrophosphate; all reactions catalyzed by enzymes.

---

the necessary temperature for the elimination of prenol triggers the two consecutive rearrangements in excellent yields. Because of the importance of citral, other synthetic procedures of industrial scale have been patented (Andrews & Hindley, 1973; Brueninger, 1980).

In comparing Figures 4–14 and 4–15, it is remarkable to see the similarity between the biogenetic pathway and the chemical synthesis; in both cases an unsaturated C_5-unit seems to fit best. This fact can be observed frequently and relies on the inherent chemical reactivity of a compound.

4.2.3.4 *The minty camphoraceous flavor note*

The minty note is described as the impact of peppermint having a sweet, fresh and cooling sensation. The main compounds are *l*-menthol, pulegone, *l*-carvylacetate, *l*-carvone and camphor, but borneol, eucalyptol (=cineol), and fenchone also belong traditionally to this odor family (Figure 4–16). The minty notes are generally used as refreshing top notes with cooling effects in special flavors and applications (chewing gum, toiletries and beverages).

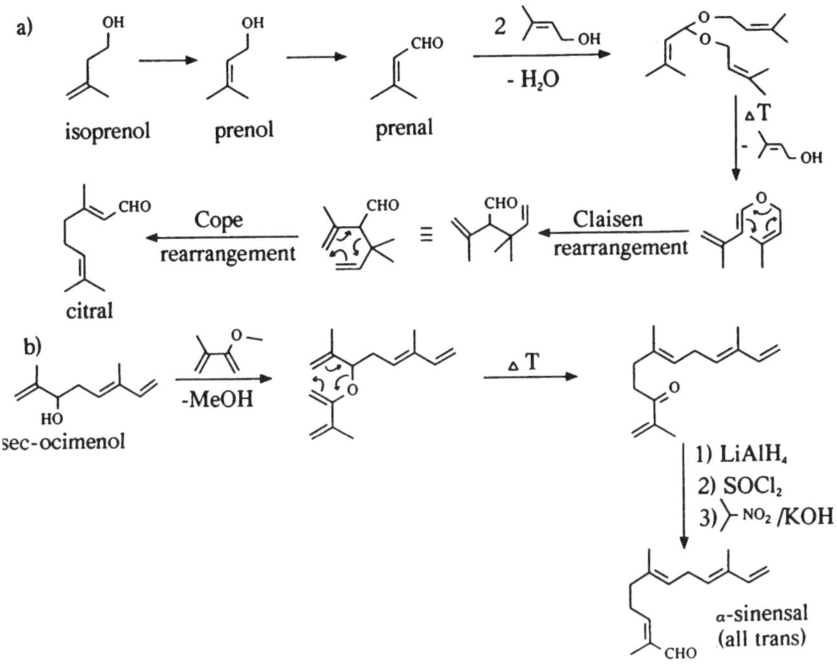

Figure 4–15 Synthesis of (a) citral (Nissen et al., 1981) and (b) α-sinensal (Helmlinger & Naegeli, 1976).

In this group *l*-menthol is economically the most important synthetic ingredient, the main quantity being used in non-food applications. Most of these minty camphoraceous compounds are oxygenated monoterpenes and are formed according to the general outline of Figure 4–2. As shown in Figure 4–17, the formation of menthol can be explained by cyclization of linalyl cation followed by adjustment of the functional groups (introduction of hydroxyl, hydrogenation of double bonds). The formation of fenchone is thought to proceed as an enzyme bound cation, because no free intermediates are involved in this conversion catalyzed by the enzyme *l*-endofenchol-synthetase (isolated from fennel) (Croteau et al., 1980). It is, however, evident that a re-arrangement of the carbon skeleton must occur, because fenchone belongs to the irregular monoterpenes. The stereochemical outcome of the reaction sequence (only one enantiomer is formed) is in accordance with the enzymatic control of each step from geranyl pyrophosphate to the end-product.

In most applications *l*-menthol is the preferred isomer of menthyl alcohols. In Figure 4–18 an industrial synthesis is shown, which produces specifically

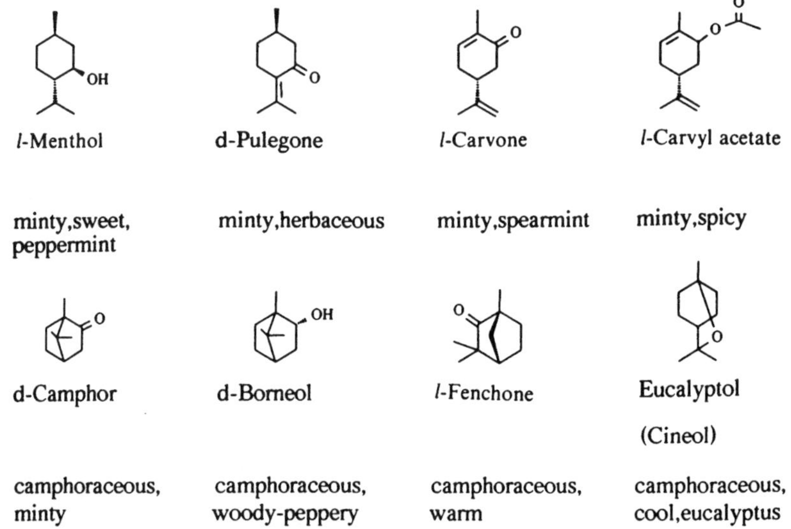

Figure 4–16 Examples of minty camphoraceous notes.

the desired isomer. The crucial step is the enantio selective isomerization of geranyl diethyl amine with an homogeneous asymmetric rhodium catalyst. Another industrial preparation includes the resolution of a mixture of menthol, neomenthol and isomenthol (Fleischer et al., 1972). Due to these and other successful syntheses *l*-menthol is a low price item, available in large quantities.

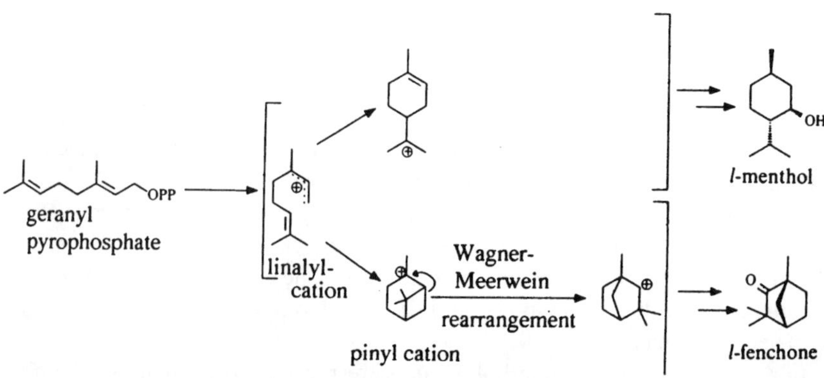

Figure 4–17 Biogenesis of menthol and fenchone.

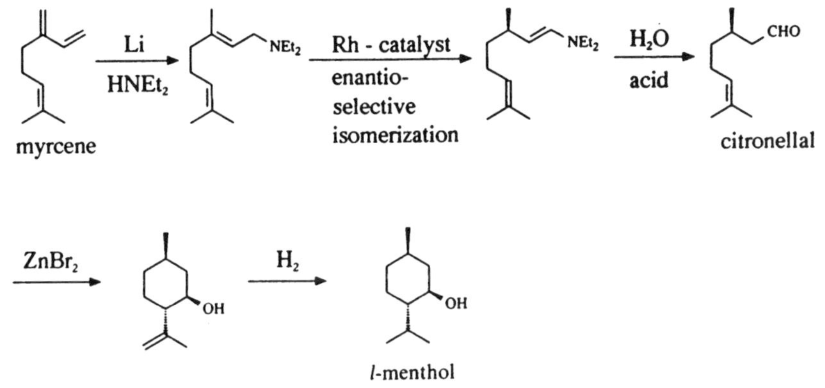

Figure 4–18 Synthesis of *l*-menthol (Noyori & Kitamura, 1989).

4.2.3.5 *The floral sweet flavor notes*

The floral note can be defined as the odors emitted by flowers and contains sweet, green, fruity and herbaceous characters. There are no typical impact ingredients, but a range of single flavor substances belonging to different chemical classes (Figure 4–19). Phenylethanol, geraniol, *β*-ionone and some esters (benzyl acetate, linalyl acetate) are important compounds in this group. All these com-

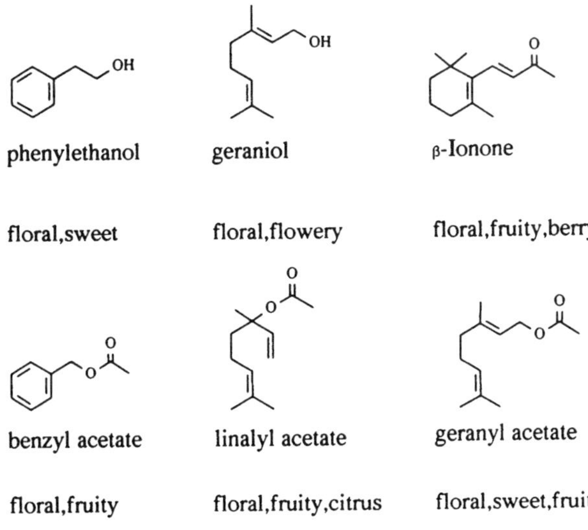

Figure 4–19 Examples of floral sweet notes.

pounds originate from plant materials. At higher concentrations many floral notes are accompanied by an unwanted perfume-like impression.

The formation of β-ionone (Figure 4–20a) includes the formation and degradation of β-carotene. The formation is yet another example of terpene biogenesis, two farnesyl pyrophosphate moieties being fused, tail-to-tail, to give the C_{30}-compound squalene. Further enzymatic steps form β-carotene, which is oxidatively split into smaller odoriferous fragments like β-ionone and other substances later in the ripening process (Schreier, *Chromatographic Studies*, 1984).

The reasons why a plant generates and metabolizes higher terpenes like β-carotene is not well understood. Oxidative degradation can happen as part of an enzymatic ripening process or as a post-harvest (photo-oxidative or auto-oxidative) event.

The formation of 2-phenylethanol (Figure 4–20b) is an example of a Strecker degradation of the amino acid phenylalanine (cf. Figure 4–1b). Tressl et al. (1975) showed that in banana tissue slices, the radioactive label of added L-[2-^{14}C]phenylalanine could be mainly found, after 5 h of incubation, in 2-phenylethanol.

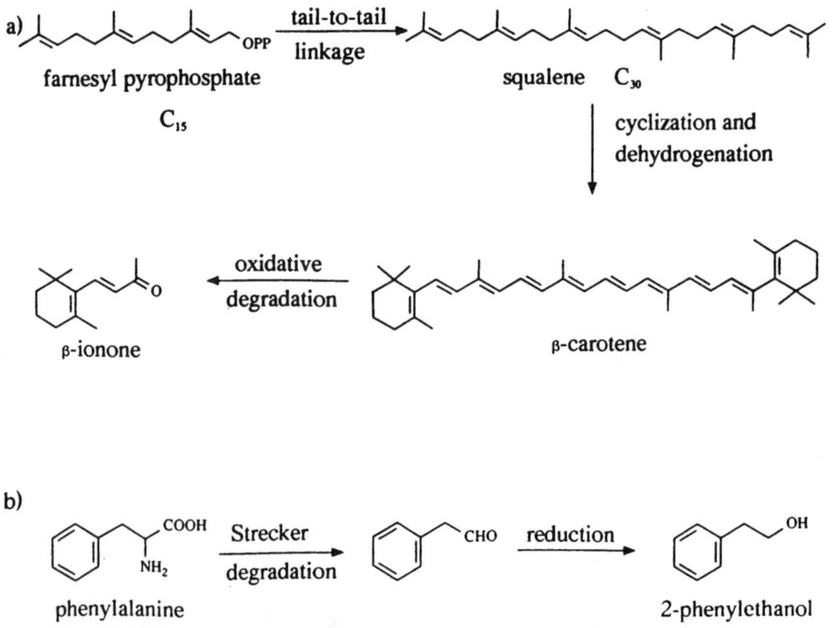

a) OPP
tail-to-tail linkage
farnesyl pyrophosphate C_{15}
squalene C_{30}
cyclization and dehydrogenation
O
oxidative degradation
β-ionone
β-carotene

b) COOH NH$_2$
Strecker degradation
CHO
reduction
OH
phenylalanine
2-phenylethanol

Figure 4–20 (a) Biosynthesis of β-carotene and degradation to β-ionone (Schreier, *Chromatographic Studies*, 1984) and (b) Generation of 2-phenylethanol (Tressl et al., 1975).

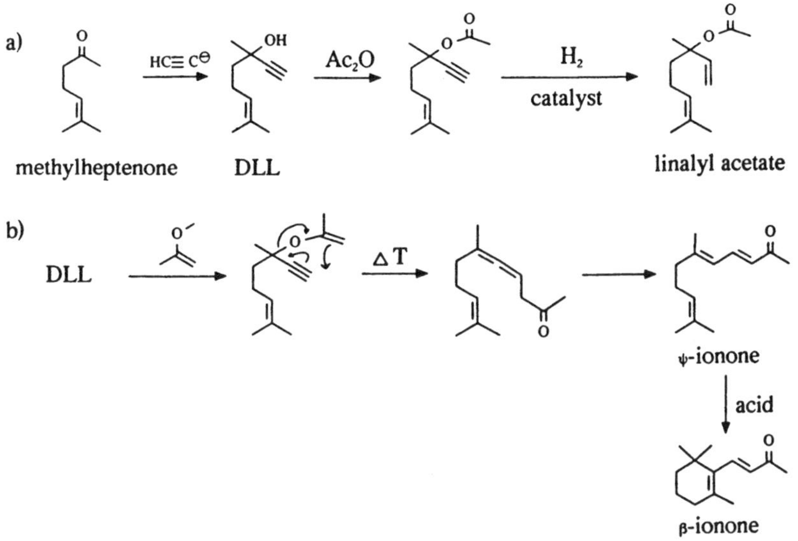

a) methylheptenone $\xrightarrow{HC{\equiv}C^{\ominus}}$ DLL $\xrightarrow{Ac_2O}$ $\xrightarrow[\text{catalyst}]{H_2}$ linalyl acetate

b) DLL $\longrightarrow$ $\xrightarrow{\Delta T}$ $\longrightarrow$ ψ-ionone $\xrightarrow{\text{acid}}$ β-ionone

Figure 4–21 Synthesis of (a) linalyl acetate (Bedoukian, 1986; Mayer & Isler, 1971) and (b) β-ionone (Saucy & Marbeth, 1967).

The examples of Figure 4–19 are the largest volume items in the flavor and fragrance industry. The consumption of linalyl acetate in the U.S. market is estimated to be about 500 tons annually (Bedoukian, 1986).

Figure 4–21 shows synthetic procedures for linalyl acetate and β-ionone, based on methylheptenone which is another very large-volume commodity. In the acetylene process, dehydrolinalool (DLL) is first produced, and then acetylated and partially hydrogenated (Figure 4–21a) (Bedoukian, 1986; Mayer & Isler, 1971). The elongation of DLL with a C_3-unit makes use of a modified Claisen rearrangement (Saucy & Marbeth, 1967). Isomerization of the allenic systems and acid catalyzed cyclization conclude the synthesis of β-ionone (Figure 4–21b).

Besides these purely synthetic approaches other manufacturers start from natural sources, e.g., pinene is transformed into linalyl acetate and geranyl acetate (Bay, 1962).

4.2.3.6 *The spicy herbaceous flavor note*

The spicy herbaceous flavor note is common to herbs and spices with all their nuances. These flavorings are certainly the oldest food ingredients used by mankind. Spices were especially desired not only as appetizers but also as food preservatives. Aromatic aldehydes, alcohols and phenolic derivatives are typical constituents with their strong flavor and physiological (e.g., bacteria-static) effects. Many are impact character chemicals: anethole (anise), cinnamaldehyde (cinnamon), estragole (estragon), eugenol (clove), *d*-carvone (dill), thymol (thyme), etc. (Figure 4–22).

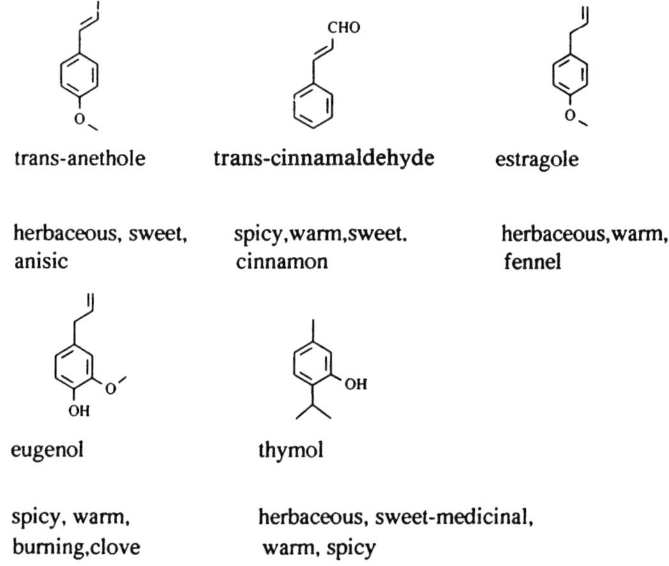

Figure 4–22 The spicy herbaceous flavor note.

Due to their strength many of these compounds are used only in small quantities. Often the spices or extracts of them are used. The range of application is very broad and includes bakery goods and alcoholic beverages as well as toothpaste and chewing gum and even fragrances of oriental type. The largest item used is cinnamaldehyde, which is also synthetically produced in considerable amounts.

The formation of cinnamaldehyde, anethole, estragole and eugenol is related to the phenylpropanoid metabolism, i.e., the transformation of phenylalanine to cinnamic acid and its hydroxylated derivatives (Figure 4–23). The formation of phenylalanine is linked with the carbohydrate metabolic pathway through shikimic acid. In higher plants, the phenylpropanoid pathways are especially important and give rise to a number of compounds in relatively high concentrations (Schreier, *Chromatographic Studies*, 1984).

Thymol has its origin in a monoterpenic *p*-menthane derivative, e.g., pulegone. Synthetic cinnamaldehyde is made by an aldol condensation of benzaldehyde with acetaldehyde, a procedure which goes back well into the last century (Figure 4–24a) (Peine, 1884).

Estragole can be synthesized from *p*-chloroanisole by a Grignard reaction with allyl chloride. Base-catalyzed isomerization of the double bond gives *trans*-anethole (Figure 4–24b) (unpublished results).

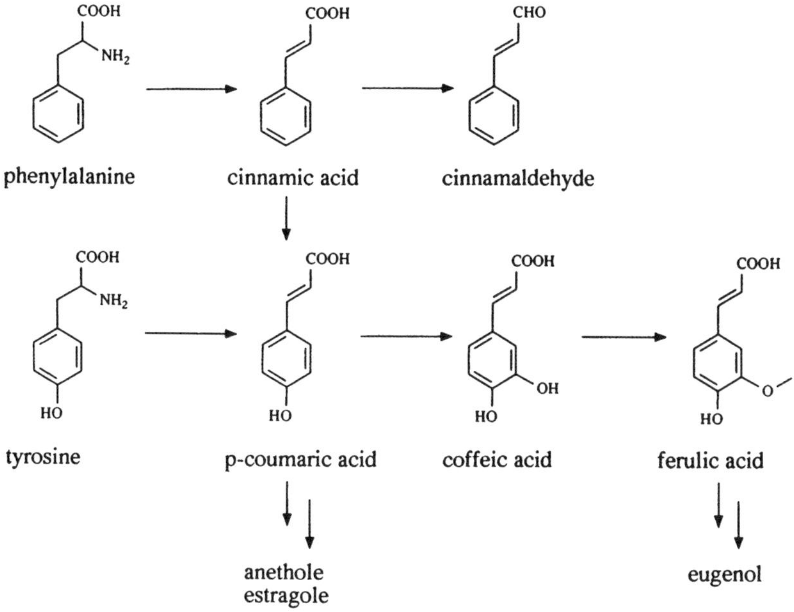

phenylalanine → cinnamic acid → cinnamaldehyde

tyrosine → p-coumaric acid → coffeic acid → ferulic acid

anethole
estragole

eugenol

Figure 4–23 Biogenesis of cinnamaldehyde and related phenylpropanoid structures (Schreier, *Chromatographic Studies*, 1984).

4.2.3.7 *The woody smoky flavor note*

Woody smoky flavor notes are characterized by such compounds as substituted phenols (guaiacol, etc.), methylated ionone derivatives (methylionone), and, exceptionally, by some aldehydes (*trans*-2-nonenal in very low concentration), which have respectively warm, woody, sweet and smoky odors (Figure 4–25).

Methylionones are examples of synthetic ingredients that have not been found in food although they are structurally very similar to the carotene degradation products, the ionones. The methylionones are regarded as safe (Chemical Source Association, 1989) and may be used in food flavors in many countries.

The woody and smoky notes are generally considered not to belong to the native flavor compounds, rather they are formed during storage or heat treatment. These compounds can only be used in such combination with flavor notes where an improvement of mouthfeel and body is needed. In higher concentrations they are perceived as off-flavors.

Although the phenols are generally available from processed flavor preparations (e.g., by dry sawdust distillation), ionone derivatives are individually synthesized. The formation of the phenolic derivatives can be explained by thermal degradation of lignin (cf. smoking processes), or by fermentative oxidations (cf.

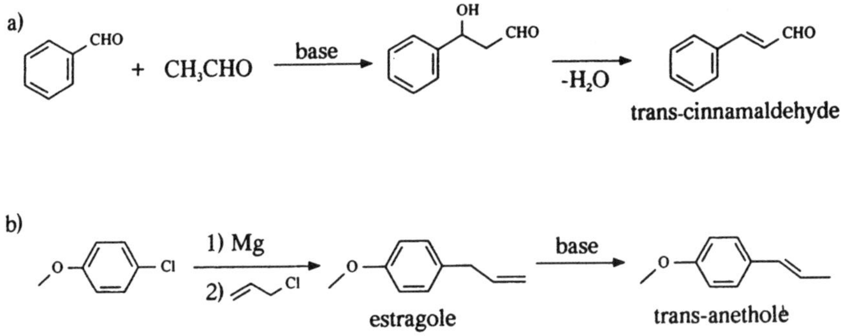

Figure 4–24 Synthesis of (a) cinnamaldehyde (Peine, 1884) and (b) estragole and anethole (unpublished).

curing of vanilla beans) of phenylpropanoid precursors (Figure 4–26, see also Figure 4–23) (Belitz & Grosch, 1987). The synthesis of furfurylpyrrole starts from furfuryl amine and 2.5-dimethoxy-tetrahydrofuran in a widely applicable procedure (Figure 4–27a) (Elming & Clauson-Kaas, 1952). Condensation of cyclocitral with butanone gives a mixture of methylionones. The olfactory value depends on the composition of the mixture, which can be adjusted by variation of

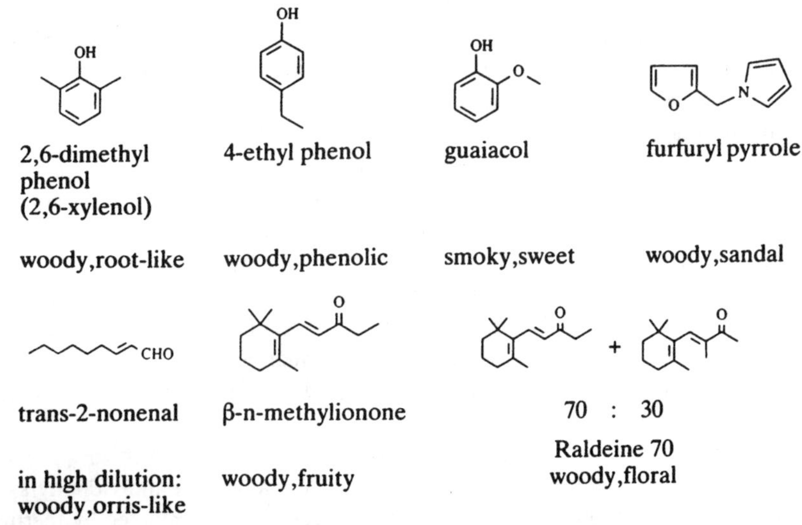

Figure 4–25 The woody, smoky flavor note.

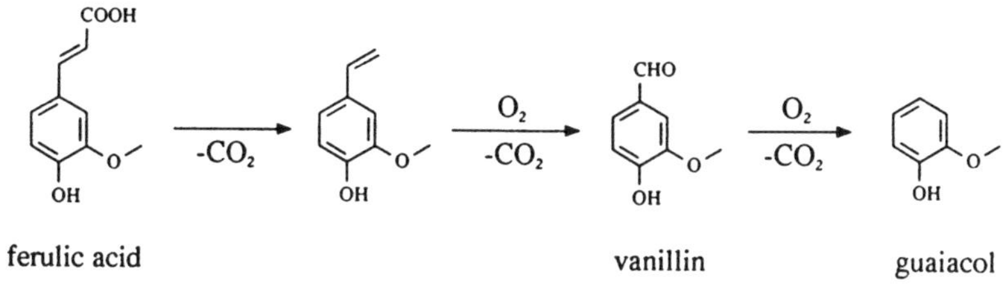

ferulic acid $\xrightarrow{-CO_2}$ $\xrightarrow[-CO_2]{O_2}$ vanillin $\xrightarrow[-CO_2]{O_2}$ guaiacol

Figure 4–26 Formation of guaiacol by oxidative degradation (Belitz & Grosch, 1987).

basic reaction conditions (Figure 4–27b) (*Givaudan Index*, 1962). The synthesis of *trans*-2-nonenal is shown in Figure 4–42.

4.2.3.8 *The roasty burnt flavor note*

The roasty burnt flavor note is generally associated with the term "pyrazine," which is broadly true because pyrazines are able to cover the whole range of this flavor type. They are a predominant chemical class in roasted products. Different substitution by alkyl, acyl or alkoxy and combinations thereof induces a great diversity of flavor impressions; so burnt, roasted, green, earthy and musty notes can result. Within this group of the roasty burnt flavor notes, alkyl and acetyl substituted pyrazines are the most important (Figure 4–28).

Several mechanisms and precursors have been proposed to explain the multitude of pyrazines formed in heat treated food (Figure 4–29) (Rizzi, 1988). Condensation of two amino acids leads to a 2,5-diketopiperazine, which is further dehydrated to a pyrazine. Another condensation starts from α-aminoketones, a "side product" of the Strecker degradation of amino acids (cf. Figure 4–1b).

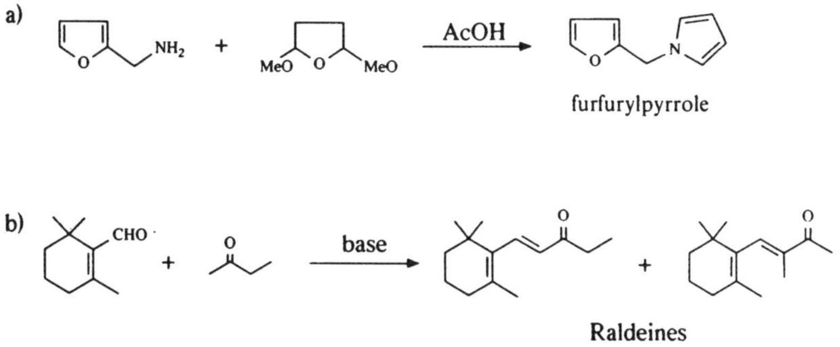

a) + $\xrightarrow{AcOH}$ furfurylpyrrole

b) + $\xrightarrow{base}$ + Raldeines

Figure 4–27 Synthesis of (a) furfurylpyrrole (Elming & Clauson-Kaas, 1952) and (b) methylionones (*Givaudan Index*, 1962).

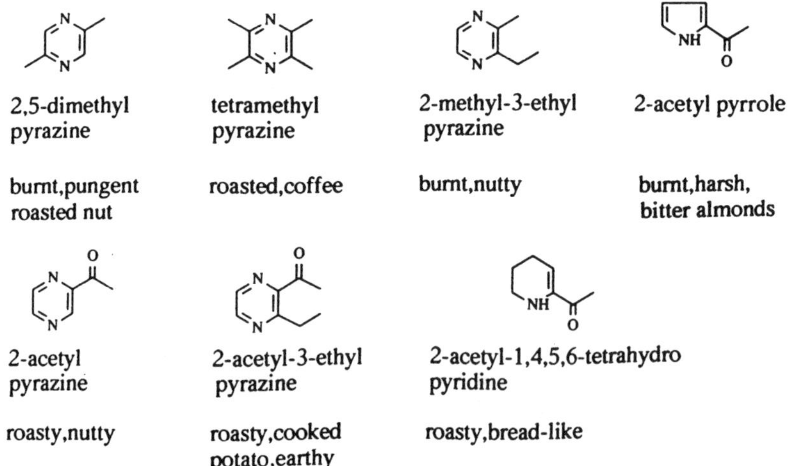

Figure 4–28 Examples of roasty burnt flavor notes.

A cyclodehydration reaction of two 1,2-aminoalcohol derivates (e.g., threonine) was also proposed to form pyrazines.

The predominant role of pyrazines as flavor chemicals is reflected by numerous publications or patents on analytical findings (Baltes & Bochmann, 1987) and synthesis (Masuda et al., 1981; JA Pat. 6 0258 168, 1984; Flament, 1979). Two examples of synthetic procedures for alkyl and acetyl substituted pyrazines are shown in Figure 4–30 (Masuda & Mihora, 1986; Mookherjee et al., 1974).

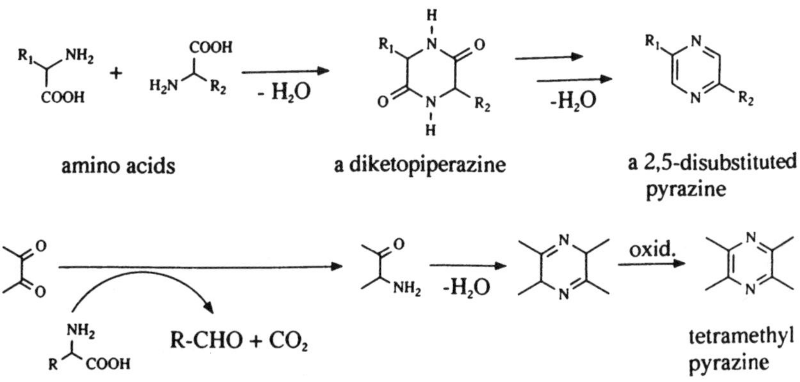

---

amino acids — $-H_2O$ → a diketopiperazine — $-H_2O$ → a 2,5-disubstituted pyrazine

R-CHO + CO_2 — $-H_2O$ — oxid. → tetramethyl pyrazine

Figure 4–29 Formation of pyrazines (Rizzi, 1988).

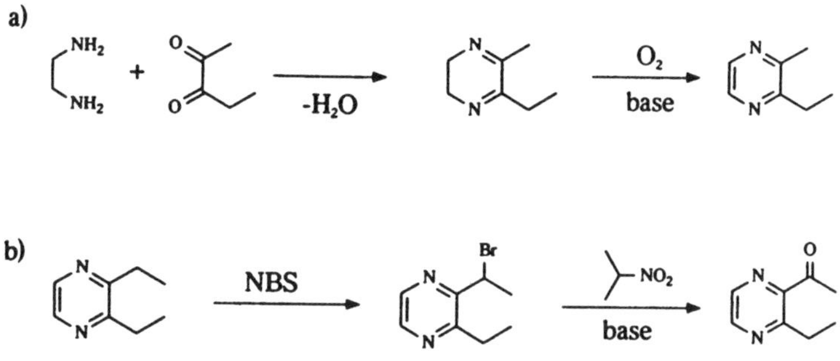

Figure 4–30 Synthesis of (a) 2-methy-3-ethyl pyrazine (Masuda & Mihora, 1986) and (b) 2-acetyl-3-ethyl pyrazine (Mookherjee et al., 1974). NBS, *N*-bromosuccinimide.

4.2.3.9 *The caramel nutty flavor note*

The caramel nutty flavor note is mostly described by heat processed sugar-containing food products. The slightly bitter and burnt odor of roasted nuts represents the complement of this flavor type. Beside corylone, maltol, furonol, etc. (Figure 4–31), a special range of components like vanillin, ethylvanillin, benzaldehyde, phenylacetic acid, cinnamic alcohol, dehydrocoumarin, trimethyl-

---

2-hydroxy-3-methyl
2-cyclopenten-1-one
(corylone)

nutty,maple-like

2-hydroxy-3,4-dimethyl-
2-cyclopenten-1-one

caramel,roasty

2-hydroxy-3,5-dimethyl-
2-cyclopenten-1-one

caramel,butterscotch-like

2,5-dimethyl-4-hydroxy-
3[2H]-furanone
(furonol)

caramel,sweet

2-ethyl-5-methyl-4-hydroxy-
3[2H]-furanone
(homofuronol)

caramel,sweet,bread-like

2-methyl-3-hydroxy-
4-pyranone
(maltol)

caramel,burnt sugar

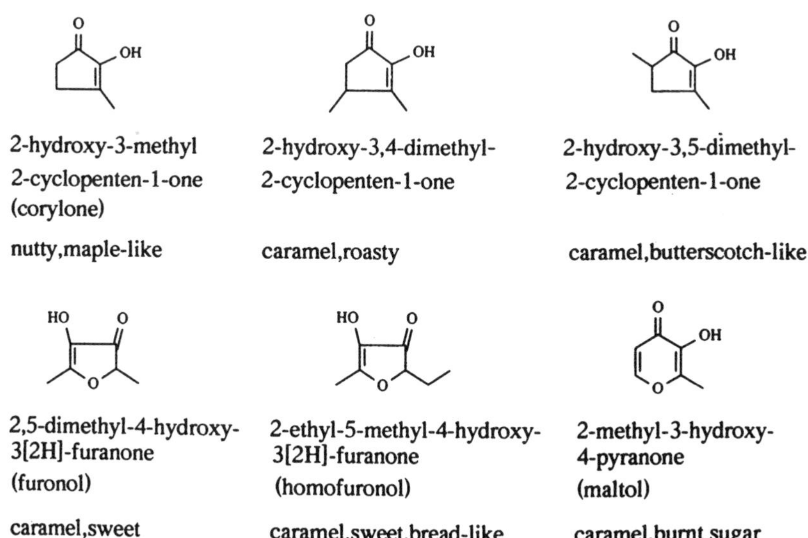

Figure 4–31 Examples of nutty caramel flavor notes.

pyrazine also belong to this group. The examples of Figure 4–31 are characterized by a common structural unit, namely a cyclic enolone system, and also by a common flavor profile. These cyclic enolones are most valuable compounds for their strong impact character and for their distinct flavor enhancing effect. These compounds can be applied in all sorts of caramel flavors and also in roasted and fruit applications.

A comparison of the structures of 18 cyclic enolones and their threshold values confirms an interesting structure-activity relationship. A higher substitution pattern, i.e., more methyl groups or ethyl instead of methyl, lowers the threshold value significantly (Wild, 1988).

The formation of corylone, maltol or furonol occurs during heat treatment in a large range of foodstuffs containing carbohydrates. In Figure 4–32, the anticipated formation of furonol from rhamnose is outlined (Hodge, 1963). This example is a somewhat special case of a Maillard reaction, because the amino acid is replaced by piperidine and acetic acid, and the yield of the end-product is, at around 20%, exceptionally high.

The synthetic routes for homofuronol and corylone as shown in Figure 4–33 are industrially feasible processes (Huber & Wild, 1977; Wild, 1986). The synthetic concept for homofuronol allows easy variation of the starting materials, so that with the same reaction sequence, other homologues can be prepared. The synthesis of corylone starts from bulk chemicals using phase transfer catalysis (PTC) conditions. Use of differently alkylated chloroacetones gives rise to higher alkylated corylones (see Figure 4–31 for examples).

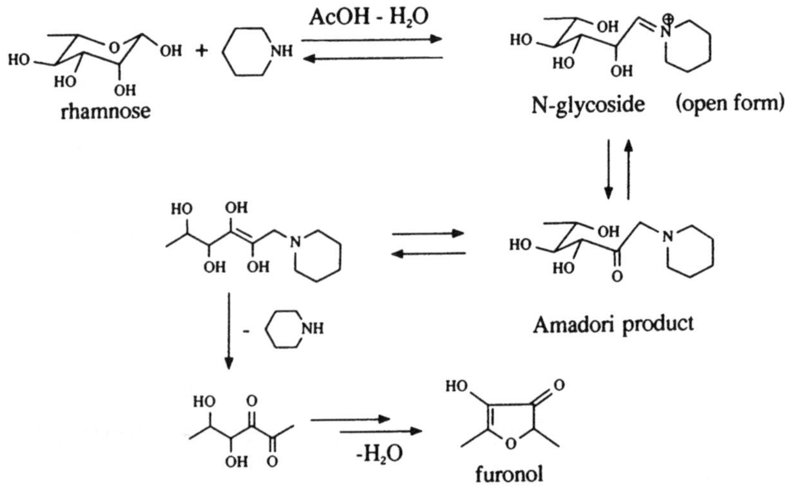

Figure 4–32 Formation of furonol from rhamnose (Hodge, 1963).

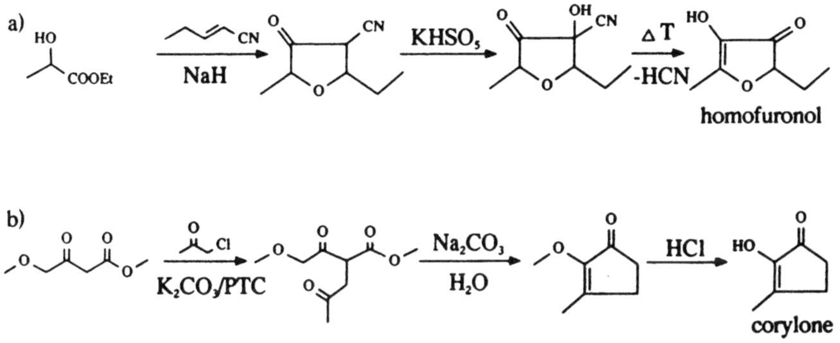

a) HO COOEt —(CN, NaH)→ O CN O —($KHSO_5$)→ O OH CN O —(ΔT, -HCN)→ HO O O homofuronol

b) O O O O —(O Cl, K_2CO_3/PTC)→ O O O O O O —(Na_2CO_3, H_2O)→ O O —(HCl)→ HO O corylone

Figure 4–33 Synthesis of (a) homofuronol (Huber & Wild, 1977) and (b) corylone (Wild, 1986). PTC, phase transfer catalysis.

4.2.3.10 *The bouillon HVP flavor note*

The bouillon HVP flavor is not a distinct flavor note based on typical representants, but described as a rather diffuse warm, salty and spicy sensation, associated with enhanced meat extracts. Single flavor ingredients are less important than the appropriate global reaction flavors, which are responsible for the basic taste perception (so-called "body"). Nevertheless some components are helpful for enhancing and topping (cf. Figure 4–34).

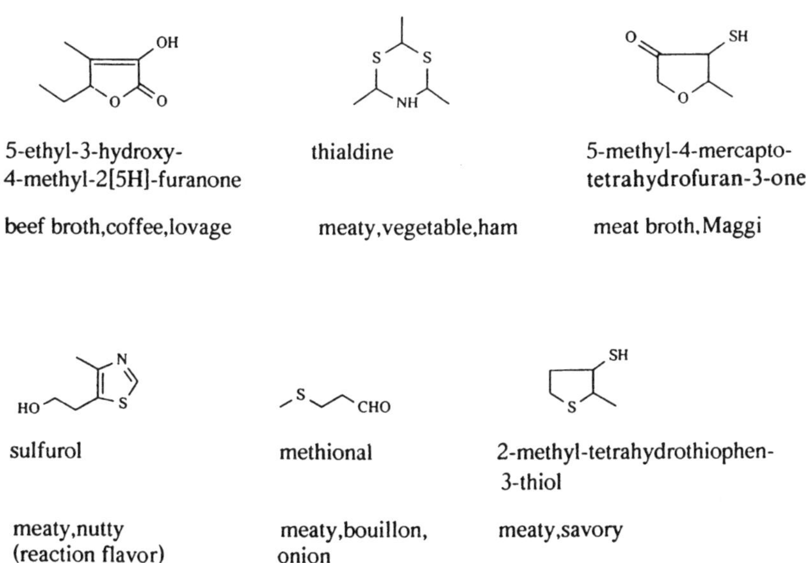

Figure 4–34 Examples of bouillon HVP-like flavor notes.

As top notes, 5-methyl-4-mercapto-tetrahydrofuran-3-one and 2-methyl-tetrahydrothiophen-3-thiol are described (Ouweland & Peer, 1977), two compounds which are chemically related to sulfurol degradation products (see also Figure 4–38). The examples of Figure 4–34 exhibit different chemical classes used in this flavor note with one representative of each. 5-Ethyl-4-methyl-3-hydroxy-2-[5*H*]-furanone ("emoxyfurone") was originally found in HVP. Its formation from threonine can be explained as shown in Figure 4–35a (Sulser et al., 1967). The odor of this hydroxybutenolide is reminiscent of lovage, where its precursor α-ketobutyric acid was identified. 4-Methyl-5-hydroxyethyl thiazole (sulfurol) is described as a thiamine degradation product (Figure 4–35b) (Hartman et al., 1984).

Sulfurol is an example of a curious observation which often irritates flavorists and flavor chemists. Whereas strictly pure sulfurol is a disappointingly weak compound, "aged" samples display a rich and strong odor of beef broth. Even today the "impurities" responsible for the odoriferous part cannot be instrumentally detected.

Figure 4–36 shows practical syntheses of emoxyfurone (Roedel & Hempel, 1974) and thialdine (Hwang et al., 1986).

4.2.3.11 *The meaty animalic flavor note* (see Chapter 9)

The meaty animalic flavor note is one of the most complex to be described. Roast beef meat flavor differs, for example, from that of barbecued or simply

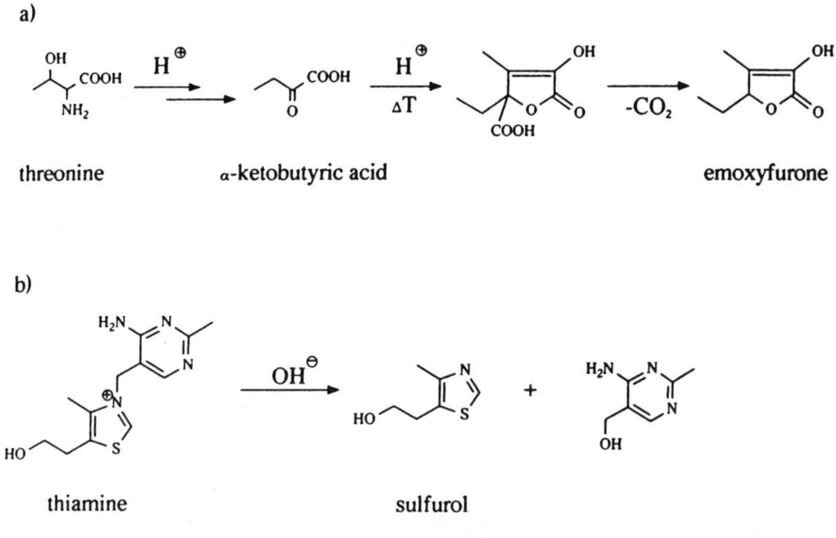

a)

threonine → α-ketobutyric acid → emoxyfurone

$H^{\oplus}$; $H^{\oplus}$, ΔT ; $-CO_2$

b)

thiamine → ($OH^{\ominus}$) sulfurol +

Figure 4–35 Formation of (a) hydroxybutenolide in HVP (Sulser et al., 1967) and (b) sulfurol from thiamine (Hartman et al., 1984).

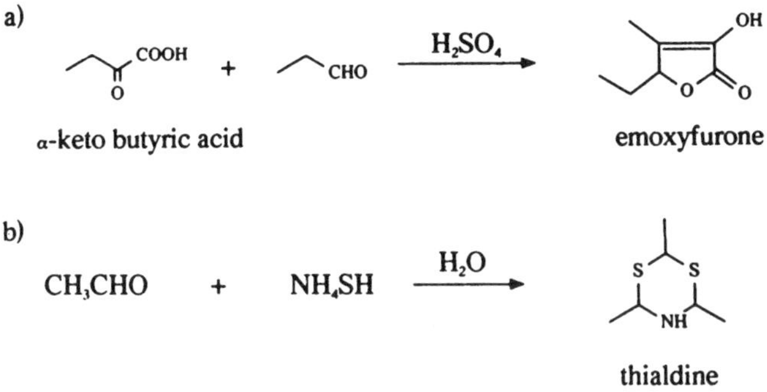

a) α-keto butyric acid + CHO $\xrightarrow{H_2SO_4}$ emoxyfurone

b) CH_3CHO + NH_4SH $\xrightarrow{H_2O}$ thialdine

Figure 4–36 Synthesis of (a) emoxyfurone (Roedel & Hempel, 1974) and (b) thialdine (Hwang et al., 1986).

boiled meat. Crude uncooked (bloody animalic) meat is poorly flavored and reminiscent of a salty, amine taste. The amine flavor note can be evoked by low concentrations of *n*-butyl amine or piperidine. The desired flavor compounds are generated only during processing at the beginning of the Maillard reaction. Sulfur containing components (mercaptans, thiazoles, thiophenes, etc.) as well as nitrogen heterocycles (pyrazines, pyrroles, pyridines, oxazoles, etc.) predominate the flavor effect. Figure 4–37 shows some flavor components.

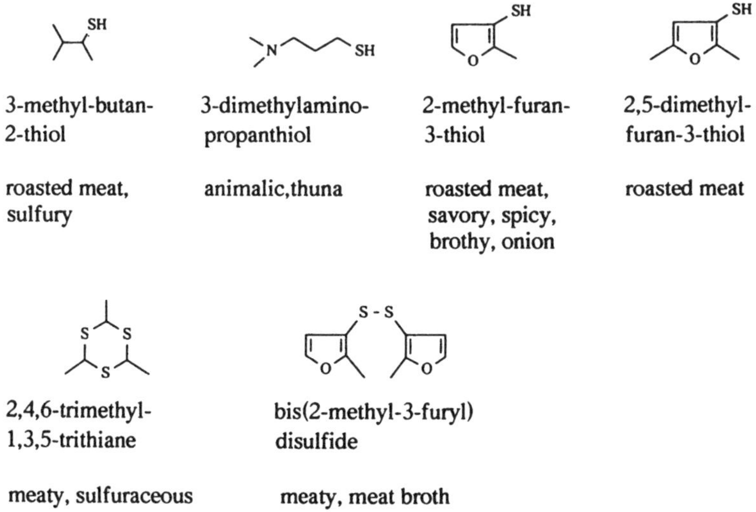

Figure 4–37 Examples of meaty animalic flavor notes.

Depending on the kind and content of fat incorporated or used, a large number of aliphatic carbonyls (aldehydes and ketones) develop and influence the resulting flavor effects. Non-volatile derivatives of nucleotides and peptides as well as minerals are presumably responsible for typical meat mouthfeel. Investigations using model systems have shown that a multitude of components are formed, but only a small range of them are really flavor significant in low concentration (MacLeod, 1986). A perfect reconstitution of meat flavor with all its nuances seems to be practically impossible. That is why processed meat flavor "bases" or extracts are preferred as building blocks for composition.

Bis-(2-methyl-3-furyl)-disulfide is described as having an extremely low threshold value (Buttery et al., 1984). A possible formation can be explained by thiamine decomposition (Figure 4–38) (Hartman et al., 1984). Investigations on the mechanism of thiamine degradation and analysis of the reaction products is still an active research area (Sugimoto & Hirai, 1987). Figure 4–39 shows the chemical synthesis of 2-methyl-3-mercapto furan (Evers et al., 1976) and 3-dimethylamino propanethiole (Huber, 1977), representing quite different meat flavor perceptions.

4.2.3.12 *The fatty rancid flavor note*

The most potent examples of fatty rancid flavor notes are butyric and isobutyric acids, which are described as unpleasant, sour and repulsive. Caproic acid is weaker and rather cheesy rancid. Medium chain methyl branched fatty acids like 4-methyl octanoic and nonanoic acids are responsible for the off-flavor of mutton fat.

Long chain fatty acids are perceived as oily, waxy and soapy. The aldehydes cited in Figure 4–40 with their individual specific notes contribute to the fatty rancid perception depending on their concentration in different applications. The same concentration of *cis*-4-heptenol can be differently evaluated in fat and fruit products.

a) thiamine $\xrightarrow[\text{acid,heat}]{H_2O}$ + + HCOOH

-H_2O oxid. bis-(2-methyl-3-furyl) disulfide

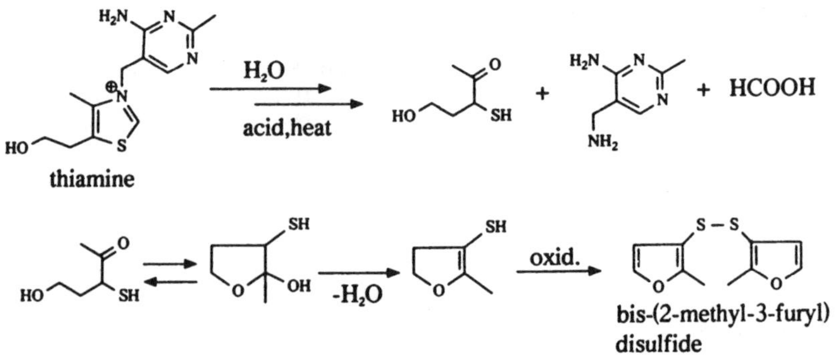

Figure 4–38 Formation of bis-(2-methyl-3-furyl)-disulfide by degradation of thiamine (Hartman et al., 1984).

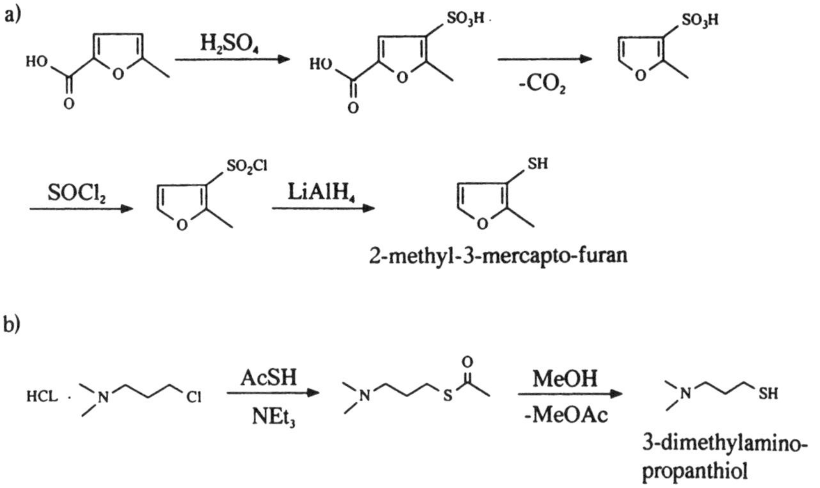
a)
HO
O
H2SO4
SO3H
-CO2
SOCl2
SO2Cl
LiAlH4
SH
2-methyl-3-mercapto-furan
b)
HCL
N
Cl
AcSH
NEt3
S
MeOH
-MeOAc
SH
3-dimethylamino-propanthiol

Figure 4–39 Synthesis of (a) 2-methyl-3-mercapto-furan (Evers et al., 1976) and (b) 3-dimethylamino-propanthiol (Huber, 1977).

Figure 4–41 shows the formation of some aldehydes and fatty acids generated by degradative fat oxidation of linoleic and linolenic acid, respectively (Tressl et al., 1981). The same products can be formed enzymatically, e.g., in fruits or vegetables, where they have desired flavor effects. In Figure 4–42 a standard synthesis of an α,β-unsaturated aldehyde *trans*-2-nonenal is outlined.

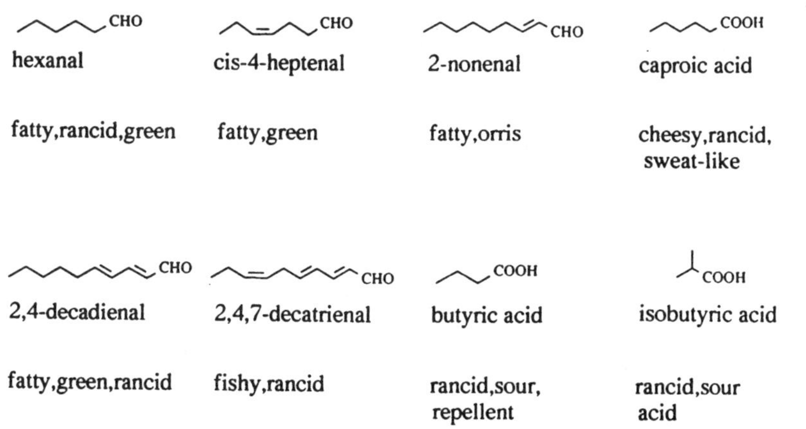
CHO
hexanal
fatty,rancid,green
CHO
cis-4-heptenal
fatty,green
CHO
2-nonenal
fatty,orris
COOH
caproic acid
cheesy,rancid, sweat-like
CHO
2,4-decadienal
fatty,green,rancid
CHO
2,4,7-decatrienal
fishy,rancid
COOH
butyric acid
rancid,sour, repellent
COOH
isobutyric acid
rancid,sour acid

Figure 4–40 Examples of fatty rancid flavor notes.

linoleic acid —oxidation→ hexanal, 2-heptenal, 2,4-decadienal (E,Z and E,E)
caproic acid, caprylic acid
9-oxo-nonanoic acid, 10-oxo-decenoic acid

linolenic acid —oxidation→ 2-butenal, 2-hexenal, 2,4-heptadienal (E,Z and E,E),
2,4,7-decatrienal (E,Z,Z; E,E,Z; E,E,E),
caprylic acid, 12-oxo-10-dodecenoic acid,
13-oxo-9,11-tridecadienoic acid

Figure 4–41 Formation of acids and aldehydes by autoxidation (Tressl et al., 1981).

4.2.3.13 *The dairy buttery flavor note*

The dairy buttery flavor note varies from buttery notes (diacetyl, acetoin, pentandione) to sweet creamy fermented notes (acetol-acetate, δ-decalactone, γ-octalactone). Figure 4–43 shows some components representing this flavor type. It is also known that several aliphatic aldehydes and acids generated by lipid oxidation reactions contribute to the full flavor sensation. Polyoxygenated long chain fatty acid precursors as well as protein degradation products are presumed to be responsible for the fatty mouthfeel that occurs when butter is processed.

Figure 4–44 shows the biogenesis of acetoin and diacetyl, resulting from the well-known fatty acid metabolism. The synthesis of δ-lactones has already been mentioned (Section 4.2.3.2). γ-Lactones can be routinely synthesized according to Figure 4–45 (Ohloff, 1969).

4.2.3.14 *The mushroom earthy flavor note*

The mushroom earthy flavor note is mainly represented by 1-octen-3-ol, which is reminiscent of a typical mushroom flavor, and by geosmin, representing the earthy part. A study of flavor constituents in mushrooms has shown that beside 1-octen-3-ol, a range of mainly C_8 components, saturated and unsaturated alcohols and carbonyls are responsible for typical mushroom flavor (Tressl et al., 1982).

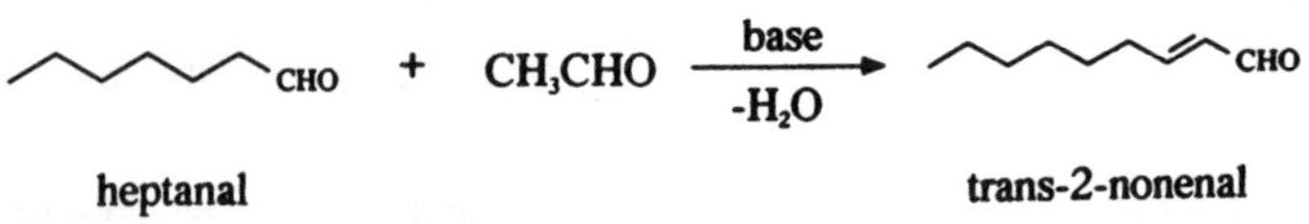

heptanal (CHO) + CH_3CHO —base, $-H_2O$→ trans-2-nonenal (CHO)

Figure 4–42 Synthesis of *trans*-2-nonenal.

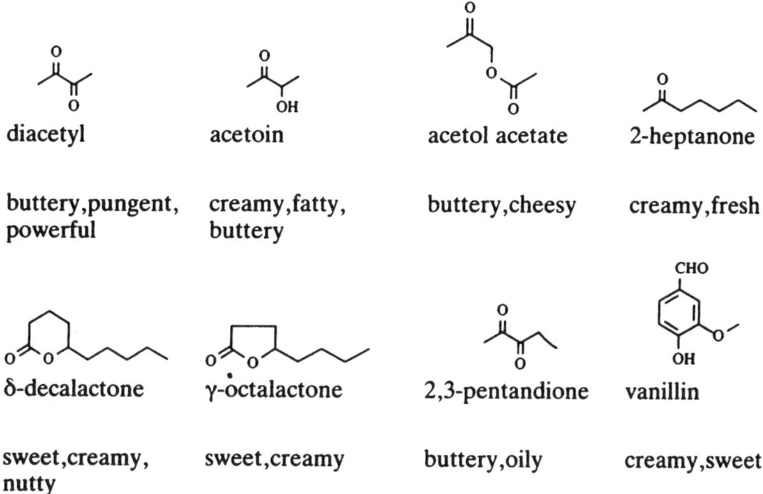

Figure 4–43 Examples of dairy buttery flavor notes.

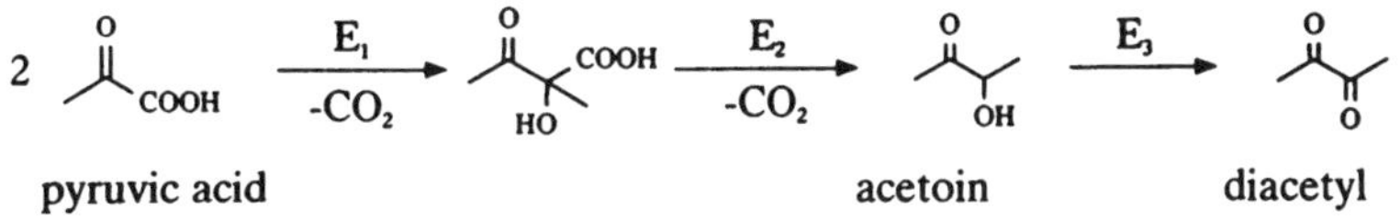

Figure 4–44 Biosynthesis of diacetyl and acetoin. E_1, E_2, . . ., different enzyme systems.

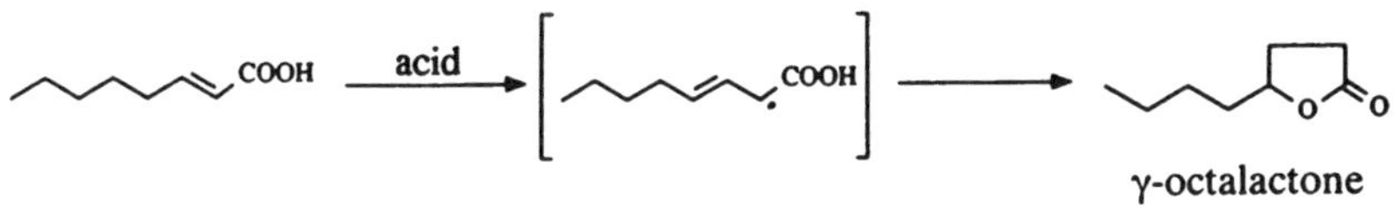

Figure 4–45 Synthesis of γ-octalactone (Ohloff, 1969).

Other synthetic ingredients like 2-octen-4-one and 1-pentyl pyrrole are known to possess mushroom-like flavor notes (Figure 4–46). 4-Terpinenol, 2-ethyl-3-methylthiopyrazine and resorcinol dimethyl ether with quite different chemical structures also have earthy flavor characters. Figure 4–47 shows the biogenetic pathway leading to 1-octen-3-ol, which has been only partially clarified (Tressl et al., 1981). The enzymatic oxidative degradation of linolic acid into C_8 and C_{10} components is observed only in mushrooms. Figure 4–48 shows a recent stereo-specific synthesis of geosmin (Kaiser & Nussbaumer, 1990).

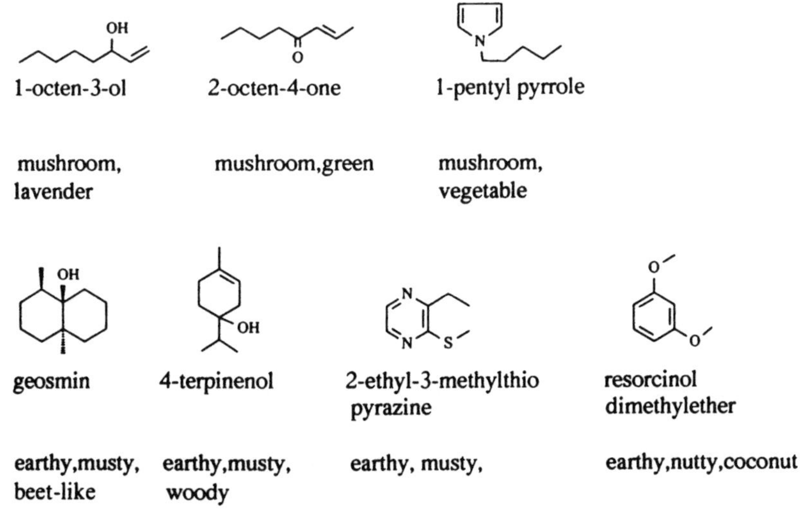

Figure 4–46 Examples of earthy mushroom flavor notes.

4.2.3.15 *The celery soupy flavor note*

The celery soupy flavor note is described as the warm spicy rooty odor, reminiscent of concentrated soup. Typical components are shown in Figure 4–49 (butylidene phthalide, butyl phthalide, 4,5-dimethyl-3-hydroxy-2[5*H*]-furanone, dihydrojasmone and *cis*-jasmone). These notes differ from those described in Section 4.2.3.10. Both have some persistent enhancer character. The distinction to be made is that the celery notes are associated with vegetable applications, whereas the bouillon notes are appropriate in meat flavors.

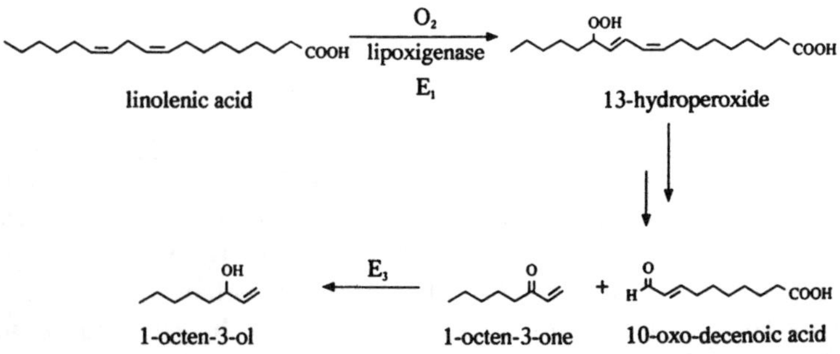

COOH
O_2
lipoxigenase
E_1
linolenic acid
OOH
COOH
13-hydroperoxide
OH
E_3
O
O
H
COOH
1-octen-3-ol
1-octen-3-one
10-oxo-decenoic acid

Figure 4–47 Biogenesis of 1-octen-3-ol (Tressl et al., 1982). E_1, E_2 . . ., different enzyme systems.

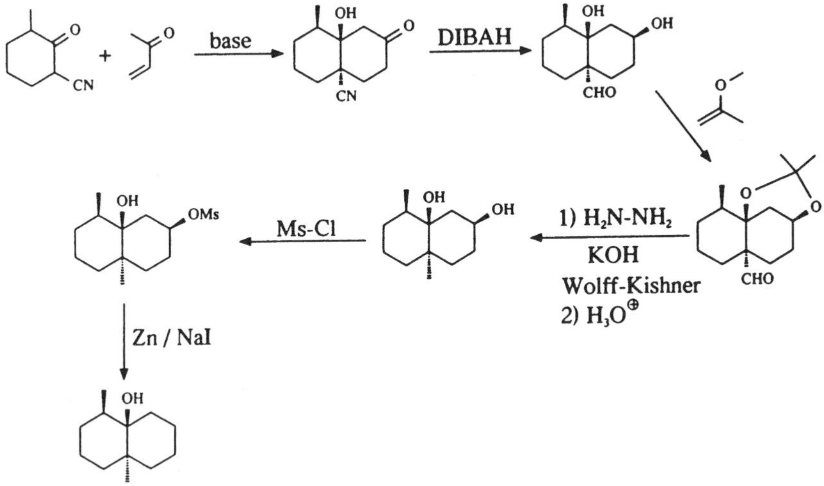

base

DIBAH

1) $H_2N\text{-}NH_2$
KOH
Wolff-Kishner
2) $H_3O^{\oplus}$

Ms-Cl

Zn / NaI

geosmin

Figure 4–48 Synthesis of geosmin (Kaiser & Nussbaumer, 1990). Ms, $Ch_3SO_2^-$(mesyl); DIBAH, $(i\text{-}Bu)_2A1H$.

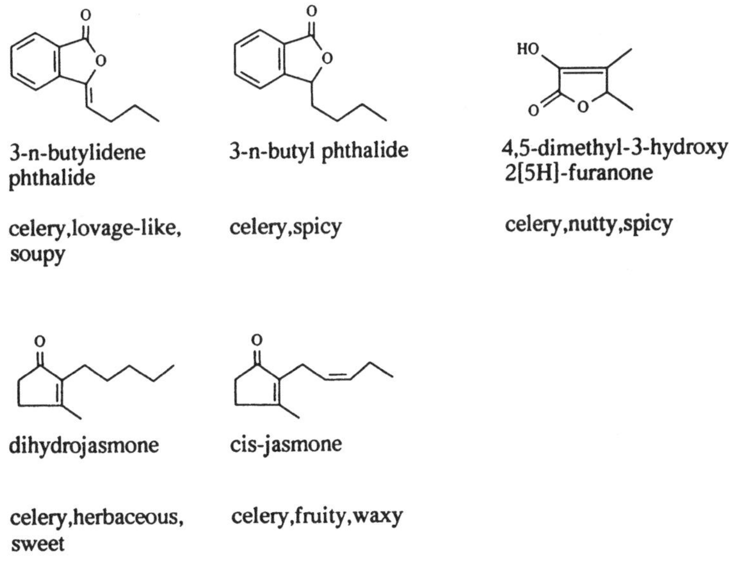

Figure 4–49 Examples of celery soupy flavor notes.

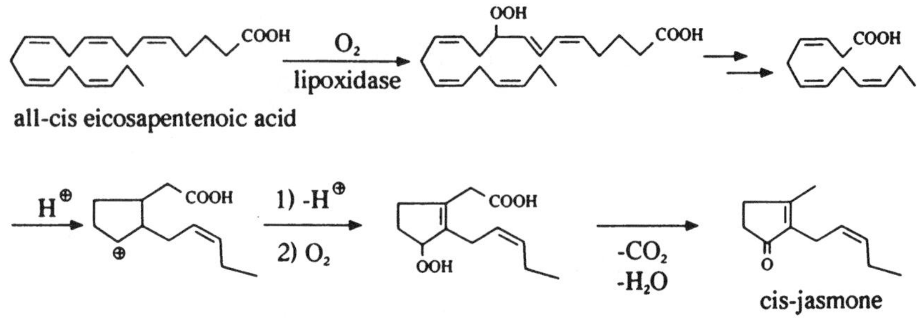

Figure 4–50 Biogenesis of *cis*-jasmone (van der Gen, 1972).

4,5-Dimethyl-3-hydroxy-2[5*H*]-furanone is in its enhancing character comparable to 5-ethyl-4-methyl-3-hydroxy-2[5*H*]-furanone (Figure 4–49). *cis*-jasmone and dihydrojasmone represent more the fruity sweet part of celery flavor. Figure 4–50 shows one of the first proposals for biogenetic formation of *cis*-jasmone (van der Gen, 1972). Investigations led to the idea that a suitable polyunsaturated fatty acid is first cyclized via an allene oxide as an activated intermediate and secondly shortened to the proper length by β-oxidation (Hamberg, 1989). Figure 4–51 shows synthetic routes to dihydrojasmone and 3-*n*-butyl phthalide according to Ho (1974) and Canone et al. (1988).

4.2.3.16 *The sulfurous alliaceous flavor note*

The sulfurous alliaceous flavor notes are vegetable specific notes, which are generally easy to recognize. The odor effect ranges from the simple unpleasant mercaptans (methyl mercaptan) through unsaturated short-chain garlic and onion

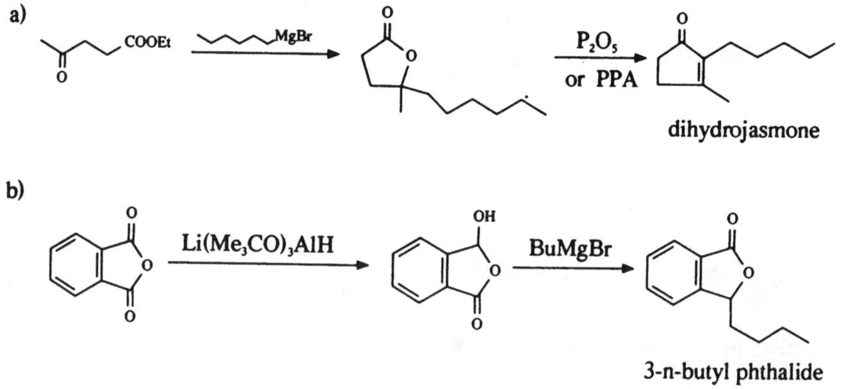

Figure 4–51 Synthesis of (a) dihydrojasmone (Ho, 1974) and (b) butyl phthalide (Canone et al., 1988). PPA, Polyphosphoric acid.

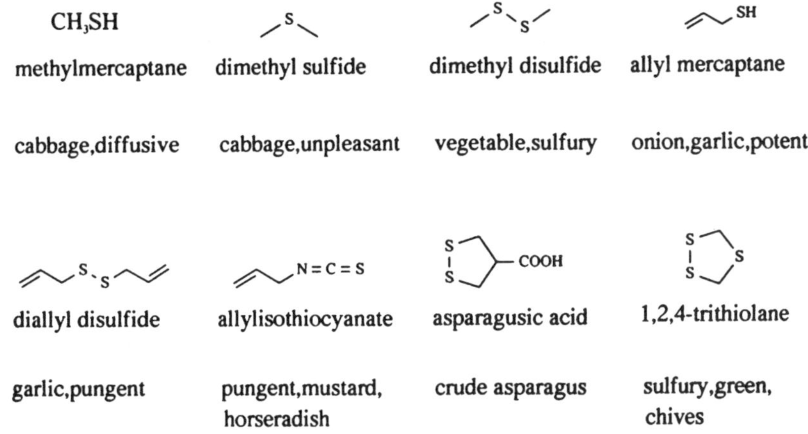

Figure 4–52 Examples of sulfurous alliaceous flavor notes.

compounds (allyl mercaptan, diallyl disulfide) to pleasant distinctly nuanced heterocyclic compounds. Examples therefore are 1,2,4-trithiolane for asparagus, lenthionine for mushroom (shiitake) and 2-methyl-4-propyl-1,3-oxathianes for passion fruit (Figure 4–52).

Sulfur containing components generally represent a very important class of flavorings with extremely low threshold values. They need very special handling because they often show instability (e.g., oxidation) and interactive reactions with other compounds in the flavor mixture. Sufficient knowledge about their behavior is crucial for their correct application. The use of corresponding precursors is, therefore, often preferred. Sulfur containing flavor components are mostly formed during heat processing of plants and food materials. Figure 4–53 shows the formation of aliphatic and cyclic sulfur components. Strecker degradation of methionine leads to methional and by heat treatment to methylmercaptan and related derivatives (Figure 4–53a).

The biogenetic precursor of asparagusic acid has been proved to be isobutyric acid (Parry et al., 1985). 1,2-Dithiole and 1,2,3-trithiane-5-carbonic acid, which contribute to the asparagus flavor, are presumed to be formed by thermal degradation (Figure 4–53b). The synthesis of methional, diallyldisulfide and allylisothiocyanate are outlined in Figure 4–54 as typical examples of organic sulfur chemistry (Pedersen & Becher, 1988).

4.3 SYNTHETIC FLAVOR INGREDIENTS AND THE FUTURE

Synthetic flavor ingredients will certainly remain indispensable and will continue to be used, for many reasons, more in the future than the past. Figure 4–55

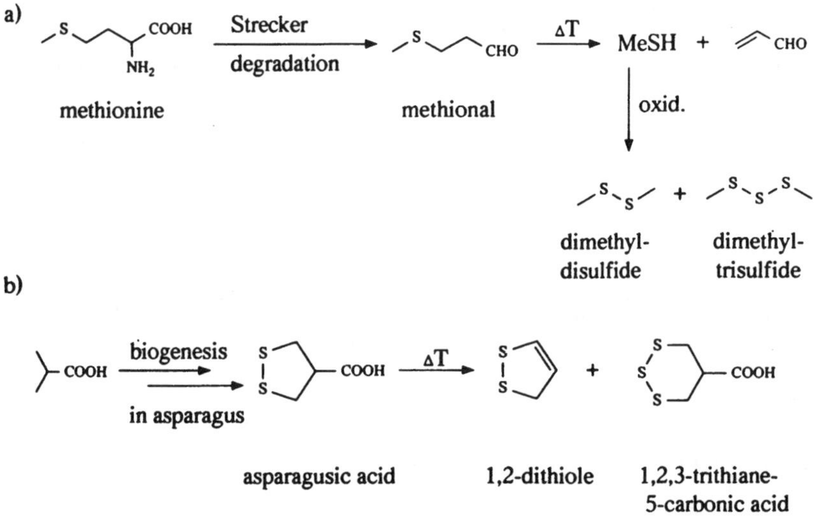

a) methionine —Strecker degradation→ methional —ΔT→ MeSH + CHO; MeSH —oxid.→ dimethyl-disulfide + dimethyl-trisulfide

b) COOH —biogenesis in asparagus→ asparagusic acid —ΔT→ 1,2-dithiole + 1,2,3-trithiane-5-carbonic acid

Figure 4–53 Formation of (a) aliphatic and (b) cyclic sulfur compounds (Parry et al., 1985).

---

shows the increase in number of single flavor components identified in foods and beverages in the last 40 years. The largest increase is observed between 1965 and 1985 due to the progress in performance of new analytical techniques like GC and GC-MS-DS, HPLC, etc.

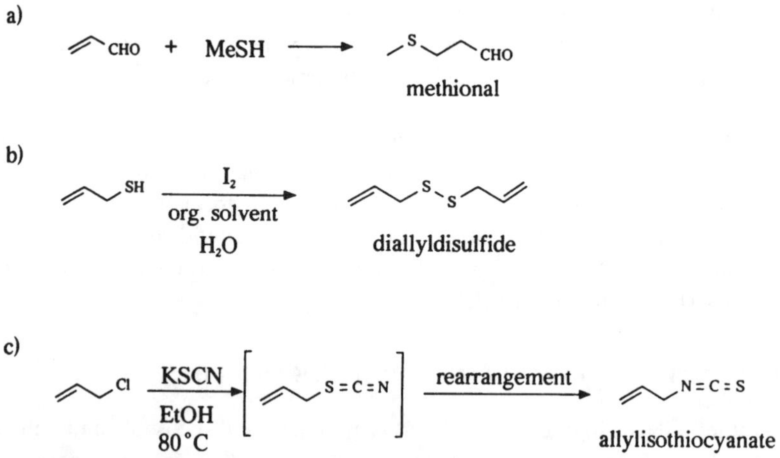

a) CHO + MeSH → methional

b) SH —I_2, org. solvent, H_2O→ diallyldisulfide

c) Cl —KSCN, EtOH, 80°C→ [S=C=N] —rearrangement→ N=C=S allylisothiocyanate

Figure 4–54 Synthesis of methional, diallyldisulfide and allylisothiocyanate.

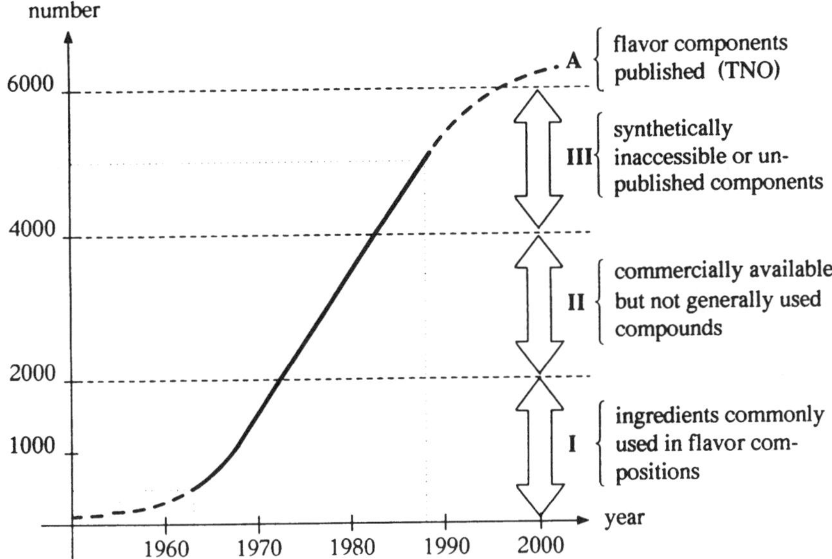

Figure 4–55 Development of flavor ingredients identified in food.

The use of these compounds (curve A) as potential flavoring agents depends mainly on the knowledge of their sensory properties, their commercial availability and toxicological behavior. This is one of the reasons why only a small number of them are expected to be used in the flavor industry (section I). Although their number is growing constantly, they will never reach the potential maximum (section II) which represents approximately two thirds of the whole range of flavor substances actually known. A better understanding of their behavior and interactions (single or in mixtures, including odor and taste threshold, stability, synergism, etc.) appears crucial and essential for the future.

At the same time a precise descriptive language using common food flavor specific terms (considering not only odors but also taste descriptions and threshold values) to describe flavor profiles is needed world-wide to facilitate the dialogue between those involved in creation, application and marketing of flavorings and their customers.

A large number of interesting flavor substances, however, will remain synthetically and economically inaccessible or forbidden for safety reasons. Some of them can be considered as unpublished, company specific know-how (section III).

Flavor science and research will concentrate on special topics like non-volatile flavor precursors (glycosinolates) biotechnologically and Maillard generated flavor compounds (or "bases") which can be used as single ingredient or as building blocks to replace or improve inaccessible flavor notes.

Flavor creation is a dynamic process which constantly evolves with progress in flavor chemistry. Considerable efforts have to be made today to detect and identify one single character impact substance occurring in the sub-ppb range. Sulfur containing flavor compounds with very low odor thresholds remain the most interesting field for future flavor compositions. To use them optimally we have to understand not only their biogenesis but also their physiological effects on perception and organoleptic sensations. In this field, much work remains to be done because we are at present just scratching the surface. Very little is known about relationships between flavor chemicals and how stimulation occurs. Different model systems and molecular mechanisms (Cagan & Care, 1981) are proposed but not yet confirmed. Selected synthetic flavor ingredients will one day be able to play the role they are designed for, being used in quantity and quality, in the same way as if they were formed naturally.

The age of flavor design or flavor engineering postulated in the early 1960s after the appearance of gas chromatography can only begin when all the preliminary work is complete (some well-known companies are already well on the way to achieving this end). Computer aided selection of synthetic ingredients can help the flavorist in his or her composition or formulations to save time and money avoiding "hit or miss" trials which are often caused by overhurried marketing demands to satisfy the needs of new short-life products observed on today's and future markets.

Even if the current trend of preference for natural ingredients continues, synthetically compounded flavors will continuously improve with regard to quality, commercial and ecological aspects. More investigations on minor compounds have to be carried out to avoid using global building blocks with all the risk they may include. A clarified and simplified legislation concerning synthetic flavor ingredients will be helpful in the future. First steps in this direction have already been initiated in Europe with a new EEC flavor directive (Muerman, 1989), but corresponding basic regulations are requested world-wide to harmonize the needs of all consumers and the flavor and food industries.

REFERENCES

J. Adda, ed., in *Development in Food Science,* 10. *Progress in Flavour Research,* Elsevier, Amsterdam (1985).

J.E. Amoore, Odor theory and odor classification in fragrance chemistry, in *The Science of the Sense of Smell,* ed. E.T. Theimer, Academic Press, New York (1982) 27–76.

D.A. Andrews and C.N. Hindley, US Pat. 3 994 936 (assigned to Hoffmann-La Roche, 1973).

S. Arctander, in *Perfume and Flavor Chemicals,* Vols. 1, 2, published by the author, Montclair, NJ (1969).

W. Baltes and G. Bochmann, *Z. Lebensm.-Unters Forsch.* 184 (1987) 485.

P.G. Bay, US Pat. 3,062,874 (assigned to Glidden, 1962).

P.Z. Bedoukian, in *Perfumery and Flavoring Synthetics,* Allured Publishing, Wheaton (1986).

P.Z. Bedoukian, *Perfumer and Flavorist* 14 (1989) 2.

H.-D. Belitz and W. Grosch, in *Lehrbuch der Lebensmittelchemie,* Springer, Berlin (1987) 297.

H.-D. Belitz and W. Grosch, in *Lehrbuch der Lebensmittelchemie,* Springer, Berlin (1987) 18–19.

H.-D. Belitz and W. Grosch, in *Lehrbuch der Lebensmittelchemie,* Springer, Berlin (1987) 206–272.

J.M.H. Bemelmans and M.C. ten Noever de Brauw, Analysis of off-flavours in food, in *Proc. Int. Symp. Aroma Research,* Pudoc, Wageningen (1975) 85–93.

R.G. Berger, S. Nitz and P. Schreier, eds., *Topics in Flavor Research,* H. Eichhorn Verlag, Marzling-Hangenham (1985).

G.G. Birch and M.G. Lindley, in *Developments in Food Flavours,* Elsevier, Amsterdam (1986).

M.W. Breuninger, Eur. Pat. 51 229 (assigned to Hoffmann-La Roche, 1980).

G. Buechi and H. Wuest, *J. Am. Chem. Soc.* 96 (1974) 7573.

R.G. Buttery, W.F. Haddon, R.M. Seifert and J.G. Turnbaugh, *J. Agric. Food Chem.* 32 (1984) 674.

R.H. Cagan and R.M. Care, eds., in *Biochemistry of Taste and Olfaction,* Academic Press, New York (1981).

P. Canone, J. Plamondon and M. Akssira, *Tetrahedron* 44 (1988) 2903.

Chemical Abstracts Service (editorial office), *Chemical Abstracts,* Vol. 111, American Chemical Society, Columbus, OH (1989).

Chemical Source Association, *Flavor and Fragrance Materials—1889,* Allured Publishing Co., Wheaton (1989).

F.M. Clydesdale, ed., in *Critical Reviews in Foods Science and Nutrition,* Vol. 28, CRC Press, Boca Raton, FL (1989).

R. Croteau, M. Felton and R. Ronald, *Arch. Biochem. Biophys.* 200 (1980) 534.

R. Croteau, *Chem. Rev.* 87 (1987) 929.

P. Dubs and R. Stüssi, *Synthesis* (1976) 696.

N. Elming and N. Clauson-Kaas, *Acta Chem. Scand.* 6 (1952) 867.

C. Eriksson, ed., in *Maillard Reaction in Food. Chemical Physiological and Technological Aspects,* Pergamon Press, Oxford (1981).

W.J. Evers, H.H. Heinsohn, B.J. Mayers and A. Sanderson, *Am. Chem. Soc. Symp. Ser.* 26 (1976) 184.

G. Fenaroli, in *Fenaroli's Handbook of Flavor Ingredients,* Vols. 1, 2, CRC Press, Boca Raton, FL (1975).

I. Flament, DOS 3 048 031 (assigned to Firmenich, 1979).

J. Fleischer, K. Bauer and R. Hopp, DOS 2 109 456 (assigned to Haarman and Reimer, 1972).

E. Gildemeister and F. Hoffmann, eds. in *Die ätherischen Oele,* Akademie Verlag, Berlin (1968).

The *Givaudan Index,* Givaudan-Delawanna Inc., New York (1962) 252.

E. Guenther, in *The Essential Oils,* Vols. 1–6, D. van Nostrand, New York (1952).

M. Hamberg, *J. Am. Oil Chem. Soc.* 66 (1989) 1445.

G.H. Hartman, J.T. Carlin, J.D. Scheide and C.-T. Ho, *J. Agric. Food Chem.* 32 (1984) 1015.

H.B. Heath, in *Source Book of Flavours,* AVI Publishing Co., Westport, CT (1981).

H.B. Heath and G.A. Reineccius, in *Flavor Chemistry and Technology,* AVI Publishing Co., Westport, CT (1986).

H.B. Heath, in *Flavor Technology: Profiles, Products, Application,* AVI Publishing Co., Westport, CT (1978).

D. Helmlinger and P. Naegeli, US Pat. 3 943 177 (assigned to Givaudan, 1976).

J.E. Hodge, *Am. Soc. Brew. Proc.* (1963) 84.

Y. Houminer, *Thermal Degradation of Carbohydrates* in *Molecular Structure and Function of Food Carbohydrate,* eds. G.G. Birch and L.F. Green, Applied Science Publishers, London (1973) 133–154.

T.-L. Ho, *Synth. Commun.* 4 (1974) 265.

U. Huber and H.J. Wild, US Pats. 4 181 666 and 4 208 338 (assigned to Givaudan, 1977).

U. Huber, US Pat, 4 285 984 (assigned to Givaudan, 1977).

S.-S. Hwang, J.T. Carlin, Y. Bao, G.J. Hartman and C.-T. Ho, *J. Agric. Food Chem.* 34 (1986) 538.

A. Ijima and K. Takahashi, *Chem. Pharm. Bull.* 21 (1973) 215.

J.A. Pat. 6 0258 168 (assigned to Koei Chem. Ind. KK, 1984).

W.G. Jennings and R. Tressl, *Chem. Mikrobiol. Technol. Lebensm.* 3 (1974) 52.

R. Kaiser and C. Nussbaumer, *Helv. Chim. Acta* 73(1) (1990) 133–139.

J.M. Kerr and C.J. Suckling, *Tetrahedron Lett.* 29 (1988) 5545.

L.B. Kier, *J. Pharm. Sci.* 61 (1972) 394.

F. Ledl, G. Fritul, H. Hiebl, O. Pachmays and T. Severin, Degradation of Maillard products, in *Amino-Carbonyl Reactions in Food and Biological Systems,* eds. M. Fujimaki, M. Namiki and H. Kato, Elsevier, Amsterdam (1986) 173.

H. Maarse and R. Belz, in *Isolation, Separation and Identification of Volatile Compounds in Aroma Research,* Akademie Verlag, Berlin (1985).

A.J. Macleod, *Instrumental Methods in Food Analysis,* Elek Science, London (1973).

G. MacLeod, The scientific and technological basis of meat flavors, in *Developments in Food Flavors,* eds. G.G. Birch and M.G. Lindley, Elsevier, London (1986) 191–223.

L.C. Maillard, *C.R. Seances Acad. Sci.* 164 (1912) 66.

P. Manitto, The isoprenoids, in *Biosynthesis of Natural Products,* Ellis Horwood, Chichester (1981) pp. 215–297.

H. Masuda, M. Yoshida and T. Shibomoto, *J. Agric. Food Chem.* 29 (1981) 944.

H. Masuda and S. Mihora, *J. Agric. Food Chem.* 34 (1986) 377.

H. Mayer and O. Isler, in *Total Synthesis in Carotenoids,* ed. O. Isler, Birkhaüser, Basel (1971) 328–575.

F.D. Mills and J.E. Hodge, *Carbohydrate Res.* 51 (1976) 9.

B.D. Mookherjee, C. Ciacino, E.A. Karoll and M.H. Vock, DOS 2 166 323 (assigned to IFF, 1974).

M. Moo-Young, ed., in *Comprehensive Biotechnology,* Vols. 1–4, Pergamon Press, Oxford (1985).

H.E. Muerman, *Perfumer and Flavorist* 14 (1989) 1.

A. Nissen, W. Rebafka and W. Aquila, Eur. Pat. 21 074 (assigned to BASF, 1981).

R. Noyori and M. Kitamura, Enantioselective catalysis with metal complexes, in *Modern Synthetic Methods 1989,* ed. R. Scheffold, Springer, Berlin (1989) 181.

G. Ohloff, Chemie der Geruchs-u. Geschmacksstofe, in *Fortschritte der chemischen Forschung,* ed. F. Boschke, Springer, Berlin (1969) 185–251.

G.A.M. Ouweland and H.G. Peer, Austrian Pat. 340 748 (assigned to Unilever, 1977).

P. Oberhänsli, CH Pat. 576 416 (assigned to Givaudan, 1976).

R.J. Parry, E.E. Mizusawa, I.C. Chiu, M.V. Naidu and M. Ricciardone, *J. Am. Chem. Soc.* 107 (1985) 2585.

C.T. Pedersen and J. Becher, eds., in *Developments in the Organic Chemistry of Sulfur,* Gorden and Breach, New York (1988).

G. Peine, *Chem. Ber.* 17 (1884) 2109.

N.A. Porter, L.S. Lehmann, B.A. Weber and K.J. Smith, *J. Am. Chem. Soc.* 103 (1981) 6447.

H.-J. Rehm and G. Reed, in *Biotechnology,* Vols. 1–8, VCH Verlagsges., Weinhem (1989).

G.P. Rizzi, *Food Reviews Internat.* 4(3) (1988) 375.

W. Roedel and U. Hempel, *Die Nahrung* 18 (1974) 133.

L. Ruzicka, A. Eschenmoser and H. Heusser, *Experientia* 9 (1953) 357.

G. Saucy and R. Marbeth, *Helv. Chim. Acta* 50 (1967) 1158.

P. Schreier, ed., *Analysis of Volatiles,* de Gruyter, Berlin (1984).

P. Schreier, in *Chromatographic Studies of Biogenesis of Plant Volatiles,* A. Hüthig Verlag, Heidelberg (1984).

P. Schreier, in *Chromatographic Studies of Biogenesis of Plant Volatiles,* A Hüthig Verlag, Heidelberg (1984) 52–126.

P. Schreier, in *Techniques of Analysis,* in *Chromatographic Studies of Biogenesis of Plant Volatiles,* A. Hüthig Verlag, Heidelberg (1984) 1–51.

H.W. Schultz, E.A. Day and L.M. Libbey, in *The Chemistry and Physiology of Flavors,* AVI Publishing, Westport, CT (1967) 331.

T.W.G. Solomons, Biosynthesis of isoprenoids, in *Organic Chemistry,* John Wiley, New York (1988) pp. 1080–1085.

H. Sugimoto and K. Hirai, *Heterocycles* 26 (1987) 13.

H. Sulser, J. de Pizzol and W. Büchi, *J. Food Sci.* 32 (1967) 611.

R. Tressl, D. Bahri and K.-H. Engel, *J. Agric. Food Chem.* 30 (1982) 89.

R. Tressl, D. Bahri and K.-H. Engel, Lipid oxidation in fruits and vegetables, *Am. Chem. Soc. Symp. Series* 170 (1981) 213.

R. Tressl, M. Holzer and M. Apetz, Biogenesis of volatiles in fruits and vegetables, in *Aroma Research,* eds. H. Maarse and P.J. Groenen, Pudoc, Wageningen (1975) 41.

R. Tressl, F. Drawert, W. Heimann and R. Emberger, *Z. Lebensm.-Unters. Forsch.* 144 (1970) 4.

A. van der Gen, *Perf. Cosm. Sav. France* 2 (1972) 356.

S. van Straten and H. Maarse, ed., in *Volatile Compounds in Food,* 5th edn., TNO, Zeist (1988) Suppl. 5.

O. Wallach, in *Terpene und Campher,* Vit., Leipzig (1914).

G.R. Waller and M.S. Feather, eds., in *The Maillard Reaction in Foods and Nutrition, ACS Symp, Series* 215, American Chem. Soc., Washington, DC (1983).

H.J. Wild, *Chem. Ind.* (1988) 580.

H.J. Wild, WO Pat. 87'03 287 (assigned to Givaudan, 1986).

E. Ziegler, in *Die natürlichen und künstlichen Aromen,* A. Hüthig Verl., Heidelberg (1982).

Chapter 5

Quality Control of Flavorings and Their Raw Materials

Günter Matheis

5.1 INTRODUCTION

5.1.1 *The Sensogram of a Food*

Like every other creature, a human being perceives the physical environment through his or her senses, that is to say through impressions which the sensory organs receive from the surroundings, register and compare with earlier impressions. There is no unanimous opinion of the number of human senses. According to Marks (1987), humans have eight senses: taste, smell, sight, hearing, pain, touch, cold and warmth. However, if one groups together touch, pain and perception of warmth and cold as a single sense (somatosensory perception) (Burdach, 1988), humans have five senses.

The first contact of a human being with a food is usually through one of the senses of sight, smell (with the breath through the nose), hearing (e.g., a steak sizzling in the pan) or touch (e.g., taking hold of an apple), or from two or three of these perceptions at the same time. The next impressions are usually touch (with the lips and in the mouth; when cold, warmth and pain can also be felt) and renewed hearing (the sound of chewing). Immediately after this come tasting and renewed smelling, although this time—as will be seen later—it takes place indirectly in the retronasal area. All these sensations affect our overall judgment of the food. The sensory perception that is frequently called "taste" in everyday speech is in fact very complex. This complex perception, shown in Figure 5–1 as the sensogram of a food, does not take account of the way the individual perceptions are staggered over a period of time.

5.1.1.1 *Taste*

Taste is perceived mainly on the tongue, but also in the oral cavity (on the soft palate, the rear wall of the pharynx and on the epiglottis). Receptors on the tongue register the four classic types of taste, sweet, sour, salty and bitter, chiefly

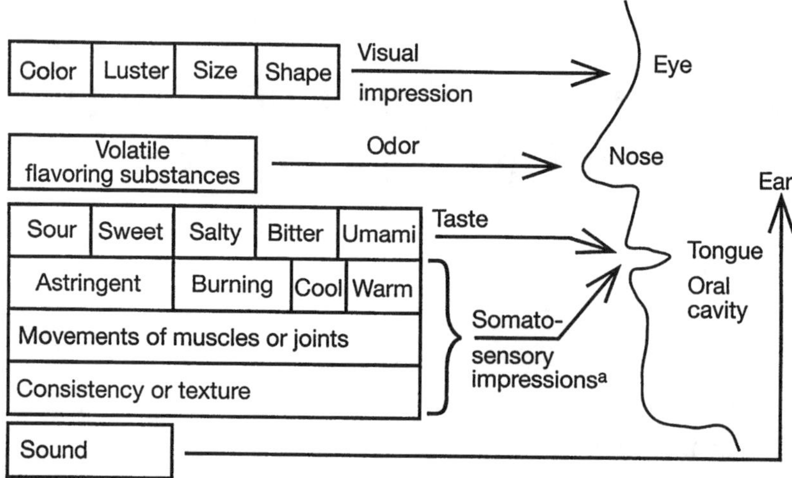

Figure 5–1 Sensogram of a food. *Source:* G. Matheis, Taste, Odor, Aroma, and Flavor, *Dragoco Report*, Vol. 39, pp. 50–65, © 1994, Dragoco.

on specific areas of the tongue (Figure 5–2). Sour and bitter are also perceived to a certain extent on the roof of the mouth.

Recently, the use of the term "umami" has become more widespread in connection with the description of taste impressions. The word comes from the Japanese, and can best be translated as palatability. Often, umami is referred to as the fifth basic taste impression. It is still not known whereabouts on the tongue the umami receptors are situated.

An adult human being has four to six thousand taste receptors, older people only two to three thousand, whereas newborn babies possess between eight and twelve thousand (Reher & Stahl-Biskup, 1987). They are constantly being renewed by a process of "moulting," and they live for about 10 days.

Sour taste is generated by H^+ ions (hydrogen ions, protons) or, to be chemically more precise, by the hydronium ions (H_3O^+ ions) of acids. However, this characteristic alone does not constitute the sour taste impression. Sensorily perceived acidity is not always directly proportional to the acidity measured chemically (pH value) and the structure of the acid molecule is of some significance in the sensory impression "sour" (Heath, 1981).

In many foods (e.g., fruits, processed foods and soft drinks), the taste is brought about by organic acids (e.g., citric, lactic, tartaric, or acetic acids). Phosphoric acid is the only inorganic acid which is of any significance as a souring agent in foodstuffs (e.g., in the soft drinks industry).

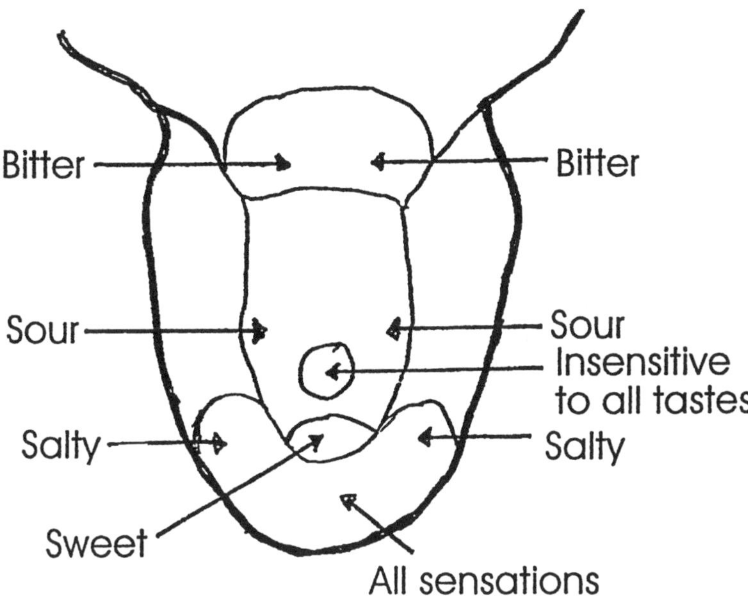

Figure 5–2 Schematic diagram of the tongue showing the areas of maximum sensitivity to the four basic taste impressions. *Source:* G. Lösing et al., Quality Control of Flavorings and Their Raw Materials, *Dragoco Report,* Vol. 42, pp. 93–135, © 1997, Dragoco.

Salty taste is brought about by low molecular weight inorganic salts, for instance by common salt (NaCl), potassium chloride (KCl), sodium bromide (NaBr) or sodium iodide (NaI). Common salt is the only salt which is described as "purely salty." All commonly used salt substitutes generate combinations of impressions; for instance KCl and NaBr are described as predominantly but not purely salty, and potassium bromide (KBr) as salty and bitter. Higher molecular weight salts can taste purely bitter (e.g., potassium iodide and cesium chloride) or sweet (e.g., lead acetate and beryllium chloride).

The mechanism which lies at the root of the salty impression has not been fully explained. It is generally accepted, however, that the ionic character of the salt is a prerequisite, and that the anionic part determines what taste notes arise in addition to salty, or else make the salty notes disappear altogether.

Sweet taste is as a rule automatically associated with carbohydrates such as sucrose. There are, however, a number of other compounds that taste sweet. They include polyalcohols (sorbitol, mannitol, xylitol), the familiar synthetic intense sweeteners (saccharin, cyclamate, aspartame, acesulfam K), amino acids and many others.

At the end of the 1960s a so-called AH/B structural unit was described (Shallenberger & Acree, 1967) as a typical feature of a sweet-tasting compound. Later studies (Kier, 1972; Höltje & Kier, 1974) showed that a third position in the molecule of a compound is responsible for the sweet taste; this has been designated as X, so that an AH/B/X-structure must be present. Here, A and B are electro-negative atoms (e.g., oxygen, nitrogen or chlorine), H is a hydrogen atom and X the non-polar remainder of the molecule. The sweet taste receptor on the tongue has a complementary arrangement so that hydrogen bridges can form between the AH/B structures of the sweet-tasting molecule and the receptors, whereas the non-polar remainder X fits into a kind of pocket in the receptor (shown schematically in Figure 5–3).

For the compound to taste sweet, the distance between A and B must be between 2.5 and 4.0 Å. Figure 5–4 shows the AH/B/X structures of fructose and saccharin.

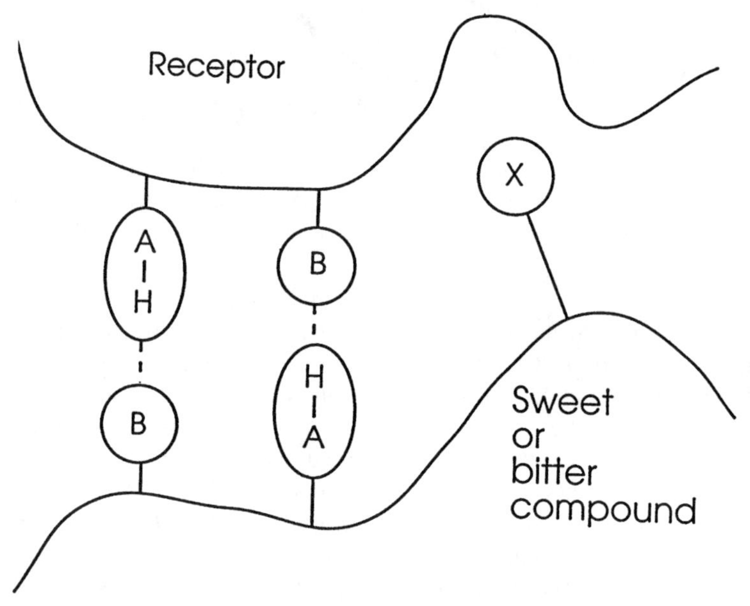

Figure 5–3 Schematic representation of the AH/B/X structure of sweet and bitter compounds and of taste receptors. Note: - - - Hydrogen bridges. Sweet compounds: Distance between A and B = 2.5 to 4 Å. Bitter compounds: Distance between A and B = 1.0 to 1.5 Å. *Source:* G. Matheis, Taste, Odor, Aroma, and Flavor, *Dragoco Report,* Vol. 39, pp. 50–65, © 1994, Dragoco.

The AH/B structure can also explain the **bitter taste** of a compound (see Figure 5–3). Here, the distance between A and B, at 1.0 to 1.5 Å, is smaller than in the case of sweet-tasting substances (Kubota & Kubo, 1969). Figure 5–5 demonstrates the AH/B structure of the bitter-tasting diterpene isodonal.

Among the most important compounds which bring about the **umami** taste impression are the salts of glutamic acid (chiefly sodium glutamate), and the disodium salts of the purine-5′-monophosphates, particulary inosine-5′-monophosphate (IMP), guanosine-5′-monophosphate (GMP) and adenosine-5′-monophosphate (AMP). A compound with umami effect has two negative charges which lie between 3 and 9 (preferably 4 to 6) carbon or other atoms apart (shown schematically in Figure 5–6) (Ney, 1987). This requirement is fully met by monosodium glutamate (MSG), IMP, GMP (Figure 5–6) and some other compounds (for example, succinic acid). AMP, on the other hand, has a negative charge only at one end of the molecule, whereas there is an amino group instead of a charge at the other end (Figure 5–6). This is reflected in the fact that the umami effect of AMP is significantly less than that of IMP, GMP and MSG.

The reader will doubtless have noticed already that the compounds mentioned above are familiar as flavor enhancers. In fact, no clear boundary can be drawn here between the fifth basic taste, umami, and the properties of flavor enhancement.

5.1.1.2 *Trigeminal Perception*

The trigeminus (the nervus trigeminus is the fifth cranial nerve) is classed as belonging to the somato-sensory system (Burdach, 1988; Prescott, 1994). This is why the terms “trigeminal perception” and “somato-sensory impressions” are often used synonymously in the literature. Recently, the old term “common chemical sense” has come back into use, together with the newly coined term “chemesthe-

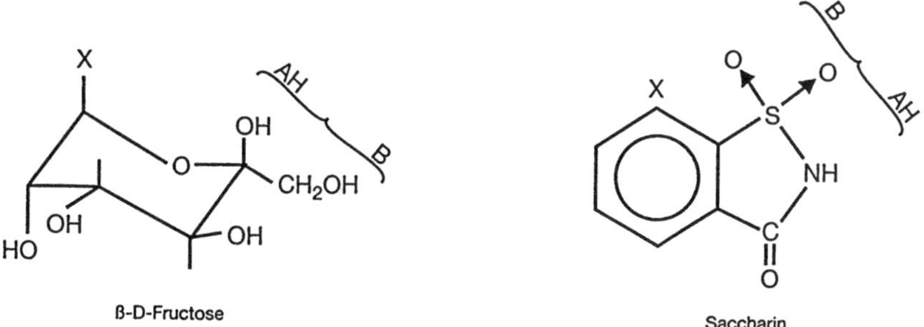

Figure 5–4 AH/B/X Structures for fructose and saccharin. *Source:* G. Matheis, Taste, Odor, Aroma, and Flavor, *Dragoco Report,* Vol. 39, pp. 50–65, © 1994, Dragoco.

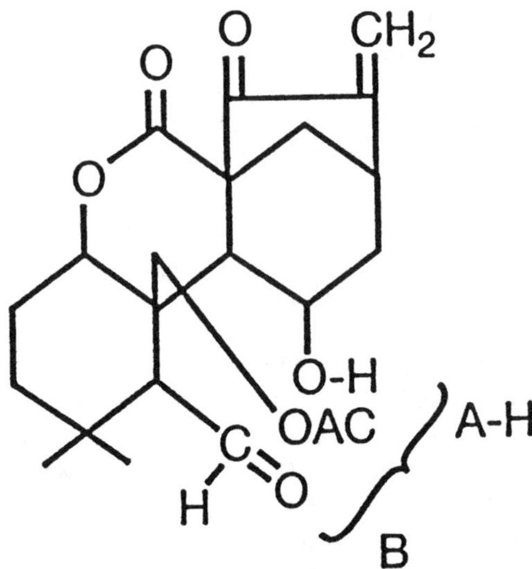

Figure 5–5 AH/B Structure of the bitter-tasting diterpene isodonal. *Source:* G. Matheis, Taste, Odor, Aroma, and Flavor, *Dragoco Report,* Vol. 39, pp. 50–65, © 1994, Dragoco.

sis" (derived from "somesthesis," which is the sensation of touch on the skin) (Lawless & Lee, 1993; Green, 1996).

The trigeminus innervates the entire facial area, particularly the eyes, nose and mouth. It has three principal branches (trigeminus means triple), namely the ophthalmic, maxillary and mandibular branches, and has a number of sensory and motor functions. We shall consider exclusively the sensory functions in the nasal and oral cavities. Trigeminal nerve endings have been identified in the mucosa of the nasal and oral cavities and on the surface of the tongue (Burdach, 1988; Silver & Finger, 1991).

Trigeminal perceptions in the nasal and oral cavities range from only very slightly stimulating through to painful (examples in Table 5–1). Substances which trigger these impressions (examples in Table 5–1 and Figure 5–7) are of value in numerous foods and drinks. Compounds which give a **burning** impression in the mouth (notice how close this is to pain) can be divided into the four groups: o-methoxyphenols, acid amides, mustard oils (isothiocyanates) and disulfides (Figure 5–7). They all contain two centers with double bonds (marked with arrows in Figure 5–7. The chemical groups that are the carriers of these double bonds are listed in Figure 5–7.

The impression of **astringency** is brought about chiefly by two classes of compounds: hydrolyzable tannins (hydrolyzable tanning agents) and non-hydrolyz-

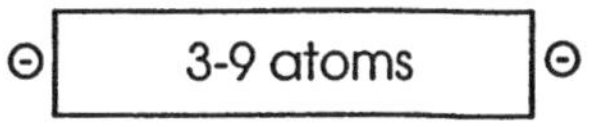

⊖ 3-9 atoms ⊖

structure of a umami compound

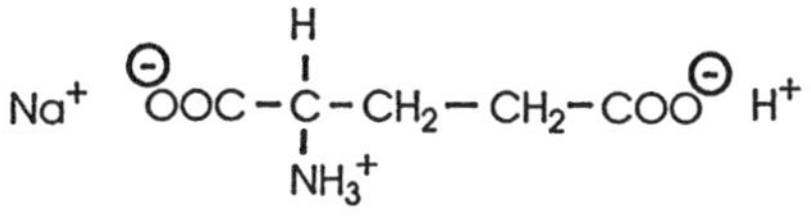

$Na^+ \ {}^{\ominus}OOC-CH(NH_3^+)-CH_2-CH_2-COO^{\ominus} \ H^+$

sodium glutamate

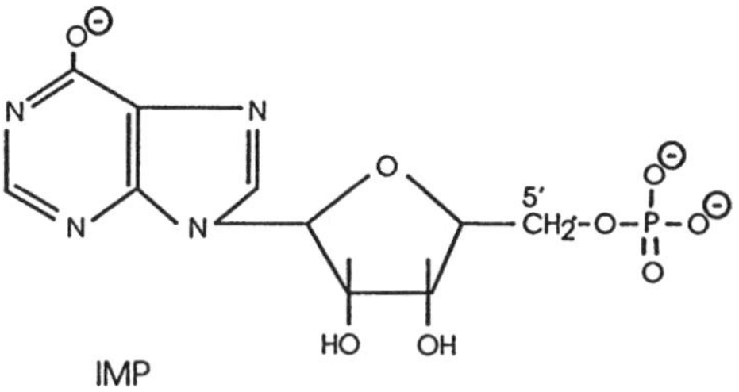

IMP

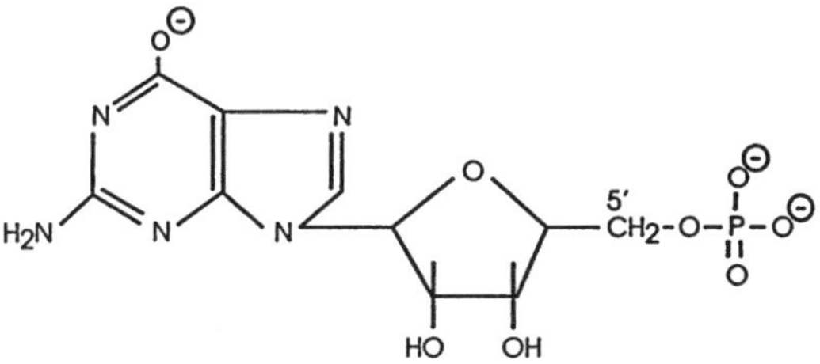

GMP

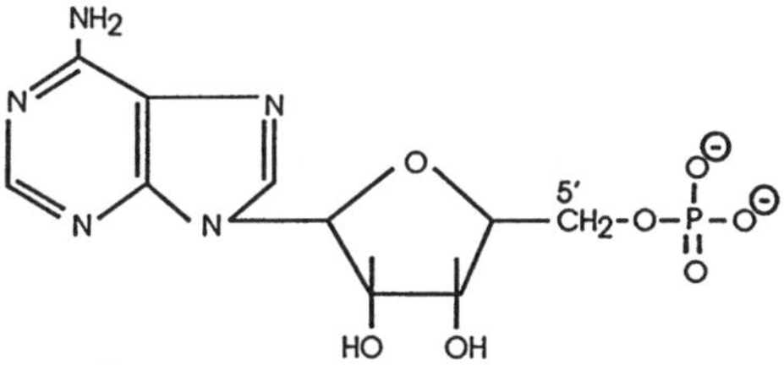

AMP

Figure 5–6 A selection of compounds which bring about the umami impression. *Source:* G. Matheis, Taste, Odor, Aroma, and Flavor, *Dragoco Report,* Vol. 39, pp. 50–65, © 1994, Dragoco.

Table 5–1 Examples of trigeminal stimuli

Nasal trigeminal stimuli	*Triggers (examples)*	*Oral trigeminal stimuli*	*Triggers (examples)*
Piquant[a]	Ethanol	Piquant[a]	Ethanol, capsaicin, piperine
Painful[a]	Ammonia, acetic acid	Painful[a]	Capsaicin, piperine
Pungent[a]	Ammonia, acetic acid	Burning[a]	Capsaicin, piperine
Tingling	Carbonic acid	Cooling	Menthol
Smell of burning	Tobacco smoke	Warming	Ethanol
		Astringent	Tannins

[a]Fleeting transitions between the individual stimuli, dependent on the concentration of the triggering substance.

Source: G. Matheis, Trigeminal Perceptions in the Nasal and Oral Cavities, *Dragoco Report*, Vol. 40, pp. 72–82, © 1995, Dragoco.

able tannins (condensed tannins) (Clifford, 1986). Both classes of compounds are polyphenols (Figure 5–8). The astringency impression arises because the tannins cause proteins and glycoproteins (i.e., proteins with a sugar moiety) in the saliva to precipitate (Figure 5–9), in such a way that the "lubricating" or "oiling" effect of the saliva proteins is lost (Clifford, 1986).

Apart from the tannins, other compounds which have no protein-precipitating properties may have an astringent effect; these include sinapin and chlorogenic acid, although there is some uncertainty because astringency and bitterness are frequently confused (Clifford, 1986). Some compounds with an astringent effect are in fact bitter as well, although only relatively few bitter-tasting substances are also astringent.

The impression of **cooling** is brought about by menthol (key compound of peppermint oil) and through melting and dissolving processes. Uncertainty still prevails about the physiological process involved in the cooling effect of menthol (Heath, 1981). Familiar examples of melting processes include coconut and palm kernel fat, which are solid at room temperature. In the mouth they melt under the warmth drawn from the oral cavity. This results in a clearly perceptible cooling effect (analogous to the cooling effect of sweating). Less familiar is the pleasant cooling effect of glucose monohydrate (glucose in which each molecule contains a molecule of water bound as water of crystallization) (Graefe, 1961). Warmth for the dissolving process is rapidly extracted from the oral cavity through the strong negative heat of solution of glucose monohydrate (−106 joules per gram at 25°C).

5.1.1.3 *Odor*

The perception of odor takes place in the upper part of the nasal cavity in the olfactory epithelium (regio olfactoria). This is an area of the nasal mucosa some 5 to

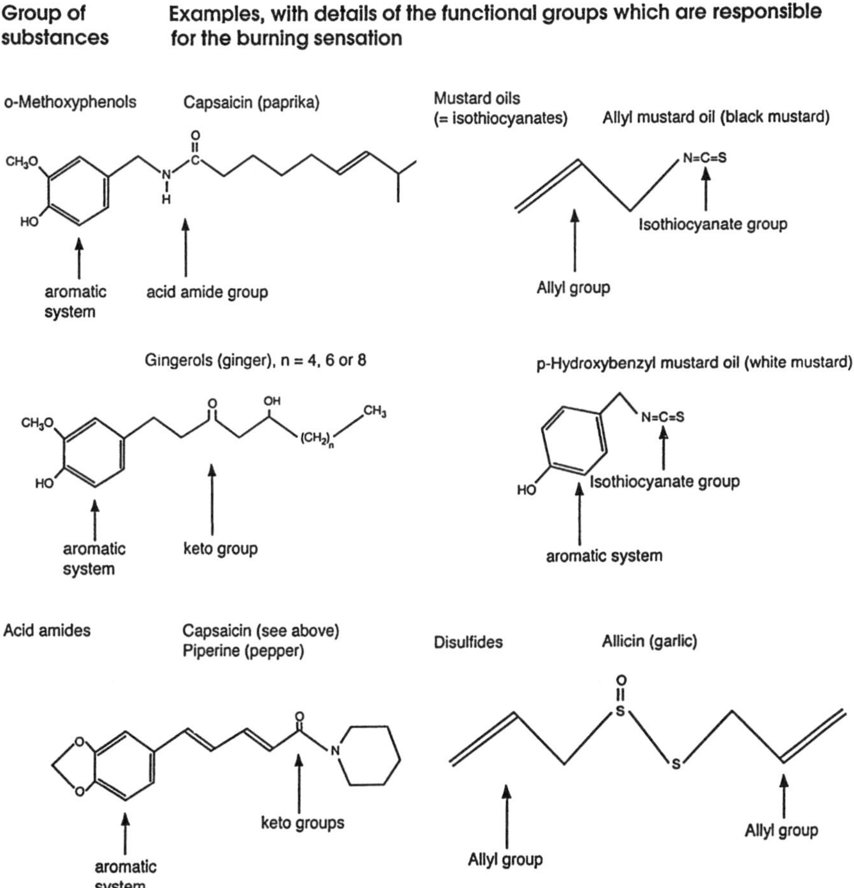

Figure 5–7 A selection of compounds which bring about a burning sensation in the mouth. *Source:* G. Matheis, Taste, Odor, Aroma, and Flavor, *Dragoco Report,* Vol. 39, pp. 50–65, © 1994, Dragoco.

---

10 cm^2, which contains between 3 and 50 million receptors for the sense of smell (Rothe, 1978; Thomson, 1986; Stahl-Biskup & Reher, 1987). Volatile aroma chemicals reach the receptors either with the breath directly through the nose or indirectly via the retronasal cavity, where they become volatile as the food is chewed, broken down, permeated by saliva and warmed (Figure 5–10).

A human being can distinguish between two and four thousand different odor impressions, although with increasing age the ability to recognize smells diminishes (Stahl-Biskup & Reher, 1987). There are more than 17,000 odor chemicals

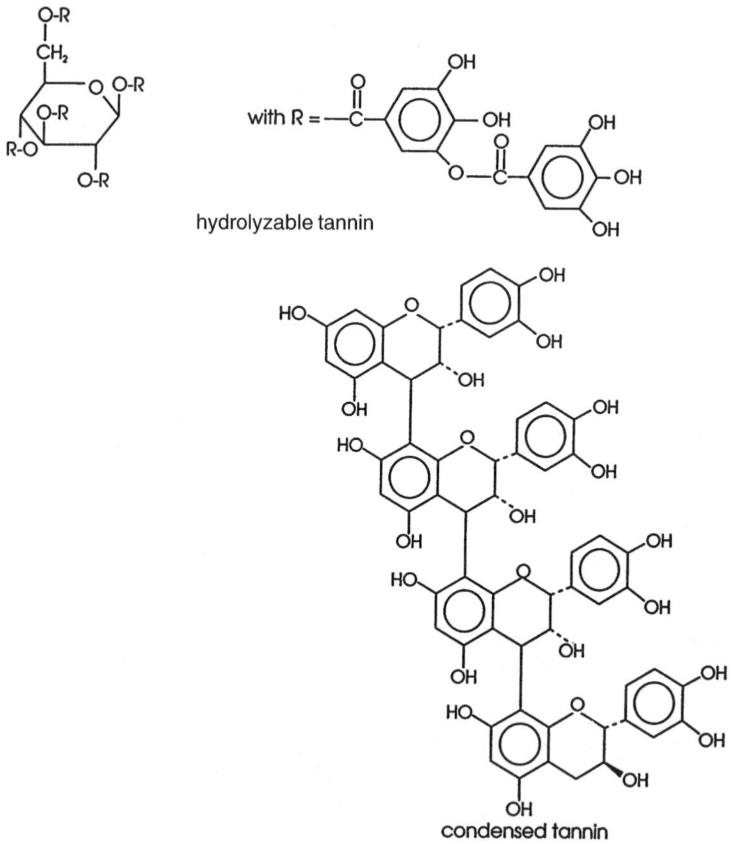

hydrolyzable tannin

condensed tannin

Figure 5–8 Examples of a hydrolyzable and a condensed tannin. *Source:* G. Matheis, Trigeminal Perceptions in the Nasal and Oral Cavities, *Dragoco Report,* Vol. 40, pp. 72–82, © 1995, Dragoco.

(Thomson, 1986). There are at least 1,000 types of odor receptor proteins and only one or very few types of receptor proteins exist in each receptor cell (Laing & Jinks, 1996).

5.1.2 *The Flavor of a Food*

The flavor of a food is a sensory impression, namely the combined effect of taste, odor and trigeminal impressions in the oral and nasal cavities (Figures 5–11 and 5–12).

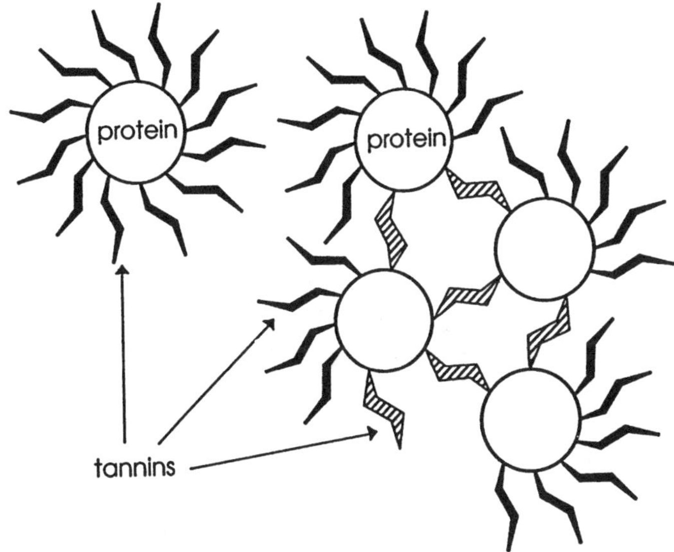

Figure 5–9 Interaction between tannins and saliva proteins which leads to precipitation of the proteins. *Source:* G. Matheis, Taste, Odor, Aroma, and Flavor, *Dragoco Report,* Vol. 39, pp. 50–65, © 1994, Dragoco.

5.1.3 *The Aroma of a Food*

The word *aroma* comes from the Greek, and originally meant "spice." Later all odor impressions that were attributable to pleasantly-smelling herbaceous plants were grouped together under *aroma.* At the present time, the term is used in many language areas (for instance in German-, English-, French-, Italian- and Spanish-speaking regions). It is not, however, used uniformly.

In the past, in German-speaking regions, *aroma* has been understood to mean principally seasonings, or combinations of them, whose effective constituents were essential oils (e.g., pepper, nutmeg, citrus fruits). At the present time *aroma* has two meanings in German. Firstly, an aroma is the pleasant odor impression of a food, regardless of whether it contains essential oil or not. We do not only speak of orange or pepper aroma, but also of the aroma of meat or bread (not determined by essential oils). This definition relates to a sensory impression, that of a pleasant smell; no one would speak of the "aroma" of bad eggs. Aroma as a pleasant odor impression is also the definition of *aroma* in English (Heath, 1981).

Secondly, in the German-speaking area, *aroma* is also a more or less complex mixture of aromatizing substances; a definition derived from legislative regula-

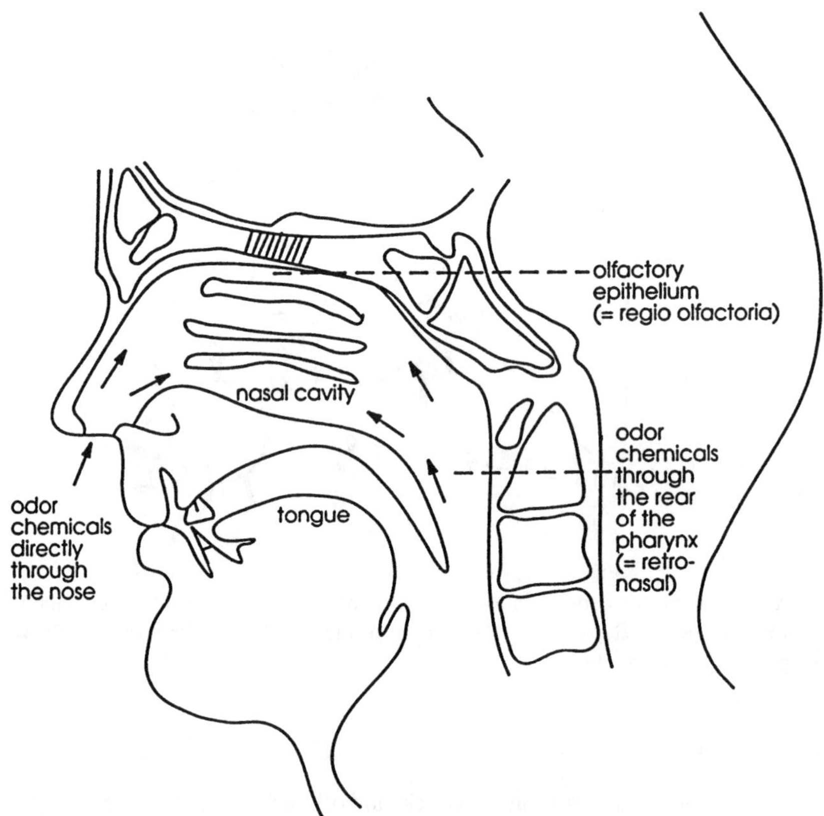

Figure 5–10 Diagram showing the human olfactory organ. *Source:* G. Matheis, Taste, Odor, Aroma, and Flavor, *Dragoco Report,* Vol. 39, pp. 50–65, © 1994, Dragoco.

tions. This is not a sensory impression, but rather a product of the flavor industry, made available to the food industry for aromatizing foods. The English term for such a product is "flavoring."

5.1.4 *Flavoring*

A flavoring is a product whose primary purpose is to impart flavor to foods and other preparations. From the definitions of the International Organization of the Flavour Industry (IOFI) and of the Council of the European Communities (CEC) (Table 5–2) it is obvious that flavorings consist of ingredients that contribute to the flavor of foods and of ingredients that do not contribute to the fla-

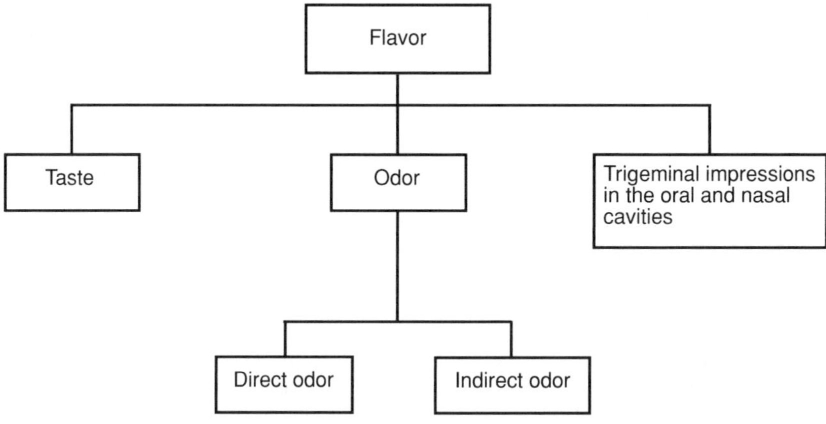

Figure 5–11 The flavor of a food.

vor. The latter are flavor adjuncts and include, for example, solvents, carriers and preservatives.

5.1.5 *Summary of Definitions*

Table 5–3 summarizes the definitions of taste, odor, aroma and flavor. We commonly understand by the term *taste* more than what is defined in Table 5–3. We

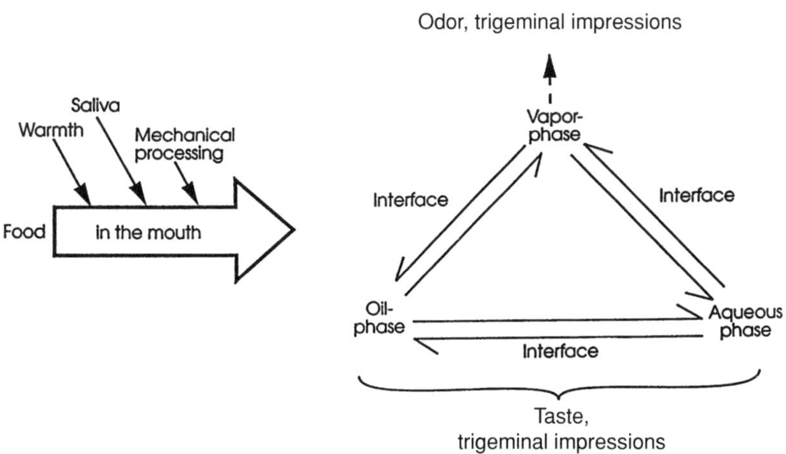

Figure 5–12 Processes in the mouth which lead to the perception of the flavor.

Table 5–2 Definitions of flavoring according to the International Organization of the Flavour Industry (IOFI, 1990) and the Council of the European Communities (CEC, 1988)

IOFI	*CEC*
Concentrated preparation, with or without flavor adjuncts,[a] used to impart flavor, with the exception of only salty, sweet or acid tastes. It is not intended to be consumed as such.	Flavoring means flavoring substances, flavoring preparations, process flavorings, smoke flavorings or mixtures thereof. Flavorings may contain foodstuffs as well as other substances.[b]
[a]Food additives and food ingredients necessary for the production, storage and application of flavorings as far as they are nonfunctional in the finished food.	[b]Additives necessary for the storage and use of flavorings, products used for dissolving and diluting flavorings, and additives for the production of flavorings (processing aids) where such additives are not covered by other community provisions.

say, for example, that when we have a head cold our food no longer "tastes" right. In fact, we do still taste sweet, sour, salty, bitter and umami, it is just that we can no longer smell them retronasally (because the mucosa of the nose are congested). Of course nobody says "I can no longer smell retronasally" but simply "I can no longer taste my food" because the layperson does not differentiate between taste and retronasal odor.

5.2 IMPORTANCE AND COMPLEXITY OF QUALITY CONTROL

Quality control is a very important function in the flavor industry. Flavor houses are expected to provide consistent, high-quality products to their customers. In addition, the flavorings must also meet legal restrictions on ingredients and their qualities and, in some causes, ingredient restrictions to meet customer guidelines.

Quality control in the flavor industry refers to the four areas: physico-chemical, biotechnology-based, microbiological (if relevant), and sensory analysis. It is carried out on all raw materials and intermediate products before they are used to make a flavoring, and on all finished flavorings before they go to the customer. Typically, more than 2,000 raw materials are used in the manufacture of approximately 10,000 different flavorings. Thus, the complexity associated with quality control of flavorings is greater than in the majority of other food-based industries.

For all raw materials, specifications and analysis protocols must be developed on first receipt to ensure the compliance of subsequent deliveries. The complexity continues with intermediate and finished products. All products are distinct and require individual attention. Satisfactory results of the tests laid down in the analysis of raw materials and intermediate products are required before their re-

Table 5–3 Summary of the definitions of taste, odor, aroma and flavor

Designation		
English	*German*	*Definition*
Taste	Geschmack	Sensory impression perceived by the taste receptors[1]
Odor	Geruch	Sensory impression perceived by the olfactory receptors (directly through the nose, and retronasally)
Aroma	Aroma	Pleasant odor impression
Flavoring	Aroma	More or less complex mixture of aromatizing and other substances
Flavor	Flavor	Combined effect of the sensory impressions of taste, odor, and of a somato-sensory nature in the oral cavity (touch, and impressions of pain, cold and warmth)

[1]In everyday speech, taste is usually taken to mean more.

Source: G. Matheis, Taste, Odor, Aroma, and Flavor, *Dragoco Report*, Vol. 39, pp. 50–65, © 1994, Dragoco.

lease for use in finished products. Likewise, all finished flavorings must pass the specified tests before dispatch to the customer. Because of the vast quantity of information that must be stored, the flavor industry has increasingly turned to computer technology.

The main objectives of quality control in the flavor industry are tests for:

- Identity (a wrong raw material might have been received from a supplier; a wrong product might have been produced)
- Purity (unacceptable impurities might be present in a raw material or a product)
- Contamination (e.g., heavy metals, pesticides, mycotoxins and microorganisms)
- Adulteration (a raw material might be adulterated)
- Restricted components (some components are restricted by legislation)
- Spoilage (age or improper storage may have altered the quality of a raw material or product)
- Authenticity (a raw material labeled "natural" might be synthetic or provenance of a raw material might be different from the labeled provenance)

5.3 PHYSICO-CHEMICAL ANALYSIS

Traditionally, physico-chemical analysis is divided into physical and chemical analysis (Figure 5–13). Although strictly incorrect, it is a useful division.

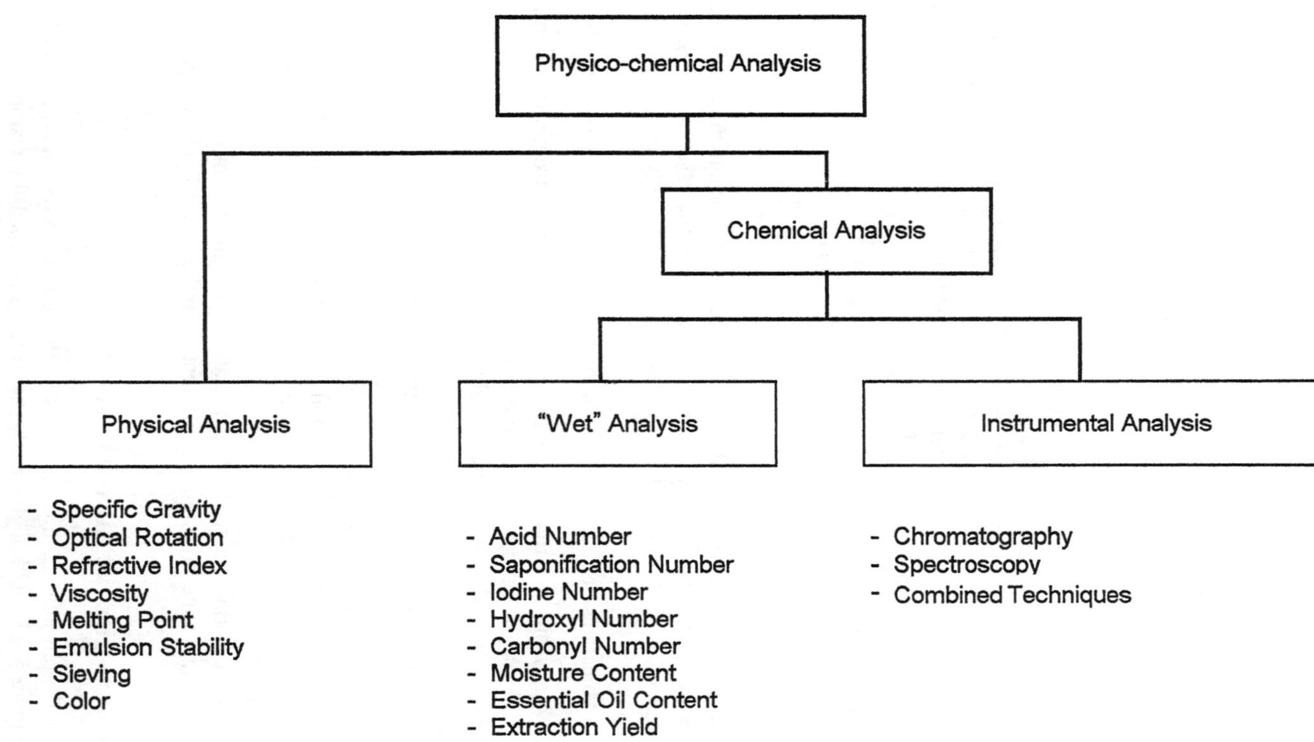

Figure 5–13 Examples of physico-chemical analysis. *Source:* G. Lösing et al., Quality Control of Flavorings and Their Raw Materials, *Dragoco Report,* Vol. 42, pp. 93–135, © 1997, Dragoco.

5.3.1 *Physical Analysis*

Typical examples of physical analysis are optical rotation, refractive index, specific gravity (or density), viscosity, melting point, color, emulsion stability, and sieving (Figure 5–13) of raw materials and products. The main purpose of most of these tests is to ascertain that the correct raw material has been received or the correct product has been produced (identity test).

One of the most commonly performed physical analyses on raw materials, intermediate and finished products is measuring *specific gravity* (SG). SG may be determined using a pycnometer or hydrometer. These methods, however, are often not rapid enough. An indirect measurement of SG developed by the Austrian company Paar KG is a quick and accurate method that can be used in modern quality control (Schulz, 1989). The sample tube is electromagnetically excited to vibrate at its natural frequency. The tube is then filled with liquid sample and again excited to its new resonance frequency. The difference in frequencies is related to the mass difference. Mass and, subsequently, density can be calculated by the instrument (Paar DMA 45 Digital Density Meter).

A second physical analysis that is performed on many raw materials and products is *refractive index* (RI). RI is typically measured using an Abbé refractometer or an automated digital refractometer (e.g., Abbemat by the German company Dr. Kernchen) (Schulz, 1989). The latter enables the determination of RI of opaque samples. The RI of a flavoring is a function of all the components and their proportions. RI and SG will detect most compounding errors in the formulation.

Optical rotation (OR) is commonly used as a quality control of essential oils and oleoresins. The significant constituents in these raw materials are frequently optically active. OR can be used to quantify specific components in these raw materials. OR is carried out using either a circular polarimeter or an automatic polarimeter on the principle of magneto-optical compensation (e.g., Propol by the German company Dr. Kernchen) (Schulz, 1989). Although intensively colored material cannot be measured with commercially available circular polarimeters, equipments with magneto-optical compensation give reliable results up to a light transmission of 0.1%.

In view of growing requirement for automation we have automated the simultaneous measurement of SG, RI and OR (Figure 5–14). The sample is placed in position by a transport module and then drawn into the interconnected vessels of the three measuring instruments via a peristaltic pump. A photocell indicates the end of this operation and the pump is switched off. After having reached the required temperature, measurements of the sample are made simultaneously. The results are processed by computer and can either be printed out or digitally transferred to the central electronic data processing system. For cleaning, the vessels are rinsed successively with two solvents of different polarity and dried with compressed air.

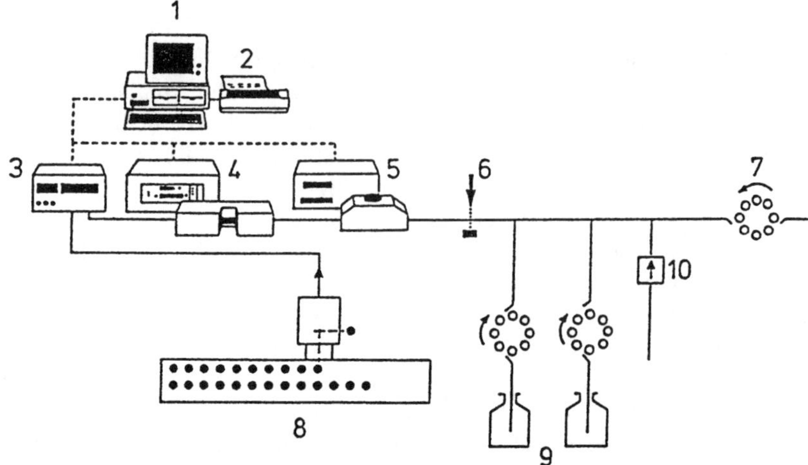

Figure 5–14 Schematic drawing of an automated system for measuring specific gravity, refractive index and optical rotation. 1 = personal computer, 2 = printer, 3 = densitometer, 4 = polarimeter, 5 = refractometer, 6 = photocell, 7 = sample pump, 8 = sample feed, 9 = rinsing pumps, 10 = two-way compressed air valve. *Source:* G. Lösing et al., Quality Control of Flavorings and Their Raw Materials, *Dragoco Report,* Vol. 42, pp. 93–135, © 1997, Dragoco.

Some products of the flavor industry are emulsions for flavoring, coloring and/or clouding purposes (Tse & Reineccius, 1995). These oil-in-water emulsions tend to separate, causing "ringing," which is very undesirable. The so-called "ringing test" is probably the most commonly used method to evaluate the *stability* of beverage flavoring *emulsions* in soft drinks (Tan & Holmes, 1988). In this test, a beverage is made up and then it is allowed to stand for a given time at either room or elevated temperature. Visible "ringing" indicates an inadequate emulsion. Alternate methods of evaluating the stability of emulsions include, for example, turbidity (Tan & Holmes, 1988) and particle size measurements. Turbidity is determined by measuring the absorbance of the diluted emulsion at a wavelength of 400 nm or at two wavelengths (800 and 400 nm), followed by forming the absorption ratio at 800 and 400 nm. There is a good relationship between absorption and mean particle size and, thus, emulsion stability. An emulsion containing weighing agent and emulsifier will typically not separate if the average particle size is in the range of 1–3 μm.

Particle size may be determined by turbidity measurement (see above), microscope, electric resistance, or light scattering techniques. The miscrocope method appears to be rather poor but it is used extensively and can work quite well if the technician is competent. Particle size evaluation by electric resistance may be carried out by using, for example, a Coulter Multisizer (Coulter Corporation, Hialeah, Florida, U.S.A.). The emulsion is converted to an electrolyte by adding salt at a very low level. The electrolyte emulsion is then drawn through a very small glass orifice. Both outside and inside the orifice is an electrode. As a particle passes through the orifice, the resistance of the electrolyte changes. In case of a large particle, there is little electrolyte and the resistance is high. If a small particle goes through the orifice, there is only a small amount of resistance. The Coulter Multisizer counts and sizes a large number of particles in a very short time. It also gives and plots the average particle size of the emulsion. Light scattering techniques, e.g., Microtrac (Leeds and Northrup, North Wales, Pennsylvania, U.S.A.) or Mastersizer (Malvern, Herrsching, Germany) are also in use. The costs of equipment for particle size evaluation by electric resistance and light scattering techniques are very significant.

Most commonly, particle size distribution in powder flavorings (e.g., spray-dried flavorings) is done by *sieving*. The powder is placed in the largest sieve of a stack and then shaken until it is distributed throughout the sieves. The powder on each sieve is then collected and weighed. Spray-dried flavorings can also be examined for particle size, using a microscope or electric resistance and light scattering techniques.

The *color* of specific raw materials and products may be determined by colorimetric measurements using either the L*a*b* or the L*C*h* system (Minolta Camera Co., 1993). These systems have been introduced by the Commission Internationale de l'Eclairage (CIE) in 1976. When using the L*a*b* system, luminosity (L*), the red/green value (a*) and the yellow/blue value (b*) are measured with, for example, a Chroma-Meter (Minolta, Japan) or a Color-View spectrophotometer (BYK-Gardner, U.S.A.) (Minolta Camera Co., 1993; Parkes, 1994).

5.3.2 *Chemical Analysis*

Traditionally, chemical analysis is divided into "wet" analysis and instrumental analysis (Figure 5–13). This is incorrect, but provides a useful basis for discussion.

5.3.2.1 *"Wet" Analysis*

Typical examples of "wet" analysis are various "chemical" parameters, moisture content, essential oil content, and extraction yield (Figure 5–13). The main purposes for determination of "chemical" parameters include tests for identity, purity and spoilage of raw materials or products.

Moisture content is of concern in dry raw materials and products. Because most raw materials and products contain volatile components in addition to water, evaporative methods for moisture determination are not recommended. Karl-Fischer Titration (Association of Official Analytical Chemists [AOAC] method

13.003) should be the method of choice. It is based on the chemical reaction of water with a pyridine/iodine complex. It has the advantage of being rapid and reasonably accurate.

The *essential oil content* of botanicals is a measure of their quality and flavor strength. The flavor industry uses various botanicals either as such or for further processing into essential oils or oleoresins. Volatile oil content is generally measured using a distillation method (AOAC methods 30.020–30.027) incorporating a Clevenger trap. Sufficient material should be placed in the sample flask to yield 10–15 ml of essential oil. After the addition of water and antifoam to the sample flask, the mixture is heated under reflux for 2–4 hours. The volume of volatile oil can be read directly from the side arm of the Clevenger trap.

Volatile oil content of spray-dried flavorings may also be determined (Anandaraman & Reineccius, 1987). Spray-dried citrus oils, mint oils or spice oils are analyzed for essential oil to determine the efficiency of the manufacturing process and the amount of flavoring present in the powder product.

Determination of *extraction yield* of botanicals that are processed into extracts is also very important. Extraction yield is typically determined by Soxhlet extraction, using a suitable solvent. After extraction, the solvent is removed under vacuum and the residue is weighed.

5.3.2.2 *Instrumental Analysis*

Instrumental analysis refers mainly to chromatography, spectroscopy and combined techniques (Figures 5–13 and 5–15). Chromatography includes, for example, gas chromatography (GC) and high performance liquid chromatography (HPLC) (Figure 5–15).

GC has grown steadily in importance as its sensitivity and separation power have improved. It is, however, not routinely used in establishing GC profiles of finished flavorings. The only time one would run GC profile on a finished product is at a customer's request. This often causes more problems than it solves. The GC profiles of two different tasting flavorings can be very similar or very different and vice versa, i.e., two flavorings may have identical GC profiles but be different in odor or taste. One must be conscious that it is more the sensory quality of the flavoring that one should be concerned about than the GC profile.

Ethanol content of finished flavorings is very readily measured by GC (AOAC method 19.001). The product may be injected directly into a gas chromatograph if it contains less than 20% of ethanol, or cut with tetrahydrofuran to dilute the sample prior to injection. Ethanol is separated from other constituents and detected using a flame ionization detector. The method is simple and takes 10 to 15 minutes.

Another use for GC is in the quality control of distillation or fractionation of essential oils (intermediate products). The use of citrus oil fractions as ingredients of flavorings has become widespread throughout the flavor industry.

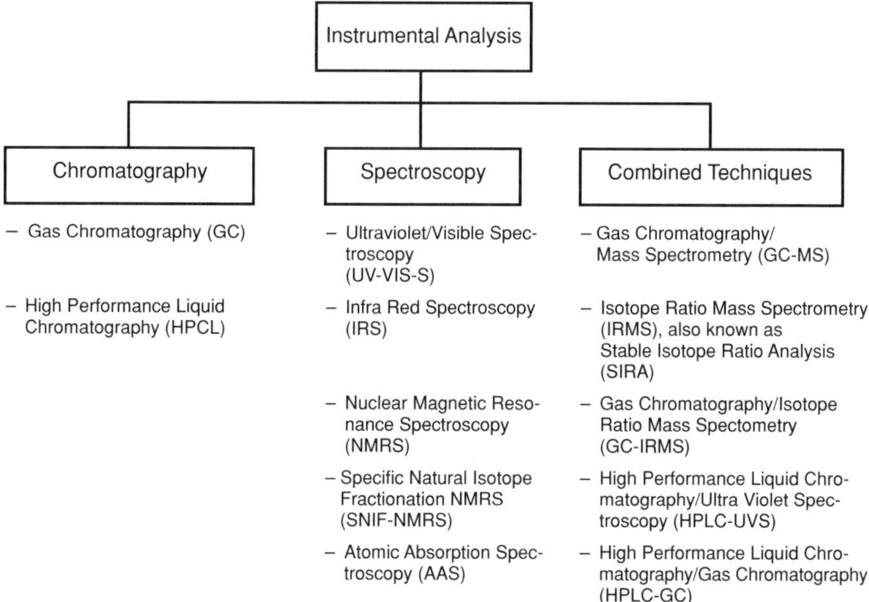

Figure 5–15 Examples of instrumental analysis. *Source:* G. Lösing et al., Quality Control of Flavorings and Their Raw Materials, *Dragoco Report,* Vol. 42, pp. 93–135, © 1997, Dragoco.

GC is very important in analysis of incoming raw materials. The purpose may be to determine composition, purity, contaminants, or adulteration. Component variations exhibited by, for example, essential oils are significant to their flavor. Reasons for these variations include geographical location of the crop, rainfall levels, plant species variation, and variations in the extraction method used.

Many flavoring ingredients are prone to deterioration during storage. Citrus oils, for example, are very susceptible to oxidative degradation. Even pure flavoring substances may undergo undesirable reactions upon storage. Benzaldehyde, for example, is readily oxidized to benzoic acid and limonene both oxidizes and polymerizes. Therefore, GC profiles may be used as one method to determine purity of raw materials.

GC is the method of choice to determine contaminants such as residual solvent (AOAC method 20.215 and 20.216, EOA method 1-1D-3-1 [EOA, 1975] or IOFI method 20 [IOFI, 1982]), and pesticides in raw materials for flavourings. Residual solvents must be determined on any flavoring material obtained by solvent extraction, e.g., oleoresins and absolutes. The oleoresin is often subjected to an initial distillation step to isolate the volatiles (including residual solvents). The distillate is then analyzed for residual solvents.

GC is also a powerful tool to analyze raw materials for potential adulteration. There is substantial incentive to dilute a more costly essential oil with a less expensive oil or other material. Examples are dilution of peppermint oil with cornmint oil and lemon oil with orange terpenes (Verzera et al., 1987). Orange terpenes do not contain carbonyl components (e.g., neral, geranial), esters (e.g., neryl and geranyl acetates), sesquiterpenes, α-terpinene and camphene. On the other hand, the orange terpenes contain δ-3-carene that is not present in lemon oil. Adulterations at even a 5% level can be detected by calculation the carene/α-terpinene and carene/camphene ratios.

Many flavoring substances are chiral, i.e., they contain one or more asymmetric carbon atoms (Figure 5–16) and thus exhibit optical activity. Chiral flavoring substances of natural origin generally have a characteristic distribution of enantiomers that is attributable to stereo selectively controlled biogenetic formation mechanisms. Thus, an excess of one or the other enantiomer occurs (Figure 5–17). When sythesized in the laboratory, the same flavoring substance will yield a racemic mixture of the optical isomers (Figure 5–18). There is substantial incentive to sell synthetic flavoring substances as natural ones or dilute natural substances with synthetic ones for economic reasons. GC (among other methods) is one method to detect such adulteration. Excellent reviews are available on chirospecific analysis of flavoring substances, using GC (Werkhoff et al., 1993; Mosandl, 1991, 1995a, 1995b; Hener et al., 1991; Hener & Mosandl, 1993; Schreier, 1993; Schurig et al., 1993; Simpkins & Harrison, 1995; Bicchi et al., 1995). In particular, enantioselective multidimensional gas chromatography (enantio-MDGC) has been demonstrated as a powerful method for the direct stereoanalysis of chiral volatiles without any further cleanup or derivatization procedures (Simpkins & Harrison, 1995).

High performance liquid chromatography (Schulz, 1988) is mainly (but not exclusively) applied in the analysis of substances that cannot be analyzed by GC at all or only with difficulties. These are mainly substances which cannot be volatized without decomposition. A prerequisite of HPLC is that the sample is completely dissolved in the mobile phase. Examples of HPLC application are

- flavor enhancers (monosodium glutamate, 5′-ribonucleotides) in raw materials and finished flavorings
- amino acids in raw materials (Figure 5–19)
- ethanol in finished flavorings
- chirospecific analysis of flavoring substances (Schurig et al., 1993; Mosandl, 1995a)
- authenticity control of citrus oils (Philipp & Isengard, 1995; Dugo et al., 1996)
- mycotoxins in raw materials (Ruhland et al., 1996)

Spectroscopic methods include ultraviolet/visible (UV-VIS-S), infrared (IRS), nuclear magnetic resonance (NMRS), specific natural isotope fractionation

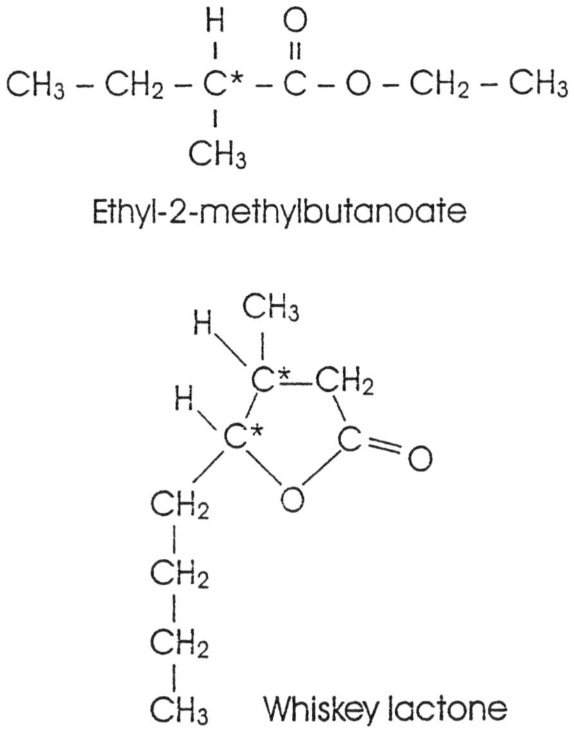

$CH_3-CH_2-C^{*}(H)(CH_3)-C(=O)-O-CH_2-CH_3$

Ethyl-2-methylbutanoate

Whiskey lactone

Figure 5–16 Examples of flavoring substances with one or two asymmetric carbon atoms. Asymmetric carbon atoms are marked with an asterisk. *Source:* G. Lösing et al., Quality Control of Flavorings and Their Raw Materials, *Dragoco Report,* Vol. 42, pp. 93–135, © 1997, Dragoco.

NMRS (SNIF-NMRS) and atomic absorption spectroscopy (AAS) (Figure 5–15). *UV-VIS-S* is commonly used for color measurement of clear colored solutions (absorbance at wavelengths from 190 to 1000 nm). Essential oils can additionally be characterized by their "CD values" (EOA, 1975; Philipp & Isengard, 1995). The CD value is obtained by the construction of a baseline which is built by holding the tangent at the absorption minima at 275 and 370 nm (points A and B). Then a vertical line is laid from the absorption maximum (point D) down to the tangent (point C). The distance between C and D, expressed as absorption units, is the CD value (Figure 5–20).

IRS is a suitable method for identity tests of flavoring substances and essential oils. Compared to other methods, both near IRS (NIRS) and middle IRS (MIRS)

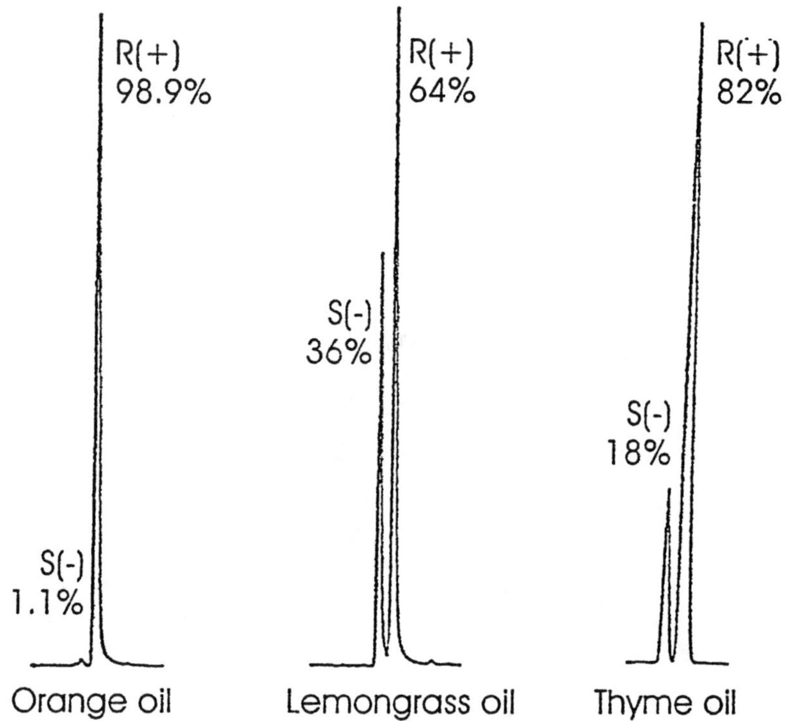

Figure 5–17 Natural enantiomer distribution of limonene in various essential oils. *Source:* G. Lösing et al., Quality Control of Flavorings and Their Raw Materials, *Dragoco Report,* Vol. 42, pp. 93–135, © 1997, Dragoco.

---

offer several advantages: short time, no reagents used, and non-destructive analysis. NIRS (Schulz & Lösing, 1995) has the additional advantage of minimal sample preparation. Typical examples of MIRS are cold pressed lime oil versus distilled lime oil, Tunisian rosemary oil versus Spanish rosemary oil, and adulterated onion oil. Distilled lime oil lacks some saturated and unsaturated carbonyl compounds that are present in cold pressed oil. Spanish rosemary oil contains verbenone which is absent in Tunisian oil. Genuine onion oil has a distinct peak of 2-hexyl-2,3-dihydro-5-methyl-furan-3-one.

SNIF-NMRS (Hener & Mosandl, 1993; Mosandl, 1995a; Martin et al., 1993) is one of the isotopic analysis methods based on measuring the abundance of stable isotopes in flavoring substances. This abundance is often different between a natural and a synthetic molecule. Isotopic analysis can be performed by isotope ratio mass spectrometry (IRMS; discussed below as a combined technique) which provides the overall molecular isotope contents. It has been reported, however, that

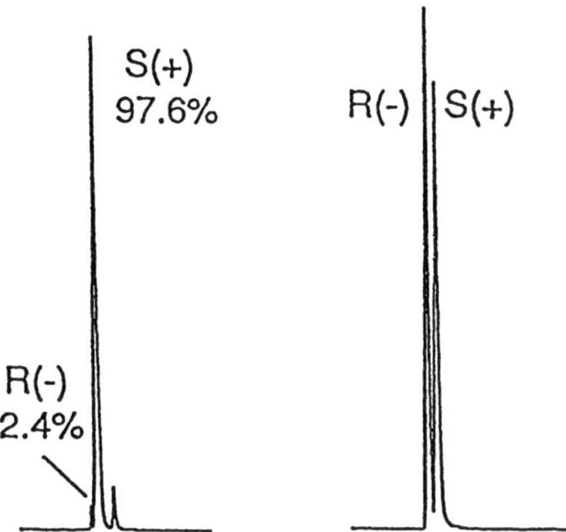

Figure 5–18 Enantiomer distribution of ethyl-2-methyl butanoate in apple (left) and synthetic material (right). *Source:* G. Lösing et al., Quality Control of Flavorings and Their Raw Materials, *Dragoco Report,* Vol. 42, pp. 93–135, © 1997, Dragoco.

such control may be circumvented by appropriate enrichment (Martin et al., 1993). SNIF-NMRS enables the direct measurement of isotopic ratios at several positions of a given molecule, thus making possible the proof of the natural origin of a flavoring substance by comparison with natural materials. For background information on natural abundance of stable isotopes of the bioelements (e.g., ^{2}H, ^{13}C, ^{15}N, ^{18}O and ^{34}S) and on isotope abundance as indicator for the origin of molecules (type of plant, biotechnology, synthesis), the reader is referred to the literature (Hener & Mosandl, 1993; Mosandl, 1995a, 1995b; Martin et al., 1993; Winkler & Schmidt, 1980; Schmidt, 1986).

Although ^{13}C measurements have, for several years, been a successful method for characterization of natural flavoring substances, ^{2}H-NMRS brought a new dimension to isotopic analyses (Martin et al., 1993). Today, SNIF-NMRS can provide isotopic fingerprints of various flavoring substances, e.g., vanillin, anethole, benzaldehyde, limonene, citral, menthol. In all cases, purification of the molecule of interest is required prior to SNIF-NMRS analysis. Because of the widespread level of adulteration, the most frequent use of SNIF-NMRS in the flavor industry has been on vanillin. Vanillin contains six monodeuterated isotopomers. Owing to fortuitous equivalence of two aromatic positions, only five signals are observed in the deuterium spectrum (^{2}H at positions 1 to 5 in Figure 5–21). Measuring the five specific isotope ratios ($^{2}H/^{1}H$) and molar fractions provides a very good means of discriminating between various sources of vanillin. A discriminant

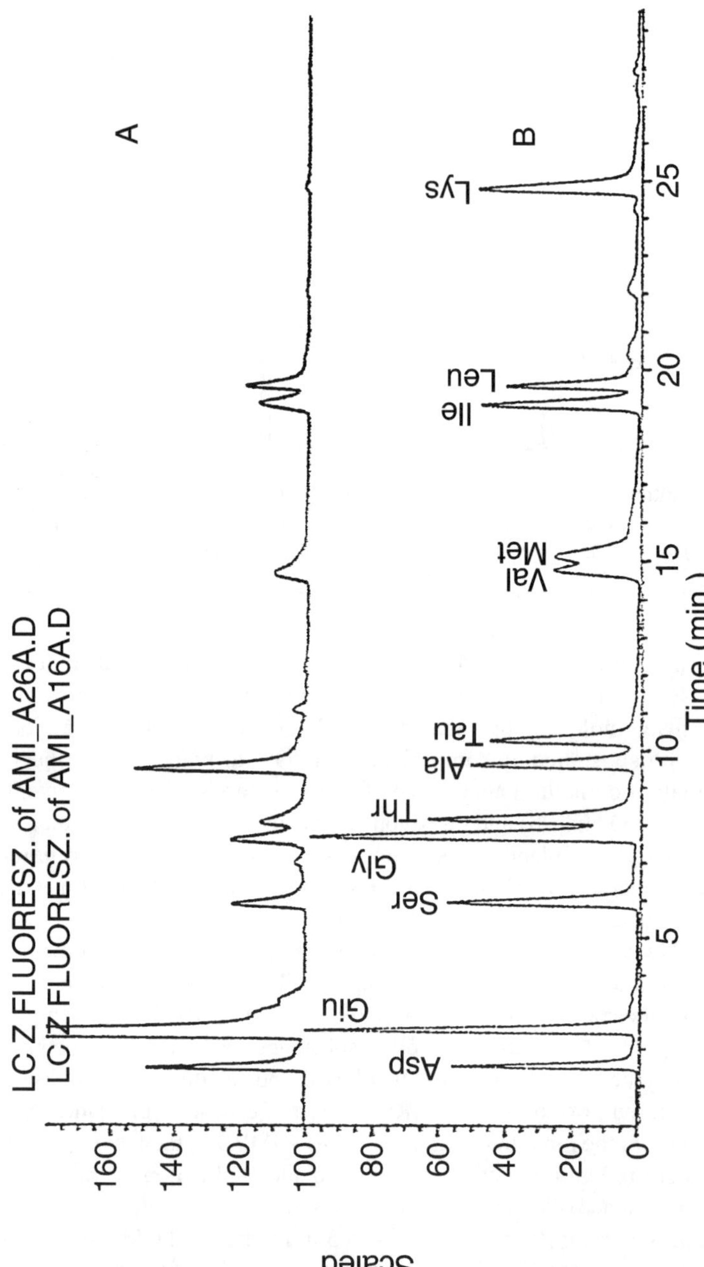

Figure 5–19 HPLC of amino acids of a hydrolyzed vegetable protein (**A**) and of standard amino acid mixture (**B**). *Source:* G. Lösing et al., Quality Control of Flavorings and Their Raw Materials, *Dragoco Report*, Vol. 42, pp. 93–135, © 1997, Dragoco.

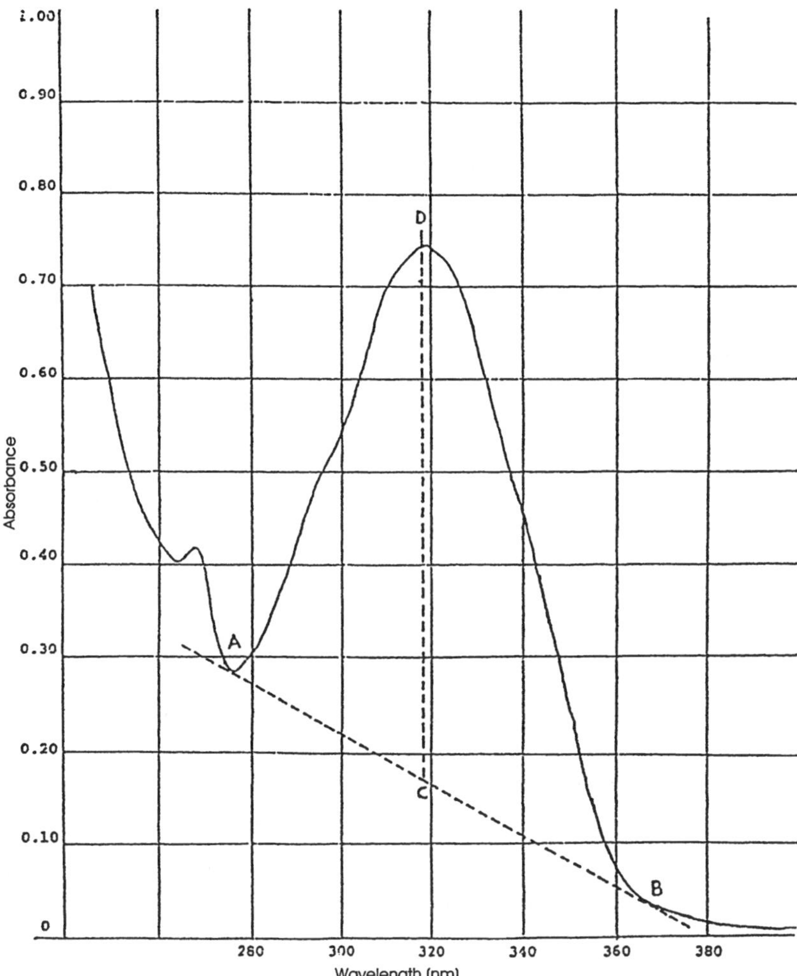

Figure 5–20 Construction of CD value. *Source:* G. Lösing et al., Quality Control of Flavorings and Their Raw Materials, *Dragoco Report,* Vol. 42, pp. 93–135, © 1997, Dragoco.

analysis plot showing the separation between vanillin from vanilla beans, lignin and guaiacol is shown in Figure 5–22.

AAS is used for the determination of heavy metals (e.g., cadmium, lead, mercury, arsenic) in raw materials.

Examples of *combined techniques* (Figure 5–15) are gas chromatography/mass spectrometry (GC-MS), isotope ratio mass spectrometry (IRMS), gas chromatography/isotope ratio mass spectrometry (GC-IRMS), high performance liquid

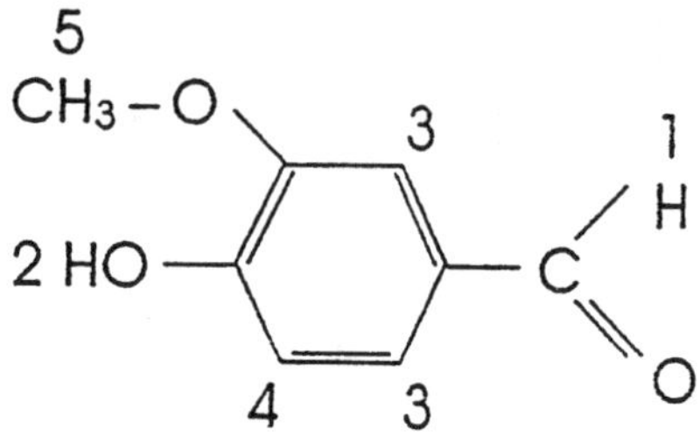

Figure 5–21 Positions in vanillin giving the five signals in the deuterium spectrum. *Source:* G. Lösing et al., Quality Control of Flavorings and Their Raw Materials, *Dragoco Report,* Vol. 42, pp. 93–135, © 1997, Dragoco.

chromatography/ultraviolet spectroscopy (HPLC-UVS), and high performance liquid chromatography/gas chromatography (HPLC-GC).

GC-MS, IRMS and GC-IRMS are used to detect adulterations of raw materials. One approach to detect whether a synthetic flavoring substance has been added to a natural raw material is to look for intermediates known to be present in chemically synthesized substances. Neroli oil, for example, contains about 40% of linalool. Synthetic linalool could be added for substantial profit. Synthesized linalool contains 0.5 to 2% of dihydrolinalool (Figure 5–23), which can serve as marker for its addition. The addition of as little as 2% of synthetic linalool can be detected by GC-MS (Frey, 1988). Another example of this approach is the addition of synthetic cinnamic aldehyde to cassia oil (Frey, 1988). In this case, phenyl pentadienal is the byproduct of cinnamic aldehyde synthesis.

A second approach is the determination of stable isotopes of a molecule by IRMS (also known as stable isotope ratio analysis or SIRA) (Hener & Mosandl, 1993; Mosandl, 1995a, 1995b; Bicchi et al., 1995). As already pointed out (see above, SNIF-NMRS), IRMS is another isotopic analysis method for measuring the abundance of stable isotopes in flavoring substances. Stable isotope ratios of special interest are $^{13}C/^{12}C$, $^{2}H/^{1}H$ and $^{15}N/^{14}N$. The use of IRMS, particularly ^{13}C, to detect adulteration of flavoring materials has been longest applied to vanilla. There is substantial incentive for adulteration of vanilla extracts with vanillin for cost reasons (McCormick, 1988). $^{13}C/^{12}C$ isotope ratios for vanillin obtained from natural and synthetic sources are shown on Table 5–4. It is apparent that natural vanillin can be readily distinguished from synthetic vanillin by this method. The financial incentive for adulteration has resulted in attempting to circumvent this method of detection. The most common approach was to enrich the methoxy group of synthetic vanillin with ^{13}C. SNIF-NMRS (see above) is able to detect this circumvention. Other examples of IRMS application are adulteration of benzaldehyde and cinnamic aldehyde (Culp & Noakes, 1990). The combination of

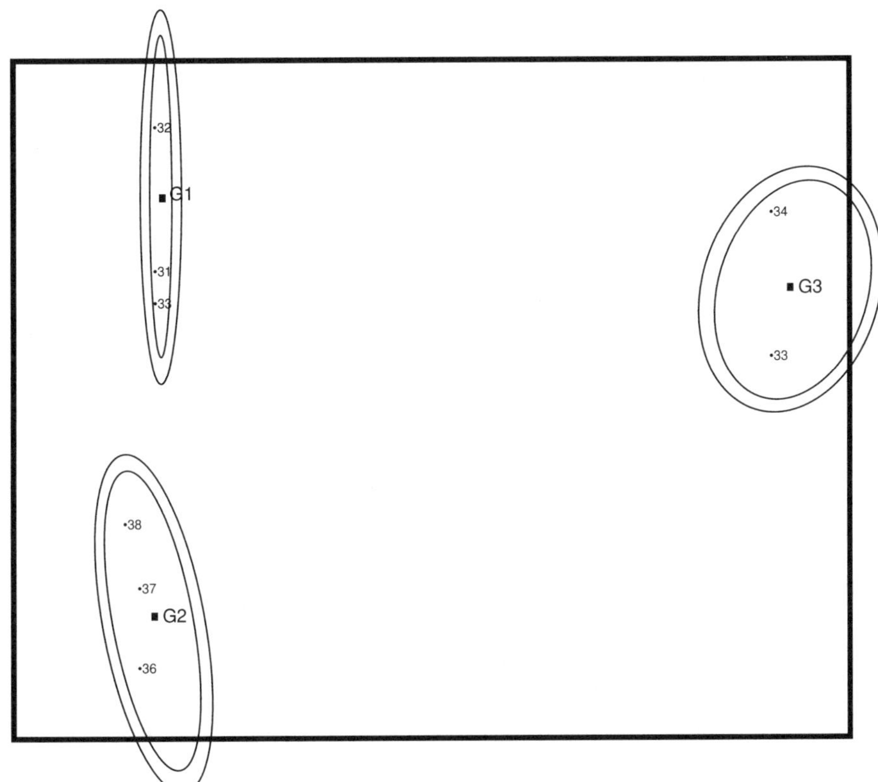

Figure 5–22 Discriminant analysis plots of three sources of vanillin obtained from the five site specific SNIF-NMRS isotopic ratios. Vanillin sources are vanilla beans (G1), lignin (G2) and guaiacol (G3). *Source:* G. Lösing et al., Quality Control of Flavorings and Their Raw Materials, *Dragoco Report,* Vol. 42, pp. 93–135, © 1997, Dragoco.

IRMS with GC (GC-IRMS), or better with MDGC (MDGC-IRMS), has been found very successful for authenticity control of flavoring substances and essential oils (Mosandl, 1995a, 1995b; Bicchi et al., 1995).

HPLC-UVS (Schulz, 1988) is used in the quantitative determination of substances that cannot be vaporized without decomposition. Examples include:

- the pungent principles of pepper (piperine and related components), paprika (capsaicin and related components [Figure 5–24]), and ginger (gingerol and related components)
- vanillin, coumarin and other phenolic substances (Figure 5–25)
- flavonoids and other phenolic substances in plant extracts (Figure 5–26).

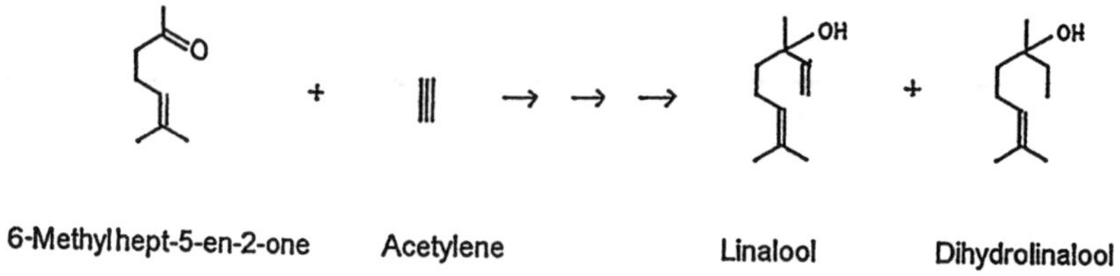

6-Methylhept-5-en-2-one + Acetylene → → → Linalool + Dihydrolinalool

Figure 5–23 Synthesis of linalool. *Source:* G. Lösing et al., Quality Control of Flavorings and Their Raw Materials, *Dragoco Report,* Vol. 42, pp. 93–135, © 1997, Dragoco.

On-line *HPLC-GC* is one of the most recently introduced combined techniques for authenticity control of flavoring substances and essential oils (Bicchi et al., 1995; Mondello et al., 1996).

5.4 BIOTECHNOLOGY-BASED ANALYSIS

Biotechnology-based techniques make use of living organisms or the products of those organisms (e.g., enzymes, antibodies) to analyze raw materials and products (Giese, 1996). Examples of enzymatic determination are acetaldehyde (in powder flavorings), amino acids (including MSG), sulfite, ethanol, guanosine-5′-monophosphate (GMP), and creatinine.

Some of the biologically active materials used in biotechnology-based techniques are part of instruments called biosensors (Schulz, 1991). One application is a biosensor for detection of MSG. Another biotechnology-based technique is the determination of mycotoxins (e.g., aflatoxins), pathogenic microorganisms

Table 5–4 δ ^{13}C Values of vanillin obtained from different sources

Vanillin source	*δ ^{13}C Value*
Bourbon	−20.2
Madagascar	−20.5
Comores	−20.0
Javan	−18.7
Mexican	−20.3
Tahitian	−16.8
Lignin	−27.0
Guaiacol	−29.5
Eugenol (clove)	−30.8

Source: G. Lösing et al., Quality Control of Flavorings and Their Raw Materials, *Dragoco Report,* Vol. 42, pp. 93–135, © 1997, Dragoco.

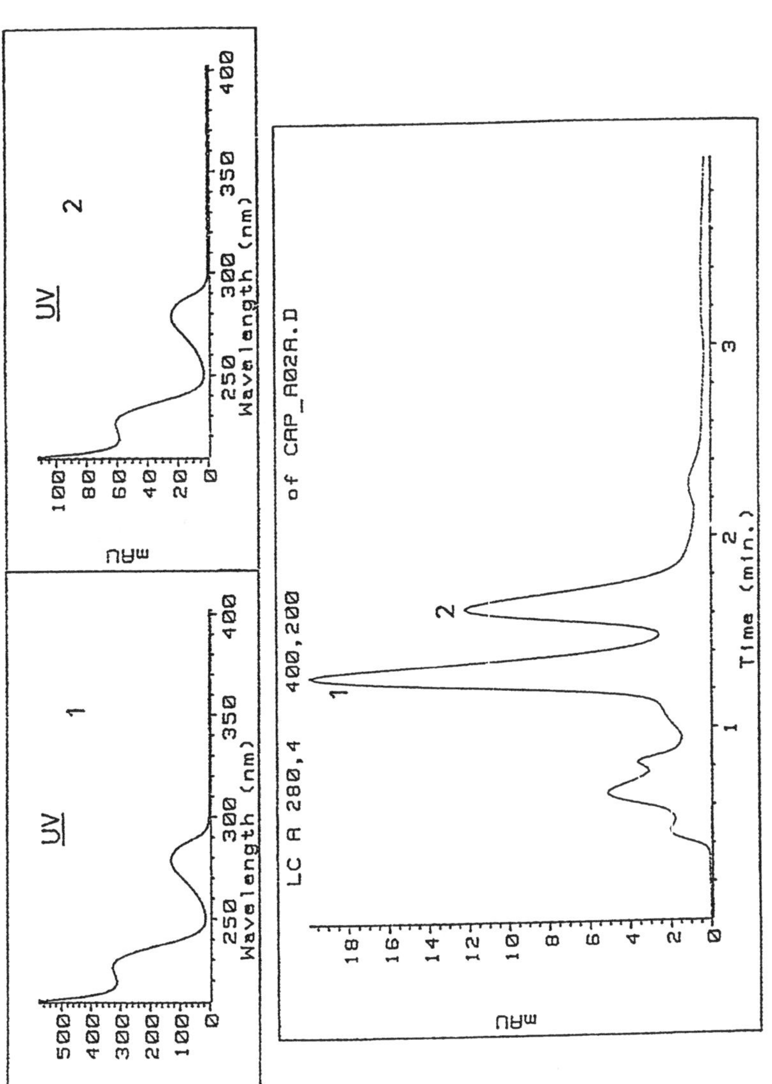

Figure 5–24 HPLC of a chilli extract and UV spectra of (1) capsaicin and (2) dihydrocapsaicin. *Source:* G. Lösing et al., Quality Control of Flavorings and Their Raw Materials, *Dragoco Report,* Vol. 42, pp. 93–135, © 1997, Dragoco.

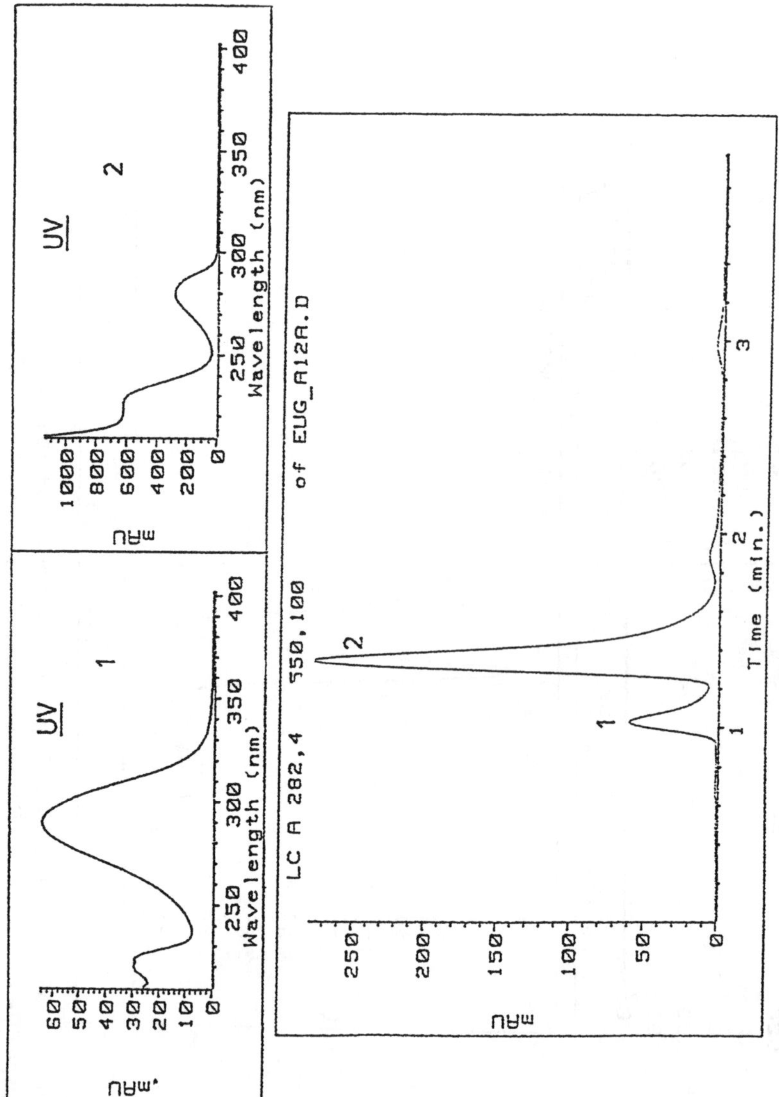

Figure 5–25 HPLC of cinnamon leaf oil and UV spectra of (1) cinnamic aldehyde and (2) eugenol. *Source:* G. Lösing et al., Quality Control of Flavorings and Their Raw Materials, *Dragoco Report,* Vol. 42, pp. 93–135, © 1997, Dragoco.

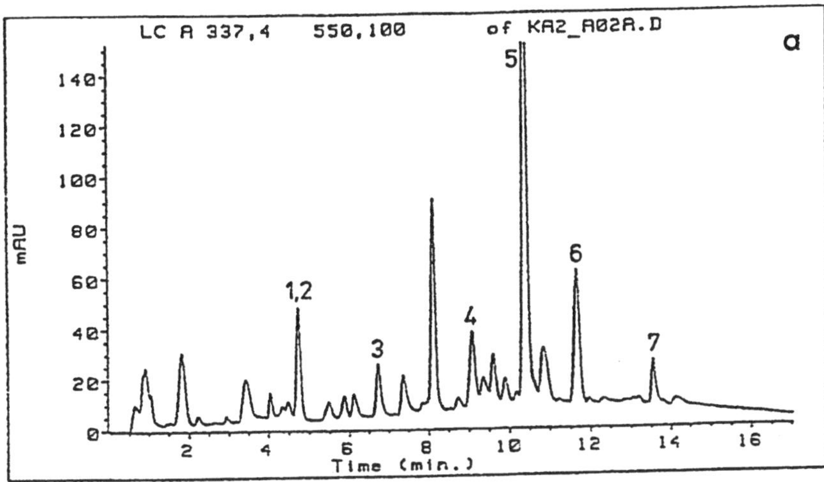

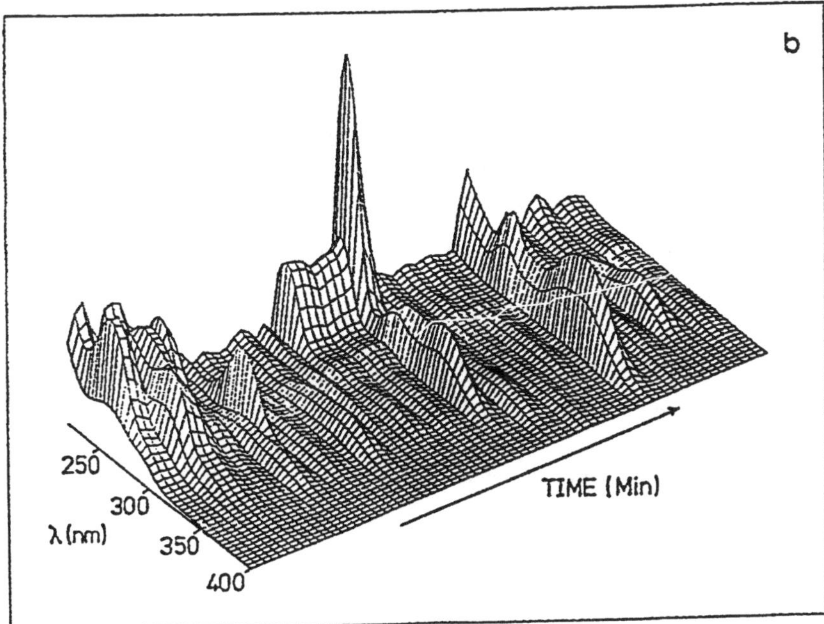

Figure 5–26 HPLC of chamomile flower extract shown as two-dimensional (**A**) and three-dimensional (**B**) graphs. 1 = chlorogenic acid, 2 = caffeic acid, 3 = umbelliferone, 4 = luteolin-7-glucoside, 5 = apigenin-7-glucoside, 6 = herniarin, 7 = apigenin. *Source:* G. Lösing et al., Quality Control of Flavorings and Their Raw Materials, *Dragoco Report,* Vol. 42, pp. 93–135, © 1997, Dragoco.

and pesticides by enzyme linked immunosorbent assay (ELISA) (Whitaker, 1984; Sweger, 1993; Chen et al., 1996). Specific antibodies form antibody/antigen complexes with the substances (antigens) to be determined. There are two versions of the ELISA technique, the "sandwich" technique (Engvall & Perlmann, 1971) and the "double-sandwich" technique (Notermans et al., 1982). In both cases, the substances to be determined are quantified by enzymatic reactions. The enzymes are covalently bound to the antibodies involved.

5.5 MICROBIOLOGICAL ANALYSIS

Increasing cases of food poisoning over the last years require careful microbiological assessment of foods and their ingredients, including flavorings and their raw materials. Only a limited number of raw materials and flavorings, however, need to be tested microbiologically. The majority of liquid flavorings contain solvents as flavor carriers (e.g., ethanol, propylene glycol, vegetable oil). The level of such solvents (normally about 70 to 90%) is such that these liquid flavorings have been found to be either bactericidal or at worst bacteriostatic. In addition, some of the flavoring substances used in the flavorings are also bactericidal or bacteriostatic. Because of this, routine microbiological testing of many raw materials and flavorings has been discontinued.

Whether or not microbiological analyses are performed depends on the origin of raw materials, manufacturing process and composition of the intermediate and finished products. Examples, where microbiological analyses are required, are:

- agricultural produces of vegetable and animal origin (e.g., spices, juice concentrates, meat, fish, and shellfish extracts)
- emulsion, paste and powder flavorings that contain these agricultural products

Typically, microbiological analysis of raw materials and products includes:

- total viable count (TVC)
- yeast and molds
- coliform bacteria or enterobacteria
- *Escherichia coli*

Depending on the type of raw material or product and on the application of the product, additional tests may be done on pathogenic and spoilage microrganisms. Examples are:

- aerobic and anaerobic spore forming bacteria (e.g., *Bacillus cereus* and *Clostridium perfringens,* respectively)
- *Staphylococcus aureus*
- *Salmonella* spp.
- *Listeria* spp.

Because the classical microbiological methods are rather time-consuming, the development of rapid methods and automation is currently in progress. Available rapid methods are not generally applicable for all materials. They offer, however, practicable solutions to specific problems, such as analysis for pathogens.

There is considerable variation within the flavor industry as to how microbiological analysis is done. In some companies, little microbiology is done in-house. Most of the work is contracted out to consulting laboratories. If anything is done in-house, it is usually total plate count, yeasts and molds. Other flavor houses (including our company) do most of the work in-house.

Microbiological analysis also includes hygiene control of plant cleanliness on production sites and is part of the hazard analysis critical control point (HACCP) concept. In our company, hygiene monitoring is performed (among other methods) by using the adenosine triphosphate (ATP) bioluminiscence method (Stannard & Gibbs, 1986; Ehrenfeld, 1995; Baumgart, 1996; Ogden, 1993; Griffiths, 1996), a rapid test method. Results are available in a very short time. Traditional microbiological methods require at least 2 days before the data are available. The ATP bioluminescence technique makes use of the fact that all living cells contain ATP, which is the universal energy donor for metabolic reactions. An enzyme-substrate complex (luciferase/luciferin), present in firefly tails, converts the chemical energy associated with ATP into light by a stoichiometric reaction. Thus, the amount of light emitted is proportional to the concentration of ATP present and can be quantified using a luminometer.

HACCP (Savage, 1995) is a food process control system developed in the early 1970s. Since then, it has evolved into a recognized means to assure the safety of food and food ingredients throughout the industry world-wide. It is an important aspect of a Quality Management System (e.g., DIN EN ISO 9001) in order to prevent any harm occurring to consumers from biological, chemical and physical hazards. According to the EU Directive on food hygiene (93/43 EWG) issued in 1993, food (and food ingredients) manufacturers are obliged to implement a system according to the principles of HACCP. In our company, an HACPP system has been established.

5.6 SENSORY ANALYSIS

Sensory analysis is without doubt a very important facet of quality control in the flavor industry. The term "organoleptic" should be avoided because it is obsolete and incomplete.

5.6.1 *Test Panels*

Sensory analysis is carried out by human subjects (panelists) and refers to odor, taste, trigeminal perceptions, and visual appearance (Figure 5–27). Al-

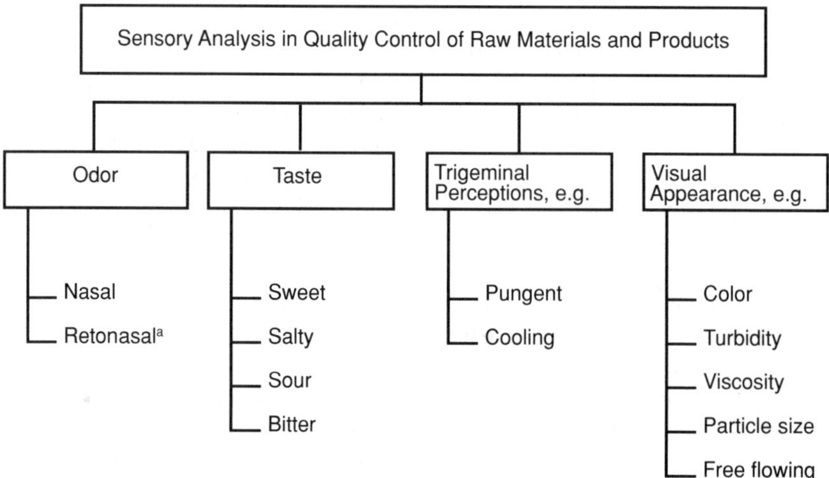

[a] = Generally, taste is used for both taste and retronasal odor.

Figure 5–27 Areas of sensory analysis in quality control of raw materials and products. *Source:* G. Lösing et al., Quality Control of Flavorings and Their Raw Materials, *Dragoco Report,* Vol. 42, pp. 93–135, © 1997, Dragoco.

though it is scientifically incorrect, the term "taste" is generally used for both taste and retronasal odor.

The senses of the panelists are the instruments of sensory analysis. Sensory analysis in quality control is carried out by trained panels. The subjects can be employees of the company or non-employees. Several factors can influence the sensory response of subjects, e.g., aptitude, motivation, communication, health, availability, and willingness for regular training (Gilette, 1994). Training (Gilette, 1994; Lawless & Claassen, 1933; ASTM, 1981; Fliedner & Wilhelmi, 1989) and proper experimental design will minimize or rule out some of these factors. Motivation is very important. The panelists must realize that sensory testing is a serious matter and not just a welcome change from daily routine.

5.6.2 *Test Facilities*

Several factors can influence the panelists' sensory response, e.g., odor, noise, paint, light, size, ambient temperature, humidity and ease (Fliedner & Wilhelmi, 1989; ASTM, 1986). Facilities must be free from mental and sensory distractions. Panelists must feel at ease. Our company has very good experience with test cabins and Figure 5–28 shows essential features of the test

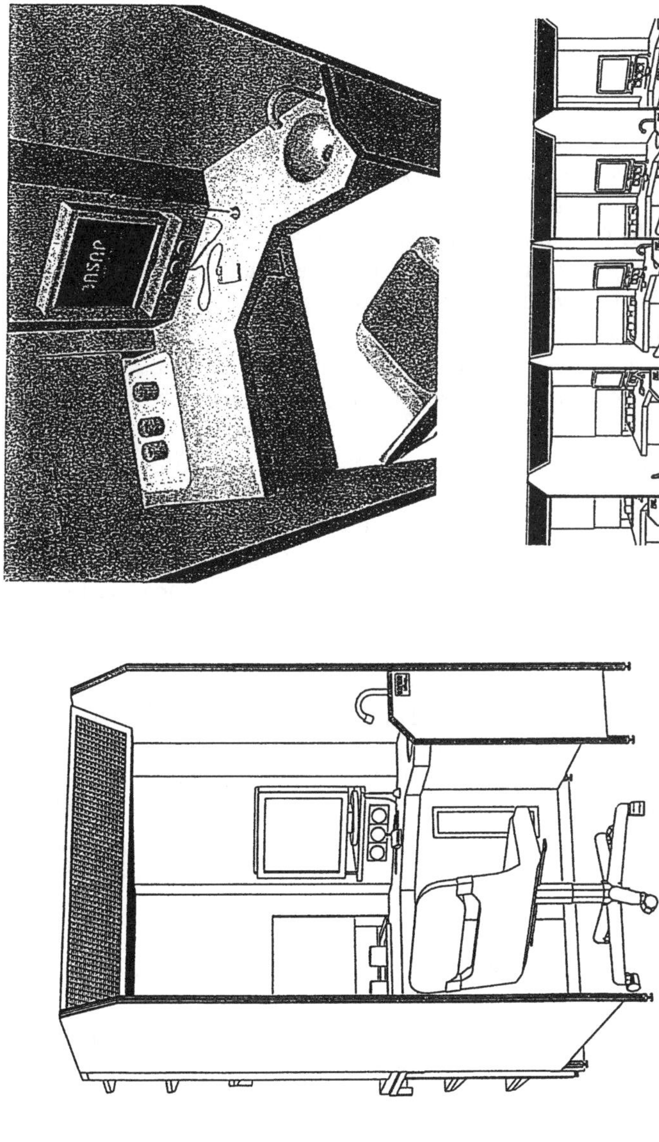

Figure 5–28 Essential features of test cabins. *Source:* G. Lösing et al., Quality Control of Flavorings and Their Raw Materials, *Dragoco Report*, Vol. 42, pp. 93–135, © 1997, Dragoco.

cabins. They are intended to be equipped with hard- and software for computer aided sensory analysis (CASA) (Arbogast, 1996).

5.6.3 *Test Media*

Raw materials are presented to panelists as such and/or diluted in test media. Examples of test media for vanilla beans, herbs and spices are shown in Figure 5–29. Finished flavorings are usually presented to test panels as 0.001 to 0.5% dilutions in test media, depending on the strength of the flavoring. Typical test media for flavorings and raw materials are shown in Figure 5–30.

The reason for using different test media is to put the flavorings into a medium that is similar to the food in which it is to be used. A sugar/citric acid solution, for example, simulates the character of fruits; a flavoring, no matter how good, would taste flat and unnatural in the absence of sugar and acid. Vanilla flavorings for dairy application may be tested in milk. A fondant may be used for evaluation of candy flavorings. Meat, cheese and vegetable flavorings are best evaluated in salt solution, and vegetable oil may be used for oil-soluble butter flavorings.

5.6.4 *Test Methods in Sensory Evaluation*

Sensory evaluation may be divided into perception, analytical testing and affective testing (Figure 5–31) (Gilette, 1990, 1994; Lawless & Claassen, 1933; O'Mahony, 1995). Perception includes threshold detection. There are four types of thresholds (Table 5–5). Threshold detection tests are used in selection and training of panelists.

Analytical testing measures objective facts and is performed with trained panelists. Affective sensory evaluation deals with subjective measurements (e.g., acceptance or degree of liking). It is performed with consumer panels. Because affective testing is not used in quality control of flavorings and their raw materials, it will not be further discussed here.

Analytical testing includes difference (or discrimination) tests, descriptive analysis, scaling tests, and time/intensity procedure (Figures 5–31 and 5–32). Typical difference tests are paired comparison, triangle test and duo-trio test. Paired comparison is a two-sample test in which the panelists are asked whether the samples are identical or not. In the triangle test, the subjects are presented with three samples. Two are identical and one is different. The panelists have to determine which of the samples is different from the other two. The duo-trio test is a multi-sample test. The subjects are presented with a sample marked as "standard" and one or more pairs of samples. These pairs of samples also contain the standard. The panelists have either to determine which is (are) the sample(s) that is (are) different or which is (are) the sample(s) that is (are) identical with the sample marked as "standard."

Vanilla beans or herbs are mixed with boiling test solution (8% aqueous sucrose). After 1 h, the suspension is filtered.

Herbs or spices are mixed with boiling test solution (0.5% aqueous sodium chloride). After 1 h, the suspension is filtered.

Figure 5–29 Examples of test media for raw materials. *Source:* G. Lösing et al., Quality Control of Flavorings and Their Raw Materials, *Dragoco Report,* Vol. 42, pp. 93–135, © 1997, Dragoco.

Examples of descriptive analysis are deviation from reference (or standard), quantitative descriptive analysis (QDA) (Stone et al., 1974; Stone & Sidel, 1985) and quantitative flavor profiling (QFP) (Stampanoni, 1993). QFP is a further development of QDA. It describes and quantifies selected sensory characteristics of a raw material or product.

A typical example of a scaling test is ranking. For ranking, three or more samples are presented to the panelists. The subjects are asked to order the samples from low to high intensity (odor, taste or color). Scaling tests are used in selection and training of panelists.

Aqueous solution of 8% sucrose

Aqueous solution of 8% sucrose and 0.08% citric acid

Aqueous solution of 0.5% sodium chloride

Aqueous solution of 20 vol% ethanol

Aqueous solution of 20 vol% ethanol and 14% sucrose

Fat filling

Milk

Vegetable oil

Fondant

Figure 5–30 Examples of test media for products and raw materials. *Source:* G. Lösing et al., Quality Control of Flavorings and Their Raw Materials, *Dragoco Report,* Vol. 42, pp. 93–135, © 1997, Dragoco.

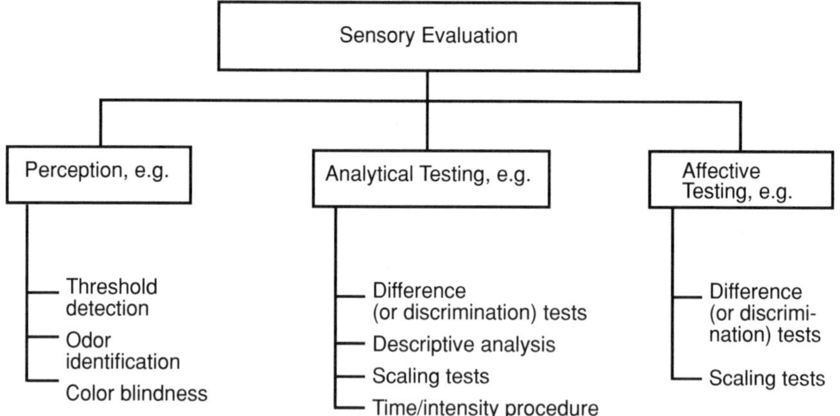

Figure 5–31 Classification of sensory evaluation. *Source:* G. Lösing et al., Quality Control of Flavorings and Their Raw Materials, *Dragoco Report,* Vol. 42, pp. 93–135, © 1997, Dragoco.

5.6.5 *Test Methods in Quality Control*

Typically, paired comparisons and/or triangle tests (both are difference tests) are used, often in combination with deviation from reference (a descriptive analysis) (Table 5–6). The same tests are also used for selection and training of panelists. In addition, ranking, threshold detection and odor identification tests are used for selection and training of panelists (Table 5–6). For odor identification, a series of odorants is given to the subjects to determine their ability to identify, or

Table 5–5 Definition of the four types of thresholds

Type	*Definition*
Detection threshold	Minimum physical intensity detectable where the subject is not required to identify the stimulus
Difference threshold	Smallest change in concentration of a substance required to give a perceptible change
Recognition threshold	Lowest concentration at which a substance is correctly identified
Terminal threshold	The concentration of a substance above which changes in concentration are not perceptible

Source: G. Lösing et al., Quality Control of Flavorings and Their Raw Materials, *Dragoco Report,* Vol. 42, pp. 93–135, © 1997, Dragoco.

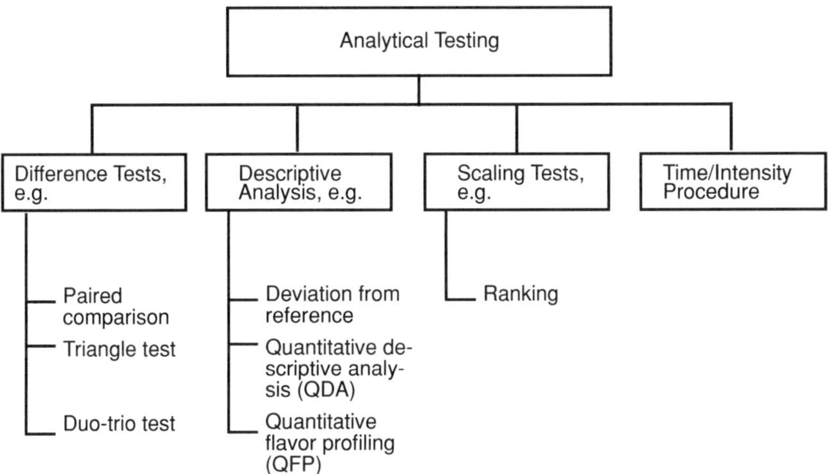

Figure 5–32 Examples of analytical testing. *Source:* G. Lösing et al., Quality Control of Flavorings and Their Raw Materials, *Dragoco Report,* Vol. 42, pp. 93–135, © 1997, Dragoco.

---

at least describe, commonly occurring odorants (Table 5–7). Odorants can be delivered by dropping a few drops onto a cotton ball inside a bottle, labeled with a code and tightly capped.

There is considerable variation within the flavor industry as to how sensory analysis for quality control of raw materials and products is done. In our company, the majority of products are evaluated by panelists in a sequential combination of difference and descriptive tests (also known as extended difference tests). As a first step, new production batches are evaluated in paired comparison against the standard. Both samples are coded. Three or more subjects are asked (extended paired comparison test):

- Are both samples identical?
- If not, describe how the right-hand sample differs from the left.

If at least one of the panelists describes the samples as different, a triangle test is performed by at least 10 subjects as a second step. In the triangle test, the subjects are presented with three coded samples and asked the following tasks (extended triangle test):

- In front of you are three samples. Two are the same, one is different. Identify the sample that is different from the other two.
- Describe the difference.

Table 5–6 Examples of typical test methods used in quality control and selection and training of panelists

Quality control	*Selection and training of panelists*	*Type of test*
Paired comparison	Paired comparison	Difference test
Triangle test	Triangle test	Difference test
Deviation from reference	Deviation from reference	Descriptive analysis
	Ranking (color, taste, odor)	Scaling test
	Threshold detection (taste, off-flavor)	Perception
	Odor identification test	Perception
	Color blindness test	Perception

Source: G. Lösing et al., Quality Control of Flavorings and Their Raw Materials, *Dragoco Report*, Vol. 42, pp. 93–135, © 1997, Dragoco.

Evaluation is carried out in test facilities equipped with sensory analysis software package.

5.7 CONCLUSIONS

It is obvious that the quality control of flavorings and their raw materials is a highly complex field, although the list of methods for physico-chemical, biotechnology-based, microbiological and sensory analysis discussed here is by no means exhaustive. Quality control laboratories in the flavor industry may have more than 500 defined analytical procedures.

Table 5–7 Examples of odorants used in odor identification test

Odorant	*Expected response*
Isoamyl acetate	Fruity
Eugenol	Clove, dentist
Benzaldehyde	Cherry, almond
Lemon oil	Lemon, citrus
1-Octene-3-ol	Mold
Diacetyl	Buttery
cis-3-Hexenol	Cut grass
Cinnamic aldehyde	Cinnamon

Source: G. Lösing et al., Quality Control of Flavorings and Their Raw Materials, *Dragoco Report*, Vol. 42, pp. 93–135, © 1997, Dragoco.

Statistical methods for treatment of analytical results (Lahiff & Leland, 1994), which are not covered in this chapter, make major contributions to quality control in general and to sensory/instrumental correlations in particular. Personal computers are becoming more and more powerful to carry out statistical analyses that were once only feasible on centralized main frame systems.

REFERENCES

American Society for Testing and Materials (ASTM). 1981. *Guidelines for the Selection and Training of Sensory Panel Members.* Special Technical Publication No. 758. ASTM, Philadelphia.

American Society for Testing and Materials (ASTM). 1986. *Physical Requirement Guidelines for Sensory Evaluation Laboratories.* Special Technical Publication No. 913, ASTM, Philadelphia.

S. Anandaraman and G.A. Reineccius. 1987. Analysis of Encapsulated Orange Peel Oil. *Perf. Flav.* 12 (2), 33–39.

A. Arbogast. 1996. Features and Trends in Sensory Analysis Packages. *Food Technol. Eur.* 3(1), 74–78.

Association of Official Analytical Chemists (AOAC), 22 Wilson Blvd., Suite 400, Arlington, VA 22201, U.S.A.

J. Baumgart. 1996. Schnellmethoden und Automatisierung in der Lebensmittelmikrobiologie. *Fleischwirtschaft* 76, 124–130.

C. Bicchi, V. Manzin, A. D'Amato and P. Rubiolo. 1995. Cyclodextrin Derivatives in GC Separation of Enantimers of Essential Oil, Aroma and Flavour Compounds. *Flav. Fragr. J.* 10, 127–137.

K.J. Burdach. 1988. *Geschmack und Geruch.* Hans Huber, Bern.

T. Chen, C. Dwyre-Gygax, S.T. Hadfield, C. Willets and C. Breuil. 1996. Development of an Enzyme-Linked Immunosorbent Assay for a Broad Spectrum Triazole Fungicide: Hexaconazole. *J. Agric. Food Chem.* 44, 1332–1356.

M.M. Clifford. 1986. Phenol-Protein Interactions and their Possible Significance for Astringency. In: *Interactions of Food Components* (G.G. Birch and M.G. Lindley, eds.) Elsevier, London, 143–163.

Council of the European Communities. 1988. Council Directive of 22 June 1988 on the Approximation of the Laws of the Member States Relating to Flavorings for Use in Foodstuffs and to Source Materials for Their Production. *Off. J. Eur. Comm.* L184, 61–67.

R.A. Culp and J.E. Noakes. 1990. Identification of Isotopically Manipulated Cinnamic Aldehyde and Benzaldehyde. *J. Agric. Food Chem.* 38, 1249–1255.

P. Dugo, L. Mondello, E. Cogliandro, A. Verzera and G. Dugo. 1996. On the Genuineness of Citrus Essential Oils. 51. Oxygen Heterocyclic Compounds of Bitter Orange Oil *(Citrus aurantium* L.). *J. Agric. Food Chem.* 44, 544–549.

E.E. Ehrenfeld. 1995. ATP Bioluminescence: Real-Time Testing of Cleaning Effectiveness. *Food Technol. Eur.* 1995 (Sept./Oct.), 151–153.

E. Engvall and P. Perlmann. 1971. Enzyme-Linked Immunosorbent Assay (ELISA). Quantitative Assay of Immunoglobulin G. *Immunochemistry* 8, 871–874.

Essential Oil Association (EOA). 1975. *EOA Book of Specifications and Standards.* Supplement 1. EOA of U.S.A., Washington, DC.

I. Fliedner and F. Wilhelmi. 1989. *Grundlagen und Prüfverfahren der Lebensmittel sensorik.* Behr's Verlag, Hamburg.

K. Frey. 1988. Detection of Synthetic Flavorant Addition to Some Essential Oils by Selected Ion Monitoring GC/MS. In: *Developments in Food Service. Flavours and Fragrances: A World Perspective* (B.M. Lawrence, B.D. Mookherjee and B.J. Willis, eds.). Elsevier, Amsterdam, pp. 517–524.

J. Giese. 1996. Instruments for Food Chemistry. *Food Technol.* 50(2), 72–77.

M. Gilette. 1990. Sensory Evaluation: Analytical and Affective Testing. *Perf. Flav.* 15(3), 33–40.

M.H. Gilette. 1994. Sensory Analysis. In: *Source Book of Flavors,* 2nd ed. (G. Reineccius, ed.). Chapman & Hall, New York, pp. 817–837.

G. Graefe. 1961. Über Traubenzucker-Schokolade. *Süßwaren* 5, 1116–1119.

B.G. Green. 1996. Chemesthesis: Pungency as a Component of Flavour. *Tr. Food Sci. Technol.* 7, 415–420.

M.W. Griffiths. 1996. The Role of ATP Bioluminescence in the Food Industry: New Light on Old Problems. *Food Technol.* 62(6), 62–72.

H.B. Heath. 1981. *Source Book of Flavors.* Avi Publishing, Westport, CT.

U. Hener, P. Kreis and A. Mosandl. 1991. MDGC-Analyse chiraler Monoterpene—Grundlagen und Anwendungen. *Lebensmittelchemie* 45, 10.

U. Hener and A. Mosandl. 1993. Zur Herkunfstbeurteilung von Aromen. *Dtsch. Lebensm. Rdsch.* 89, 307–312.

H.-D. Höltje and L.B. Kier. 1974. Sweet Taste Receptor Studies Using Model Interaction Energy Calculations. *J. Pharm. Sci.* 63, 1722–1725.

International Organization of the Flavor Industry. 1982. Solvent Residues in Extracts, Including Concretes, Oleoresins, Resionoids, etc.-Determination by Headspace Chromatographic Analysis. Recommended Method 20 (1981). *Z. Lebensm. Unters. Forsch.* 174, 402.

International Organization of the Flavour Industry. 1990. *Code of Practice for the Flavour Industry.* Geneva.

L.B. Kier. 1972. Molecular Theory of Sweet Taste. *J. Pharm. Sci.* 61, 1394–1397.

T. Kubota and I. Kubo. 1969. Bitterness and Chemical Structure. *Nature* 223, 97–99.

M. Lahiff and J.V. Leland. 1994. Statistical Methods. In: *Source Book of Flavors.* 2nd ed. (G. Reineccius, ed.). Chapman & Hall, New York, pp. 743–787.

D.G. Laing and A. Jinks. 1996. Flavour Perception Mechanisms. *Tr. Food Sci. Technol.* 7, 387–389, 421–423.

H.T. Lawless and C.B. Lee. 1993. Common Chemical Sense in Food Flavor. In: *Flavor Science. Sensible Principles and Techniques* (T.E. Acree and R. Teranishi, eds). American Chemical Society, Washington, pp. 33–66.

H.T. Lawless and M.R. Claassen. 1933. Application of the Central Dogma in Sensory Evaluation. *Food Technol.* 47(6), 139–146.

W.E. Lee III and R.M. Pangborn. 1986. Time-Intensity: The Temporal Aspects of Sensory Perception. *Food Technol.* 40(11), 71–78, 82.

L.E. Marks. 1987. *The Unity of the Senses.* Academic Press, New York.

G. Martin, G. Remaud and G.J. Martin. 1993. Isotopic Methods for Control of Natural Flavours Authenticity. *Flav. Fragr. J.* 8, 97–107.

R. McCormick. 1988. Vanilla and Vanillin Remain Ever-versatile Ingredients. *Prepared Foods* 157, 123.

Minolta Camera Co. 1993. *Precise Color Communication. Color Control from Feeling to Instrumentation.* Minolta Camera Co., Osaka, Japan.

L. Mondello, G. Dugo, P. Dugo and K.D. Bartle. 1996. On-Line HPLC-HRGC in the Analytical Chemistry of Citrus Essential Oils. *Perf. Flav.* 21(4), 25–49.

A. Mosandl. 1991. Neue Ergebnisse der Aromastoff-Forschung. *Lebensmittelchemie* 45, 2–7.

A. Mosandl. 1995a. Enantioselektivität und Isotopendiskriminierung als biogenetisch fixierte Parameter natürlicher Duft- und Aromastoffe. *Lebensmittelchemie* 49, 130–133.

A. Mosandl. 1995b. Enantioselective Capillary Gas Chromatography and Stable Isotope Ratio Mass Spectrometry in the Autheticity Control of Flavors and Essential Oils. *Food Rev. Int.* 11, 587–664.

T. Nagodawithana. 1994. Flavor Enhancers: Their Probable Mode of Action. *Food Technol.* 48(4), 79–85.

K.H. Ney. 1988. Sensogramme, eine methodische Erweiterung der Aromagramme. *Gordian* 88, 19.

K.H. Ney. 1987. *Lebensmittelaromen.* Behr's Verlag, Hamburg.

A.C. Noble. 1996. Taste-Aroma Interactions. *Tr. Food Sci. Technol* 7, 439–444.

S. Notermans, A.M. Hagenaars and S. Kozaki. 1982. The Enzyme-Linked Immunosorbent Assay (ELISA) for the Detection and Determination of Clostridium botulinum Toxins A, B, and E. *Methods Enzymol.* 84, 223–238.

K. Ogden. 1993. Practical Experiences of Hygiene Control Using ATP-Bioluminescence. *J. Inst. Brew.* 99, 389–393.

M. O'Mahony. 1995. Sensory Measurement in Food Science: Fitting Methods to Goals. *Food Technol.* 49(4), 72–82.

R.W. Parkes. 1994. Measurement of Colour in Food. In: *Food Technology International Europe* (A. Turner, ed.), Sterling Publications, London, pp. 175–176.

O. Philipp and H.-D. Isengard. 1995. Eine neue Methode zur Authentizitätsprüfung von Zitronenölen mit HPLC. *Z. Lebensm. Unters. Forsch.* 201, 551–554.

J. Prescott. 1994. The Hot Topic in Food Flavours. *Food Australia* 46, 74–77.

G. Reher and E. Stahl-Biskup. 1987. Geschmack und Geruch, die chemischen Sinne des Menschen und die Chemie der Geschmack-und Riechstoffe. Teil I: Geschmack. *Dtsch. Apotheker Z.* 127, 2468–2478.

G. Reineccius. 1994. Adulteration. In: *Source Book of Flavors,* 2nd ed. (G. Reineccius, ed.). Chapman & Hall, New York, pp. 731–742.

M. Rothe. 1978. *Handbuch der Aromaforschung: Einführung in die Aromaforschung.* Akademie-Verlag, Berlin.

M. Ruhland, G. Engelhard and P.R. Walnöfer. 1996. Application of Gradient-HPLC for the Separation of Ochratoxin A and Its Derivatives. *Adv. Food Sci.* (CMTL) 18, 32–34.

R.A. Savage. 1995. Hazard Analysis Critical Control Point: A Review. *Food Rev. Int.* 11, 575–595.

H.-L. Schmidt. 1986. Food Quality Control and Studies on Human Nutrition by Mass Spectrometric and Nuclear Magnetic Resonance Isotope Ratio Determination. *Fresenius Z. Anal. Chem.* 324, 760–766.

P. Schreier. 1993. Chirality Evaluation in Flavour Precursor Studies. In: *Progress in Flavour Precursor Studies* (P. Schreier and P. Winterhalter, eds.). Allured Publishing, Carol Stream, pp. 45–61.

H. Schulz. 1989. The Automation of Classical Physical Methods of Measurement in Quality Control. *Dragoco Rep.* 36, 178–186.

H. Schulz. 1988. Possible Applications of High Performance Liquid Chromatography in Quality Control. *Dragoco Rep.* 35, 3–11.

H. Schulz and G. Lösing. 1995. Anwendung der nahen Infrarotspektroskopie bei der Qualitätskontrolle etherischer Öle. *Dtsch. Lebensm. Rdsch.* 91, 239–242.

J.S. Schultz. 1991. Biosensors. *Sci. Am.* 264(8), 64–69.

V. Schurig, M. Schleimer, M. Jung, S. Mayer and A. Glausch. 1993. Enantiomer Separation by GLC, SFC and CZE on High-Resolution Capillary Columns Coated with Cyclodextrin-Derivatives. In: *Progress in Flavour Precursor Studies* (P. Schreier and P. Winterhalter, eds.). Allured Publishing, Carol Stream, pp. 64–75.

R.S. Shallenberger and T.E. Acree. 1967. Molecular Theory of Sweet Taste. *Nature* 216, 480–482.

W.L. Silver and T.E. Finger. 1991. The Trigeminal System. In: *Smell and Taste in Health and Disease* (Getchell, T.V. et al., eds.). Raven Press, New York, pp. 87–108.

W. Simpkins and M. Harrison. 1995. The State of the Art in Authenticity Testing. *Tr. Food Sci. Technol.* 6, 321–328.

E. Stahl-Biskup and G. Reher. 1987. Geschmack und Geruch. Die chemischen Sinne des Menschen und die Chemie der Geschmacks-und Riechstoffe. Teil II: Geruch. *Dtsch. Apotheker Z.* 127, 2529–2542.

C.R. Stampanoni. 1993. The "Quantitative Flavor Profiling" Technique. *Perf. Flav.* 18(6), 19–24.

C.J. Stannard and P.A. Gibbs. 1986. Rapid Microbiology: Applications of Bioluminescence in the Food Industry—A Review. *J. Biolumin. Chemilumin.* 1, 3–10.

H.Stone, J. Sidel, S. Oliver, A. Woolsey and R.C. Singleton. 1974. Sensory Evaluation by Quantitative Descriptive Analysis. *Food Technol.* 28, 24–34.

H. Stone and J.L. Sidel. 1985. *Sensory Evaluation Practices.* Academic Press, New York, pp. 194–226.

L.B. Sweger. 1993. Rapid *Salmonella* Methodology in the Chocolate Industry. *Manufact. Confect.* June 1993, 77–80.

C.-T. Tan and J.W. Holmes. 1988. Stability of Beverage Flavor Emulsions. *Perf. Flav.* 13 (1), 23–41.

D.M.H. Thomson. 1986. The Meaning of Flavour. In: *Developments in Food Flavours* (G.G. Birch and M.G. Lindley, eds.). Elsevier, London, pp. 1–21.

K.Y. Tse and G.A. Reineccius. 1995. Methods to Predict the Physical Stability of Flavor-Cloud Emulsion. *ACS Symp. Ser.* 610, 172–182.

A.Verzera, A. Cotroneo, G. Dugo and F. Salvo. 1987. On the Genuineness of Citrus Essential Oils. Part XV. Detection of Added Orange Oil Terpenes in Lemon Essential Oils. *Flav. Fragr. J.* 2, 13–16.

P. Werkhoff, S. Brennecke, W. Brettschneider, M. Güntert, R. Hopp and H. Surburg. 1993. Chirospecific Analysis in Essential Oil, Fragrance and Flavor Research. *Z. Lebensm. Unters. Forsch.* 196, 307–328.

J.R. Whitaker. 1984. Biological and Biochemical Assays in Food Analysis. In: *Modern Methods of Food Analysis* (K.K. Stewart and J.R. Whitaker, eds.). Avi Publishing, Westport, CT, pp. 187–225.

F.J. Winkler and H.-L. Schmidt. 1980. Einsatzmöglichkeiten der ^{13}C-Isotopen-Massen-Spektrometrie in der Lebensmitteluntersuchung. *Z. Lebensm. Unters. Forsch* 171, 85–94.

Chapter 6

Beverage Flavorings and Their Applications

A.C. Mathews

6.1 INTRODUCTION

Our requirement for liquid refreshment is as longstanding as the origins of the species *Homo sapiens*. An examination of Figure 6–1 shows that at relatively low levels of fluid loss, e.g., 3%, impaired performance results. A 20–30% reduction in capacity for hard muscular work occurs with a moisture/fluid loss of 4%, heat exhaustion at 5%, hallucinations at 7% and circulatory collapse and/or heat stroke at 10% fluid loss. The required fluid intake for the average person in the arid areas surrounding the Red Sea is a staggering 8 liters per day.

With the very existence of life depending on our fluid intake, why do we not simply take water alone? The first reason is that water alone is not readily taken into human body systems, because it requires a certain level of carbohydrate and salt for rapid transfer across the brush border of the gut. The second reason is that water alone is uninteresting and unquenching in taste.

The importance of fluid intake at the correct osmotic pressure (a measure of tonicity and hence compatibility with body fluids, the desired goal being isotonicity) is at the very origin of human nutrition. What was the reason for the original manufacture of local wines at alcohol contents of approximately 5% in the tropical, subtropical and arid areas of the world? It was simple to use the preservation properties of alcohol as a means of keeping grape juice from degrading to a non-potable state. In the process, rehydration fluid (water) was held in a microbiologically stable and acceptable state suitable for use over extended periods of time. This was the origin of the category of food products that has the universally accepted nomenclature, beverages.

In the early days of wine production, it was found that by using different grape juices and fruit juices in varying proportions a multitude of completely different tastes was achievable. The occasional use of only small quantities of certain fruits, vegetables, or herbs with strong characters (e.g., mango or thyme) in the blend of juices to be fermented would result in a completely different taste of the

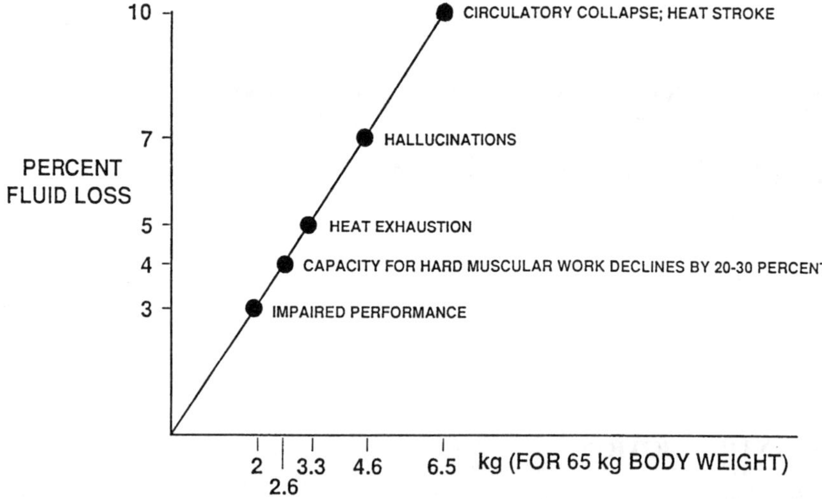

Figure 6–1 The effects of fluid loss (Bergstrom & Hultman, 1972; National Academy of Sciences, 1974).

finished wine. This was probably the first regular use of the application of food flavorings to beverages.

Any substance which, when added to a beverage in quantities lower than 50% of the total volume, causes the flavor of the whole product to assume its own character may technically be called a flavoring. However, for the purposes of this chapter, we will consider such flavorings only in passing and concentrate our discussions on intense, relatively low volume use flavors.

One of the problems when trying to discuss the composition and application of food flavorings to beverages is the number and diversity of beverage industries around the world. An overview of some of the major sectors of these industries is presented in this chapter.

6.2 CATEGORIES OF BEVERAGES

From a technical point of view this diversity of beverage type and composition is considered so that flavor solubility and miscibility are not compromised when applying flavorings to beverages. For this purpose we will consider three broad categories of basic composition:

(a) alcoholic beverages, subdividing into alcohol contents
  (i) lower than 20%
  (ii) higher than 20%

(b) aqueous and aqueous/sugar beverages subdividing into sugar contents
  (i) lower than 2.5%
  (ii) higher than 2.5% and lower than 60%
  (iii) higher than 60%
(c) mixtures of (a) (b), e.g.,
  (i) liqueur
  (ii) low alcohol wine
  (iii) shandy or beer cooler

The composition of either naturally occurring or synthetically produced flavoring material must now be examined closely for compatibility with the category of beverage that requires flavoring. For example, some of the most effective and readily available natural flavoring materials are citrus essential oils released from the peel of citrus fruits during processing for juice extraction. These oils, as obtained from the fruit are almost insoluble in aqueous and aqueous sugar-containing beverages, are generally soluble in high alcohol containing beverages and have limited solubility (but greater miscibility) in aqueous beverages with sugar contents higher than 60%. These solubility/miscibility characteristics are predictable from an examination of the composition of the oils, which shows a very high water insoluble terpene content. The composition of both lemon and orange oils is given together with information on essential extraction requirements (Bauer & Gerbe, 1985) in Appendix I.

The best beverage composition for the natural whole citrus oil which has been neither solubilized nor processed is thus a high alcohol-containing beverage with a high sugar content. This is usually called a liqueur. Many famous names in this beverage category are renowned for their strong, fresh authentic citrus characters that would be almost impossible to obtain via any other flavor route because the other methods which we shall consider for the processing of these and other such essential oils result in a diminution in authenticity of character unless this is compensated for in other ways.

6.3 TYPES OF FLAVORINGS FOR BEVERAGES

Just as there are broad compositional categories for beverages, so it is with flavorings for beverages. They result from either the form in which the concentrated natural or synthetic flavoring exists or from the method or methods used to extract or manufacture the flavoring.

When the flavorist is compounding a flavor, he or she will start with an array of flavoring components of either natural or synthetic origin. These components will have different physical performance characteristics of dispersibility and solubility as well as individual flavor effects and interactive properties with other components. For natural flavoring materials these properties will, in part, be the result of the method of extraction and solvents used and will fall into three broad categories or types which then result in two basic performance properties:

(a) oil soluble flavorings subdividing into type by solvent
  (i) alcoholic
  (ii) non-alcoholic
(b) water soluble flavorings
(c) mixtures of (a) and (b), for performance reasons and generally operating at the limits of technical acceptability of the beverage application

The synthetic components will also carry with them performance characteristics which may be categorized in a similar way.

6.4 METHODS OF EXTRACTION, SOLUBILIZATION AND CONCENTRATION OF FLAVORINGS

Because water (alone)-soluble flavorings are by far the smallest category for the flavoring of beverages it would seem sensible to consider these before oil-soluble and mixed-solvent flavorings. Their acceptability and ease of use in the aqueous non-alcoholic beverage category and industry (as defined in this chapter) would make them the preferred route for flavor application if it were not for the limitation they impose on flavor range and intensity. The biggest and most widespread extraction and use of water-soluble flavorings is instant coffee.

6.4.1 *Extraction of Coffee Flavor and Manufacture of the Instant Product*

Instantly soluble coffee is prepared in a series of column extractors in which the finely ground roast beans are extracted with counter currents of water under pressure at 175° C (350° F) for a period of about 4 h. This results in an almost complete recovery of soluble solids from beans and gives about 45–50% of extract. The initial concentrate is then either spray dried to give the familiar free-flowing powder or freeze dried and agglomerated to give a product looking more like ground coffee. To improve the "freshly perked or filtered" coffee aroma, about 0.2% of recovered coffee oil is incorporated into the product by either spraying onto the surface of the powder or by incorporation into the concentrate prior to freeze drying.The resulting products are packed under inert gas into glass jars to protect flavor freshness and intensity during storage prior to use.

In this example of what appears on the surface to be a fully water extracted and soluble flavoring system, we can already see that certain very important and significant volatile (taste and, in particular, aroma) components are lost if only the water extract alone is used. This is because when any flavor extraction is carried out at temperatures of 100°C and above (at atmospheric pressure) any essential oils and, dependent on temperature, resinous materials will be volatilized and carried over in the steam released. The effectiveness of this volatilization, which must be compensated for when formulating the finished beverage, will be dependent upon the particle size of the material under extraction because this has to be small enough to allow good steam penetration. This can be seen even more clearly in the next most commonly perceived "water-soluble or extracted" group of flavorings, concentrated fruit compounds and flavoring extracts where the flavor of the fruit is divided unequally between the juice and the volatile and oil components.

6.4.2 *Flavorings Extracted from Harvested Fruits*

Fruits are the matured ovary of the plant or tree and may be with or without seeds and sometimes with the flower still attached. The wall of the fruit developed from the wall of the ovary is called the pericarp and may be either dry or fleshy; it is this edible fleshy part which forms most of the varieties we call "fruits." Nuts are also fruits and they will be covered separately for their application to beverages and, in particular, the manufacture of cola nut extracts and flavorings. Botanically fruits may be classified as in Appendix II (Heath, 1982).

This is the type of classification and way of thinking about fruits that a beverage formulator will pursue as it follows the perception and requirements of the consumer. A flavorist, however, will be more interested in the value, diversity and intensity of the flavoring materials as library items from which to draw when compounding flavorings. He or she may be less interested in the replication of the whole fruit flavor in the finished beverage, this being the job of the applications technologist. The processing of fruit for the purpose of extracting juice is a complex area of technology involving the use of enzymes, preservatives, different expression techniques, de-activation of natural enzymes, heat processing and flavor retention, collection, and recovery techniques (Chapter 3). A very full account of fruit processing also has been given by Tressler and Joslyn (1982) and will only be covered in outline here as far as it relates to flavor extraction, formulation, beverage application and their interactions.

Very few flavorings which may be obtained from natural materials for beverages can be extracted and dissolved by water alone. In citrus fruits the juice cannot possibly be characteristic of the whole fruit because so much of the character is contained in the oil component in the peel.

In so-called berry fruits, the method of hot enzyming and extraction ensures that the volatile organic components will be lost during concentration in the steam/water phase unless steps are taken to recover them. The characteristic volatile aroma that is responsible for the flavor profile of most freshly pressed juices is present at levels typically around only 700 parts per million and considerably less in heat treated juices. In the extraction of flavoring materials from the juicing process, it is intended that no components of potential value are wasted and that the valuable nonaqueous phase materials are collected. These components are insoluble and form an unstable emulsion when applied to most liquid beverages.

6.4.3 *Extraction and Use of Oil Soluble Flavorings*

Oleoresins, tinctures and extracts have already been discussed in Chapter 2. In this section, discussion is restricted to the methods of their application to beverages and, where necessary (e.g., cola nut extract), specific methods of extraction which result in products compatible with aqueous and aqueous sugar-containing beverages, in addition to beverages containing less than 20% alcohol.

6.4.3.1 *Solvent extraction*

If normal distillation techniques are considered too harsh for the extraction of more delicate or sensitive flavorings such as floral notes or cola nuts, then solvent extraction may be employed. The collected flower or finely ground nuts are placed in tiers on perforated plates in a suitable extraction vessel and the selected solvent (usually about 60% alcohol or propylene glycol in the case of cola nuts) is allowed to percolate until the raw material is exhausted. The resulting solvent containing extract may then be used directly or reduced to a tincture or soft extract by distillation to recover solvent. This may take place at elevated temperature and an example is given in Table 6–1.

6.4.3.2 *Essential oil extraction by pressing*

This is the method used to extract essential oils from fruits with peels which have a naturally high oil content such as the citrus fruits of orange, lemon, lime, etc. It is the method used by most of the world's citrus juice processors using FMC-type extractors. Oil released from the peel during processing is washed away from the fruit with a spray of water. The oil and water are subsequently separated by centrifugation. Owing to the incompatibility of essential oils with aque-

Table 6–1 Cola nut flavor

	Quantity (g)
1. Finely divided (into small particles) cola nuts	400.00
2. Add propylene glycol	800.00
3. Heat and maintain at 30 to 35° C for at least 8 h	
4. Gently distill under vacuum to reduce distillate to	100.00
5. Remove vacuum and add sufficient water (approx. 900 g) to give total essence of	1000.00

ous phase systems (see above), the highest quality essential oils are those with minimal (or the least) contact with water.

Oils may also be extracted by steam distillation but will still require the use of solvent extraction and/or subsequent use of solvents to make the resulting flavoring practically usable in aqueous sugar beverages and beverages containing <20% alcohol.

Ethyl alcohol or isopropyl alcohol (although imparting different flavor characters in its own right) are widely used as a solvents for beverage flavorings. Both are normally compatible either used singly or in combination with each other or water with aqueous, aqueous sugar-containing and alcohol (<20%)-containing beverages. In parts of the world where the use of these or other alcohols is either undesirable or legislatively not permitted, alcohol as an extractive solvent must be removed by distillation and the resultant extract redispersed in other solvents/diluents such as propylene glycol or tri-acetin (glyceryl triacetate).

A traditional method for the preparation of flavorings from essential oils is that of "washing the oil" or oils. This involves mixing the essential oil(s) with alcohol, water and sometimes other solvents. The mixture is allowed to stand for up to 24 h for separation to occur with the subsequent removal of insoluble terpenes. The resulting essence is then further solubilized with alcohol. A typical formula is given in Table 6–2.

6.5 BEVERAGES BASED ON GINGER

An example of a flavor which has application across all three of the previously defined beverage categories (including the subdivisions) and the three flavor categories (including the two broad performance types) is ginger (Table 6–3).

Table 6–2 Lemon essence

	Quantity (g)
1. Cold pressed lemon oil[a]	150.00
2. Cold pressed lime oil	50.00
3. Alcohol, 95%	550.00
4. Water	250.00
Total	1000.00
5. Agitate and stand 24 h to separate terpenes	−150.00
Yield	850.00
6. Add alcohol, 95%	150.00
Lemon essence, terpene-free (filter if necessary)	1000.00

[a]May be partially replaced by, or supplemented with lemon grass oil which is the main source of natural citral (approx. 80% geranial and 20% neral).

Table 6–3 Examples of beverages based on ginger

Beverage[a]	*Aqueous/sugar (% sugar content)*	*Alcoholic (% alcohol content)*
American ginger ale	10–12	—
Diet American ginger ale	<2.5	—
Original ginger ale	4–8	—
Ginger beer flavor*	6–13	—
Ginger beer*	6–13	1
Ginger cordial	23–60	—
Ginger cordial	23–60	10
Ginger wine	30–40	10–20
Ginger liqueur	60	40

[a]All products listed are normally of clear appearance. However, those marked with * are often cloudy because they use an emulsion containing oil (or other optically reflective droplets) finely divided and stabilized in the aqueous phase.

Every flavorist or formulator who has worked with ginger or its derivatives knows how dependent the final product is on the origin of the natural raw material used. The flavor ranges from lemony, spicy, through aromatic, earthy, harsh, to camphoraceous and the methods of flavor extraction employed (e.g., to give sweet, soft ginger flavor through to fiery, hot and nose tingling). These different flavor effects may be achieved by the use of different proportions and combinations of raw material and also by the selection of both aqueous and non-aqueous phase extracts and by the skillful manipulation of the product formulation into which these materials are applied.

6.5.1 *Manufacture of Ginger Extract*

A formula and process for the manufacturer of a general purpose ginger extract is given in Table 6–4 (Heath, 1982). The flavor of the extract so achieved may be modified in a number of ways in order to achieve the required performance target. For a flavorist this will be to obtain the maximum number of flavoring components to use in a variety of different product applications including beverages. By use of different extraction methods to obtain very different flavor characteristics, the general extract can be widely modified, e.g., by use of oleoresin of ginger (see Chapter 2).

The flavorist's perception of the required performance criteria of the flavoring will usually be physical (appearance and compatibility) and organoleptic stability in the general flavor target area. This is a result of the single most difficult problem that has to be overcome with beverages, i.e., the compatibility of most flavorings with their varying liquid compositions.

Table 6–4 Basic ginger extract

Extraction of:

(1) 132 kg Jamaica ginger, comminuted, with:
480 kg or 580 liters heated water of about 90° C, circulate the liquid over the ginger for 10 min. then let the mixture stand for 24 h to cool

(2) To the mash of (1) is then added:
285 kg or 430 kg alcohol, 95%: the menstruum is to be circulated for 10 min to obtain uniform alcohol content of 40% strength

(3) After 3 days extraction during which the menstruum has been circulated twice daily, the liquid is drained off and yields approx.

(4) 670 kg ginger extract; it is used in (5)
226 kg remaining mash is mixed with
150 kg water, and distilled at atmospheric pressure, to yield
110 kg flavor distillate of about 25% alcohol content, it is used in (5)

(5) Mix:
670 kg ginger extract of (3) with
110 kg flavor distillate of (4), to yield

Total
1720 kg or 1000 liters basic ginger extract

An examination of how the basic ginger extract may be modified to achieve different desired flavor effects for different beverage applications will serve to demonstrate how the flavorist and applications technologist can contribute to the development of different flavorings. The two products considered are:

1. "Original" (hot) dry ginger ale (carbonated beverage) and
2. "American," "Canada," or "pale" (soft and sweet) ginger ale (carbonated beverage)

6.5.2 *"Original" (hot) Ginger Ale*

The formula required to prepare flavoring of the correct character for "original" (hot) dry ginger ale is shown in Table 6–5. It can be seen that the flavorist is achieving the "drier" and "hotter" ginger taste by the use of oleoresins of both capsicum and ginger and is "deepening" or "broadening" the resulting flavor with the use of ginger, orange, lime, mace and coriander oils. The use of rose oil in this context must be very carefully controlled because an excess will render the flavor "out of balance." This expression, "out of balance," is used a great deal in the flavor and flavor application industries and we will return to it in a different context later.

The product formulation given in Table 6–6 is typical of the type of carbonated beverage of the "original" or "dry" ginger ale category. Here it can be seen that the applications technologist is achieving the "dryness" of taste by keeping the sweet-

Table 6–5 Original ginger ale essence

Mixture of:

9.4	ml oleoresin ginger[a]
0.95	ml oleoresin capsicum
1.9	ml oil of ginger[a]
1.9	ml oil of orange, cold pressed
1.9	ml oil of lime, distilled
0.25	g oil of mace
0.25	g oil of coriander
200	ml alcohol, 95%
400	ml propylene glycol
12.5	g magnesium carbonate
	water

Procedure:

(1) Mix oleoresins with 150 ml alcohol (95%) and 220 ml propylene glycol; add 6.85 g magnesium carbonate, mix 1 h, and then let stand overnight; decant and filter

(2) Mix the oils with 66.15 ml alcohol (95%) and then with the filtered mixture of (1); afterwards add to it 20 ml propylene glycol and 20 ml water to make 1000 ml essence

(3) A few drops of oil of rose are sometimes added to the essential oil mixture to give it a distinctive character

[a]May partially be supplemented or enhanced by the use of a proportion of "basic ginger extract" shown in Table 6–4.

ness low (total sweetness of 6.9% w/v compared with 10–12% w/v for most products of this type) and the acidity at an average level (0.25% w/v as citric acid monohydrate) and hence is obtaining a low sweetness/acid ratio which will make the product taste "sharp" and relatively low in sweetness. The use of sodium saccharin in this context (where permitted legislatively and by consumer acceptance) is to give the product an even "harsher" taste because a lower sugar content used in combination with this sweetener will remove some of the perceived "body" of the beverage. The inherent "thinness" or "strident" nature of the sweetness of saccharin will further enhance the "dryness" of taste and allow the strong ginger and heat characteristics of the flavoring components to become apparent. The total effect is to intensify the "dry" and "hot" ginger character of the beverage. The use of sweetness/acid balance achieves a better "original" ginger character than by the use of flavoring alone at higher addition levels combined with higher sweetness. If this product formulation has intensified the "hot" and "dry" characters of the product too much for consumer acceptance (this information should always be gathered by correctly structured consumer taste testing/research), slight reduction of the level

Table 6–6 "Original" or "dry" ginger ale (carbonated beverage): materials and quantities for 1000 liters finished product

Material	*Quantity*	*Units*
Granulated sugar	38.9	kg
Sodium saccharin dihydrate	0.06	kg
Sodium benzoate	0.174	kg
"Original" ginger ale essence and adjust to taste	0.400	liters
American ginger ale or tonic water essence	~0.160	liters
Citric acid anhydrous (to give 0.25% w/v CAMH)	2.30	kg
Caramel MW3	0.21	liters
Water to 1000 liters	~970	liters
Carbon dioxide to 4.5 vols. Bunsen (v/v)	1000	liters

Directions/special notes

Make into a suitable syrup, dissolving the solid ingredients in water before addition; dilute the caramel 1:1 with water before addition; after adequate mixing use the syrup for bottling by diluting one volume with an appropriate volume of water, adding carbon dioxide in a suitable manner

Analytical characteristics

Refractometric Brix (% w/w at 20°C)	4.0 ± 0.3
will increase if saccharin replaced by sugar to	7.0 ± 0.3
Acidity as citric acid monohydrate (% w/v)	0.25 ± 0.03

Calculated sweetness

(1)	Due to sugar (% w/v)	3.90
	Due to saccharin (% w/v)	3.00
	Total	6.90
	or, any combination up to	
(2)	Due to sugar (% w/v)	7.00

of the "original" ginger ale essence and inclusion of a small proportion of either the "softer," more "citrus" "American" or "pale" ginger ale essence or a citrus blend such as that used in Indian tonic water can be used beneficially.

The flavorist may in this way achieve through flavoring composition the same effect as the applications technologist does by varying the composition of the beverage.

6.5.3 *"American" or "Pale" Ginger Ale*

A formula for the manufacture of a flavoring suitable for "American" or "pale" ginger ale is given in Table 6–7. An examination of the formula for the manufac-

Table 6–7 "American" or pale ginger ale essence

Mixture of:

0.5 g	oil of rose
0.5 g	phenylethyl alcohol
9.5 g	methyl nonyl acetaldehyde 50%
22.0 g	oleoresin of ginger
22.5 g	oil of ginger
27.0 g	oil of bergamot
246.0 g	oil of orange, cold pressed
300.0 g	oil of lemon, cold pressed
372.0 g	oil of lime, distilled
5000.0 g	alcohol, 95% then add
7000.0 g	distilled water and mix

Procedure:

(1) The mixture is then poured into a separator, and left covered in a cool place for 48 h

(2) After the terpenes have settled on the surface of the mixture, the clear extract is taken off

(3) The remaining terpenes are mixed well with
1500.0 g alcohol (95%) and to it added
3500.0 g distilled water
mix well for 5 min then pour the mixture into a separator, after 48 h take off clear extract and discard the terpenes

(4) Extract mixtures of (2) and (3) are united, and agitated for 5 min before the compound is poured into a separator; there the mixture is left for 48 h; the terpenes will separate to the surface; the clear extract is taken off, and the terpenes are discarded

ture of "American" or "pale" ginger ale essence shows obvious differences from that for the "original" (hot) ginger ale essence. Oil of rose is used as a small but required part of the character of the flavoring because in this type of product floral notes are necessary to obtain the "balance" of required taste. It can also be seen that a "softer," "sweeter," less "hot" product is achieved by omission of oleoresin capsicum and the use of much greater quantities of citrus oils, including bergamot oil with its much "sweeter," "softer," and "fragrant" character.

The product formulation given in Table 6–8 is typical of the type of carbonated beverage of the "American" or "pale" ginger ale category. Here it can be seen that the applications technologist achieves a "sweeter," "softer" character by keeping the usual level of sweetness for a carbonated beverage (total sweetness of 9–10% w/v compared with 7% w/v for the "original" or "dry" ginger ale) in combination with a relatively low level of acidity (0.15% w/v CAMH) and hence obtaining a high sweetness/acid ratio which will make the product taste "soft," "full-bodied"

Table 6–8 "Original" or "pale" ginger ale (carbonated beverage): materials and quantities for 1000 liters finished product

Material	*Quantity*	*Units*
Granulated sugar[a]	90.0	kg
Sodium benzoate (give 145 mg/kg benzoic acid)	0.177	kg
Sodium citrate	0.21	kg
Lemon essence	0.10	liters
American ginger ale essence	3.25	liters
Caramel MW3	0.080	kg
Citric acid anhydrous (to give 0.15% w/v CAMH)	1.37	kg
Water to 1000 liters	~940	liters
Carbon dioxide to 4.5 vols. Bunsen (v/v)	1000	liters

Directions/special notes

Make into a suitable syrup, dissolving the solid ingredients in water before addition; dilute the caramel 1:1 with water before addition; after adequate mixing, use the syrup for bottling by diluting one volume with an appropriate volume of water, adding carbon dioxide in a suitable manner

Analytical characteristics

Refractometric Brix (% w/w at 20°C)	9.0 ± 0.3
Acidity as citric acid monohydrate (% w/v)	0.15 ± 0.03

Calculated sweetness

(1) Due to sugar (% w/v)	9.00
or, any combination up to	
(2) Due to sugar (% w/v)	5.00
Due to aspartame (% w/v 200)	4.00
Total	9.00

[a]May be partially replaced by Aspartame up to 0.20 kg.

and relatively high in sweetness. The use of either all sugar or a high quality intense sweetener like aspartame helps the good perception of "body" and "roundness."

In this example, the applications technologist supplements the efforts of the flavorist in achieving the required "softer," "sweeter" taste with the choice of a different sweetness/acid ratio and also in obtaining the required citrus blend with the use of additional lemon flavoring. Addition of extra floral notes and the use of other citrus components such as orange, lime or bergamot are also at the discretion of the product formulator.

A specialist area of flavoring/product formulation interaction is that of "diet," "low calorie" or "sugar-free" carbonated beverages. The intense sweetener aspartame has increasingly gained acceptance for use in this area and is renowned for its interaction with flavorings and dependence upon shelf-life management and correct product formulation and composition.

An example of a "sugar-free" "American" type ginger ale product formulation is given in Table 6–9. The flavorings which the product formulator selects for use in such a product must be carefully evaluated and assessed for interaction with the sweetener system. In this example, it is almost impossible for a flavorist to compound the correct flavoring without the specific help and interaction of the application technologist.

An examination of the sugar-free "American" style ginger ale product formulation shows how the application technologist has compensated for loss of body and sweetness which results from the removal of 10% sugar by the use of a combination of polydextrose (provides body but no sweetness) and aspartame (provides sweetness but no body) with buffered low acid content to maintain a high

Table 6–9 "Sugar-free" "American" style ginger ale (carbonated beverage): materials and quantities for 1000 liters finished product

Material	*Quantity*	*Units*
Aspartame (NutraSweet)	0.50	kg
Sodium benzoate (give 145 mg/kg benzoic acid)	0.15	kg
Sodium citrate	0.25	kg
Polydextrose (Pfizer)	50.00	kg
American ginger ale essence	3.50	liters
Lemon essence	0.15	liters
Original ginger ale	0.15	liters
Caramel MW3	0.080	kg
Citric acid anhydrous (to give 0.15% w/v CAMH)	1.37	kg
Water to 1000 liters	q.s.	liters
Carbon dioxide to 4.5 vols. Bunsen (v/v)	1000	liters

Directions/special notes

Make into a suitable syrup, dissolving the solid ingredients in water before addition; dilute the caramel 1:1 with water before addition; after adequate mixing, use the syrup for bottling by diluting one volume with an appropriate volume of water, adding carbon dioxide in a suitable manner

Analytical characteristics

Refractometric Brix (% w/w at 20°C)	9.0 ± 0.3
Acidity as citric acid monohydrate (% w/v)	0.15 ± 0.03

Calculated sweetness

(1) Due to aspartame (% w/v × 200)	9.00

perceived sweetness/acid ratio. This combination does, however, have "dampening" effects on the taste from the flavorings used in the sugar-containing analogue of this product. It is necessary to compensate for this effect by the use of extra levels of these same flavors and by the incorporation of a low level of "original" (hot) ginger ale essence to restore the "bite" of the product taste.

If product formulator and flavorist were to work together, it may be possible to formulate a flavoring which would encompass many of these requirements for use specifically in sugar-free beverages. It is the skill of the applications technologist, however, in combining different flavorings to achieve the same result that gives many products their uniqueness and makes the beverage producer independent of any one flavor manufacturer.

6.6 FORMULATION OF BEVERAGES

6.6.1 *General Principles*

The general approach to the formulation of beverages is applicable across the wide range of products which fall into this category. These general principles apply to aqueous sugar-containing, alcohol-containing and powdered, or dried, beverages as long as the effects of the different processes employed in the manufacture of the different products are taken into account when formulating. As an example, the flavoring of milks or milk-based beverages where the high temperatures employed in the UHT process are necessary for the long-life microbiological stability of the product completely destroy the effect of certain (e.g., strawberry) natural flavorings. Nature identical versions of the sensitive flavorings are similarly, although to a lesser extent, affected by this loss of flavor. Artificial flavorings will normally withstand this type of process without loss of flavor character, intensity, or acceptance.

A strawberry natural flavoring extract is shown in Table 6–10 and the formula of a milk-based product which might use such a flavoring is shown in Table 6–11. A heat stable artificial strawberry flavoring formulation is given in Table 6–12 for comparison.

There are only a limited number of things that the applications technologist can do to try to overcome this problem such as adjustment of pH, use of anti-oxidant and an investigation of the process to see if it would be possible to add this heat sensitive natural flavoring in an aseptic way after pasteurization of the product. If the flavorist is working with the beverage formulator there is the possibility of producing a WONF (with other natural flavors) strawberry essence that could have improved stability to heat. Heat instability of a natural flavoring could, for example, be the result of the initial process employed for the extraction of the natural flavoring from the fruit. A strawberry WONF flavoring of the type that may be used is given in Table 6–13 for information.

Table 6–10 Strawberry fruit juice and flavoring extract (this procedure describes the production of a strawberry juice flavor from frozen strawberries)

Procedure	*Quantity*	*Units*
A. Production of strawberry juice and flavoring extract		
(1) Defrosted whole Marshall strawberries	1000	kg
Granulated sugar[a]	200	kg
Pectinesterase (pectinol)	2.00	kg
Ethyl alcohol (95% v/v)	200	liters
This mix is comminuted using the Fitzpatrick machine with a no. 4 sieve and then pressed		
(2) The pressed pomace is mixed with water and distilled at atmospheric pressure	100	liters
Yield of distillate at approximately 60% v/v alcohol	14	liters
(3) Juice from (1) is mixed with distillate from (2) to give flavoring extract		
Approximate yield	450	liters
This extract has an alcohol content of 19–20% v/v; it is important that the level is more than 18% v/v alcohol; below this strength the juice will ferment rapidly; 1 liter expressed juice is obtained from approximately 1 kg of strawberries without sugar (0.85 kg with sugar)		
B. Strawberry fruit flavor		
Optimum flavor is obtained by vacuum distillation of the expressed juice from no more than 2 kg fruit per liter; use of more than 2 kg per liter yields an inferior product. This formula for strawberry fruit flavor is derived from the extract above, produced without added sugar		
(1) Strawberry juice extract ((3) above) expressed from approximately 1 kg fruit per liter with alcohol content 19% v/v	200	liters
This extract is concentrated under vacuum to produce a first fraction distillate at approximately 50% v/v alcohol		
Yield of first distillate	28	liters
(2) A second fraction produces distillate at approximately 30% v/v alcohol	72	liters
This is redistilled to higher alcohol content and used in the next production of flavor		
(3) A third, non-alcoholic fraction to be used in the next production batch in place of water, where it is mixed with pressed pomace ((2) above) prior to distillation		
Yield stage three distillate	64	liters
(4) Strawberry concentrate remaining	36	liters

continues

Table 6–10 continued

Procedure	*Quantity*	*Units*
(5) Formula for strawberry fruit flavor		
Cooled strawberry concentrate ((4) above)	36	liters
Add to the concentrate in the still to avoid loss of material		
Expressed strawberry juice and flavoring extract ((3) above)	36	liters
First fraction flavor distillate ((1) above)	28	liters
Yield of finished strawberry fruit flavor containing approximately 19% v/v alcohol	100	liters

[a]Optional ingredient.

6.6.2 *Principal Components Used in the Formulation of Beverages*

When gathering together the flavorings to be used in the formulation of a beverage or preparing the brief for a flavorist to manufacture a speciality flavoring for that particular application, the product formulator will already have decided upon the general structure of the beverage formulation. This means that he or she will have already conducted an in-depth survey of the legislative framework into which the beverage will fit and will have included information on local "custom and practice" which may be different from that allowed legislatively. It will therefore be assumed for the purpose of the discussion that this most vital of the beverage formulator's tasks is completed and the actual formulation only is considered.

The beverage formulator has a range of raw materials from which is drawn in a very similar way to the modus operandi of the flavorist. These materials vary dramatically in functional and taste characteristics as well as bulk, stability and interactivity. The principal components used in beverage composition are given in Table 6–14. Different raw materials from the list in Table 6–14 can be used to give similar effects. For example, body, bulk, or mouthfeel are contributed by:

(a) alcohol
(b) water
(c) sugars
(d) fruit or other characterizing ingredient, e.g., milk
(e) artifical sweeteners
(f) emulsifiers and stabilizers
(g) acidity regulators

A further investigation would reveal that each of these contribute to the body of the beverage in different ways and it is the expertise of the product formulator to know how the materials interact and to use those most effective for a particular

Table 6–11 Strawberry milk-based beverage: materials and quantities for 1000 liters finished product

Material	*Quantity*	*Units*
Skimmed milk	400.00	liters
Strawberry purée	400.00	liters
Sugar	80.00	kg
Pectin	5.00	kg
Strawberry essence	0.40	liters
Citric acid	q.s.	to pH4
Water to 1000 liters	q.s.	liters
Total	1000	liters

Analytical characteristics

Refractometric Brix (% w/w at 20°C)	12.5 ± 0.5
pH	4.0

Process details

(1) Mix the ingredients in the order listed taking care to disperse fully all of the small addition items; carefully add the flavoring and citric acid, measuring the pH during addition

(2) Ultra high temperature process this product by heating to 132°C for a minimum of 1 s prior to cooling and filling into appropriate containers

Table 6–12 Artificial strawberry flavor (wild strawberry)

Mixture of:

0.80 g	ethyl heptylate
0.80 g	oil of sweet birch
2.10 g	aldehyde C_{14}
2.40 g	cinnamyl isobutyrate
2.60 g	ethyl vanillin
3.00 g	Corps Praline (trade name) dissolved in
3.20 g	cinnamyl isovalerate
3.40 g	dipropyl ketone
5.00 g	methyl amyl ketone
6.00 g	diacetyl
21.20 g	ethyl valerate
23.15 g	aldlehyde C_{16}
43.20 g	ethyl lactate
100.00 g	alcohol, 95%
783.15 g	propylene glycol
1000.00 g	Total

Table 6–13 Strawberry WONF (yield 1000 g)

Ingredients:

500.00 g	alcohol, 95%
2.00 g	oil of rose
2.80 g	oil of jasmin
1.00 g	oil of cassie
1.00 g	oil of wintergreen
0.50 g	oil of lovage
2.50 g	oil of valerian
0.02 g	oil of celery
0.10 g	oil of coriander
520.00 g	distilled water

Procedure

(1) Agitate extract mixture of alcohol and oils, then add 520 g distilled water and continue agitating; transfer mixture into separator and leave it covered overnight; the terpenes will separate to the suface, while the extract below will be clear

(2) Take off the clear extract, discard the separated terpenes

application. Product formulators have to work to very tightly controlled briefs which specify individual requirements including total raw material cost and this has to be considered from the outset of the formulation of a product because it is much more difficult to incorporate such factors at a later stage. To some product formulators this approach is unacceptable because the most difficult stage in the development of any new product is preparing the initial product concepts so that meaningful consumer research may follow. Some would argue that the imposition of precise cost limits prevents these initial steps being carried out correctly when the remainder of the product formulation path will follow smoothly.

An example of the diversity of just one category of principal components used in the formulation of beverages is "sugars" or, more correctly, "sugars and artificial, or intense, sweeteners." The formulator will consider both together as they contribute to the total sweetness of the product. This diversity of character and contribution of sweetness (not to mention intensity) is demonstrated in Table 6–15. A similar list is available for each of the other principal ingredients.

6.6.3 *Label Claims*

Specific claims which are to be made for the beverage and included on the label must be considered at this juncture, e.g., is the product to be low calorie, is it to offer vitamins, or contain only natural flavorings, etc. If not taken into account at an early stage, label claims may be difficult to support after the product is formulated.

Table 6–14 Principal components for the formulation of beverages

Ingredient	*Contribution*
Sugars	Flavor, sweetness, mouthfeel, body, fruitiness, nutrition, facilitate water absorption (appearance and preservation in syrups)
Fruit/extract/milk/other characterizing ingredient (e.g., glucose syrup, spring or mineral water, etc.)	Flavor, body, appearance (nutrition)
Nutrient additions including salts	Nutrition: ascorbic acid and tocopherols are anti-oxidants; also controlled absorption of sugars and water
Acids Flavors Artificial sweeteners Colorings Emulsifiers and stabilizers	Flavor, antimicrobial effect Flavor, body, appearance; carotene and riboflavin colorings are nutrients also
Anti-oxidants	Improved flavor and vitamin stability
Preservatives	Antimicrobial effect; sulfite also has antibrowning and anti-oxidant effectiveness
Acidity regulators	Improved dental safety; reduced can corrosion; body
Alcohol (e.g., beers, wines, spirits, etc.)	Body, mass, solvent, carrier, flavor, mouthfeel, bite, punch, flavor potentiator or releaser
Water	Bulk and mass; solvent carrier; thirst quenching

6.6.4 *Sweetness/Acid Ratio*

This is a very important product attribute which, in many ways, controls the basic characteristics of the beverage in conjunction with the major ingredient(s). We have already seen how the sweetness/acid ratio, or "balance," in ginger ales can dramatically affect the perception of the added flavoring in dimensions other than sweetness/sharpness alone. A flavoring may be acceptable at one ratio and unacceptable (i.e., without modification) at another or acceptable with one sweetener system and unacceptable with another (e.g., sugar and saccharin). An example of this is the formulation of carbonated lemonade which contains no fruit and uses the harshness of saccharin in its sugar/saccharin sweetener system (medium

Table 6–15 Comparison of sweeteners used in beverages

	Taste characteristics					
Carbohydrates (100% solids base)	*Sweetness intensity (10% sucrose)*	*Sweetness quality*	*Time profile*	*Associated taste*	*Mouthfeel, body*	*Enhancement of fruitiness*
Sucrose	1.0	Full rounded	Fast, slight linger	None	More than other carbohydrate	Good
Invert sugar						
50% inverted	1.0	Close to sucrose	As sucrose	None	Typical carbohydrate	Fair
100%	1.1					
Fructose	1.3	Slightly thin	Fast without linger	None	Typical carbohydrate	Good
Glucose	0.7	Slightly thin	Fast without linger	None	Typical carbohydrate	Fair
Glucose syrup						
42 DE	0.33		Fast, some linger	None	More than other carbohydrate	Fair
63 DE	0.50		Fast, some linger	None	More than other carbohydrate	Fair
Iso-glucose	1.0		Fast without linger	None	Typical carbohydrate	Fair
Intense sweetness						
Saccharin	350	Slightly chemical sweetness	Slower and persistent	Bitter/metallic aftertaste	Thin	Nil

continues

Table 6–15 continued

	Taste characteristics					
Carbohydrates (100% solids base)	*Sweetness intensity (10% sucrose)*	*Sweetness quality*	*Time profile*	*Associated taste*	*Mouthfeel, body*	*Enhancement of fruitiness*
Cyclamate	33	Slightly chemical sweeteners	Slower and lingering	Of taste at high concentrates	Good	Good
1:10 saccharin/ cyclamate	100	Sugar like	As sucrose	None	Good	Good
Aspartame	140–200	Sugar like	Slight delay, slight linger	Little	Fair to good	Good
Sucralose	450	Sugar like	Fast, slight linger	None	Thin	Nil
Stevia extract	150	Clean sweetness	As sucrose	Slight liquorice menthol/ aftertaste	Thin	Good
Alitame	2000	Clean sweetness	Slight delay, slight linger	None	Thin	Nil
Acesulphame K	100	Good quality sweetness	Fast, slight linger	Bitter at high concentrates	Thin	Nil

total sweetness/acid ratio) to give a brighter, harsher more tangy edge than a beverage utilizing a terpeneless lemon essence (see Table 6–4). Without the effect of saccharin this flavor would be much "rounder" and "sweeter" tasting. A typical carbonated lemonade formulation is given in Table 6–16. An exercise often carried out in the training of a beverage formulator is to attempt to achieve an acceptable tasting carbonated lemonade without using added flavorings.

6.6.5 *Alcoholic Components*

Alcoholic components in beverages must be considered concurrently with the sweetness/acid ratio and sweetness type because they are interactive. All alcoholic components provide body and bite but vary dramatically in other contribu-

Table 6–16 Sparkling lemonade (carbonated beverage): materials and quantities for 100 liters finished product

Material	*Quantity*	*Units*
Granulated sugar	60.00	kg
Sodium saccharin dihydrate	0.088	kg
Sodium benzoate	0.175	kg
Citric acid anhydrous	2.28	kg
Sodium citrate BP	0.38	kg
Lemon essence terpeneless	0.9625	liters
Water to 1000 liters	~961	liters
Carbon dioxide to specification		
Total	1000	liters

Directions/special notes

Add ingredients in the order listed, dissolving solid ingredients in water before addition, to make a suitable syrup; after adequate mixing, use this syrup for bottling diluting one volume of syrup with an appropriate volume of carbonated water; fill into suitable bottles at the correct vacuity, adequately rinse bottle threads and sealing surface; close and label appropriately

Analytical characteristics

Refractometric Brix (% w/w at 20° C)	6.0 ± 0.3
Acidity as citric acid monohydrate (% w/v)	0.25 ± 0.02
Sodium saccharin dihydrate (% w/w at 20° C)	0.0088 ± 0.00003

Calculated sweetness

Due to sugar (% w/v)	6.00
Due to saccharin (dihydrated) (% w/v)	4.40
Total	10.40

tions, e.g., flavor. Alcohol may be contributed in the form of almost flavorless spirit of 40% alcohol through strongly flavored red, white and fruit wines to beers (including lagers), ciders and their low alcohol varieties which all contribute different tastes. Some alcoholic beverages use no sweetener or acidulant at all and here the flavorist controls the composition of the whole beverage. An example of this is gin manufacture which is described in Table 6–17.

From the foregoing it can be seen that by this stage in the formulation of the beverage the "category of beverage" as defined in Section 6.2 is decided and may be used as a guide to choosing the correct flavor extract or extraction method as well as dispersant or solvent.

6.6.6 *Water*

Even the choice of water with different analytical parameters must be considered for interactivity with flavorings selected for the beverage. Different methods of water treatment result in varying levels of different residual chemicals. The most noteworthy of these is chlorine and its reactivity (particularly noticeable and damaging to a beverage such as carbonated lemonade) with lemon oil components producing an unpleasant taste. One of the methods used to protect such a flavor is to use low levels of sulfur dioxide (<5 ppm) in the product to "block" the action of chlorine (Schroeter, 1966; Taylor, 1958).

Table 6–17 Manufacture (distillation) of gin

(1) The distillation is undertaken at atmospheric pressure in special vessels; all the solid ingredients are finely divided prior to addition (see separate section)

8.25 kg	juniper berries (mixture of different origins including Italy, Yugoslavia and Russia)
0.875 kg	coriander
0.85 kg	angelica root
1.125 kg	bitter orange peels (of mixed origin)
0.35 kg	sweet orange peels
155.0 g	orris root
100.0 g	liquorice root
35.0 g	cassia
35.0 g	cardamom
35.0 g	calamus root
975.0 kg	alcohol, 95%
1200.0 kg	water

and distill to give first fraction of

975.0 kg	gin distillate

(2) The second fraction is redistilled and used in the next batch

The only other consideration here is the effect even these low levels of sulfur dioxide may have on the flavoring (see below). It has also been shown (P.W. Harmer, private communication) that the chlorination of water with sodium hypochlorite sterilants can cause off-flavors in orange flavored beverages prepared from fruit concentrates (containing low levels of orange oil) preserved with sulfur dioxide. It is believed that this off-flavor is caused mainly by sodium chlorate and other impurities in the hypochlorite and this method of sterilization should be avoided.

This effect of chlorine in water supplies can also be seen at the consumer end of the retail chain with the preparation of tea and coffee prepared with the new generation of plastic jug kettles. The constant re-boiling of the same quality water with appreciable levels of chlorine causes interactions with phenolic components in the material used in the construction of the kettle to produce chloro-phenol off-flavors in the products made using the boiled water. This off-note has low awareness thresholds in many people.

6.6.7 *Characterizing Ingredients*

It can be seen from Table 6–16 that the main characterizing ingredients in non-alcoholic and some alcoholic beverages are usually:

(a) fruit
(b) extract of natural plant material, etc.
(c) milk/yogurt
(d) other, e.g., glucose syrup

We have already seen for example how milk can provide both body to the mouth-feel and a less acidic taste in beverages, with the examination of the strawberry milk beverage.

6.6.7.1 *Formulating with fruit*

It is for these reasons that the beverage formulator must consider the effect and impact a particular characterizing ingredient will have on the product composition and performance. This may be directly, e.g., the taste and body contribution of fruit juices, or indirectly via the process requirements the use of this ingredient necessitates, e.g., the pasteurization and preservation requirements of beverages containing fruit. This is best exemplified via the formulation of a "comminuted" or "whole" fruit drink and a formulation for such a product in an orange flavor is given in Table 6–18. There were legal requirements for such a product in the United Kingdom (*The Soft Drinks Regulations*, 1971) as follows:

(a) minimum 10% w/v potable fruit
(b) minimum 22.5% w/v added carbohydrate

Table 6–18 Orange squash: materials and quantities for 1000 liters finished product

Material	*Quantity*	*Units*
Granulated sugar	340.88	kg
3:1 Orange base blend	60.00	liters
4.5:1 Orange juice concentrate	19.55	liters
Sodium benzoate	0.296	kg
Sodium metabisulphite	0.226	kg
Citric acid anhydrous	15.36	kg
Tri-potassium citrate	2.00	kg
Ascorbic acid (to give 18 mg/100 g)	~0.16	kg
Orange essence	5.0	liters
Water to 1000 liters	698.00	liters
Total	1000.00	liters

Directions/special notes

This product must not be held as a syrup once it has been prepared, without pasteurization

Analytical characteristics

Refractometric Brix (% w/w at 20° C)	34.0 ± 0.5
Acidity as citric acid monohydrate (% w/v)	2.00 ± 0.05

Calculated sweetness

Due to sugar (% w/v)	34.1

(c) maximum for artificial sweeteners
(d) maximum for preservatives
(e) controls on many other additives

The requirement for "comminuted" as "whole" fruit means that a certain level of oil-soluble materials will be dispersed throughout the ingredient, e.g., orange oil in orange comminute and this will impart an intensity of flavor disproportionate to the juice content (% v/v) present. If the product is to be cloudy, then, as long as physical stability of the oil is maintained in the comminute and the finished beverage, there will be no need for terpene removal. In this situation, the requirement of the product formulator from the flavoring will be the supplementation of the "juice" character of the fruit if the orange oil content in the beverage is approximately 0.06% v/v (expressed as single strength oil) and of the "peel" character of the fruit if the oil content is <0.01% v/v (expressed as single strength oil). Such a flavoring is known as a "top-note" and a formula for a suitable essence is included in Table 6–19.

Table 6–19 Terpeneless orange "top-note" essence

Mixture of:

(1)	113.00 kg	orange oil,[a] cold pressed concentrate 10-fold by vacuum distillation with removed terpenes to give
	11.30 kg	10-fold orange oil
(2)	11.30 kg	10-fold orange oil
	55.00 kg	alcohol, 95%
	70.00 kg	water

Procedure

Agitate well and allow mixture to stand in separator for 24 h; remove terpenes and waxes to give

190.00 kg	terpeneless orange essence
30.00 kg	alcohol, 95%
100.00 kg	terpeneless orange "top-note" essence

[a]The flavor taste of the "top-note" essence may be adjusted by the selection of country and variety of origin of the orange oil.

6.6.8 *Other Ingredients*

6.6.8.1 *Vitamins and nutrients*

If vitamins and nutrients are to be included in the beverage formulation, the stability of these materials and their reactivity must be evaluated under all relevant storage conditions prior to application of the flavorings with which they will almost certainly interact. The B group of vitamins is a good example of unstable composition and off-flavor development in beverages.

6.6.8.2 *Anti-oxidants*

This is one of the most important areas where the flavorist and beverage formulator must work together to achieve flavor/taste/stability in the finished beverage. If working separately, the flavorist may consider only the anti-oxidant requirements of the flavoring and the formulator may only consider the requirements of the beverage. Colorings, particularly carotenoids, flavorings and some vitamins are susceptible to oxidation both initially and during storage. Oxygen from the atmosphere is entrained in the flavoring materials during initial extraction, e.g., lemon oil from lemons, in the beverage during manufacture, in the headspace of the pack and during ullaging (if multi-use pack) by the consumer.

Oxygen can also diffuse through partially permeable plastic barriers used in packaging materials for both flavorings and beverages. Oxygen can lead to the breakdown of organic molecules in the flavoring and in the beverage and is often accelerated by sunlight and heat. Citrus oils in flavorings and fruit beverages are most susceptible to oxidation. The most common protection for the oil is the use of BHA, BHT and tocopherols.

Removal of terpenes reduces the susceptibility of the citrus oils to oxidation and flavorings based on terpeneless citrus (or "folded") oils exhibit better stability. Such flavorings will require the expertise of the beverage formulator to provide some of the flavor characteristics of the terpenes by other routes. Because this is extremely difficult, the alternative available to the formulator is to provide and maintain a reducing environment in the beverage in which the terpenes will be more stable. Ascorbic acid and sulfur dioxide may be used for this purpose but the latter causes interactive problems of its own. Ascorbic acid can be introduced into the oil phase as the palmitate.

6.6.8.3 *Preservative*

Benzoic acid is possibly the most widely used beverage preservative, probably owing to its low taste threshold, low volatility and wide antimicrobial spectrum. For maximum effectiveness of benzoic acid, the pH of the beverage must be about 3.0 or less; benzoate resistant yeasts are widespread in manufacturing plants. Benzoic acid is relatively non-interactive with flavoring materials.

Sorbic acid is used increasingly in beverages and provides good protection through its non-volatility and effectiveness against yeasts. Unfortunately sorbic acid has a lower taste threshold over a wide range and, in citrus flavors, some consumers are very sensitive to its particular flavor.

Sulfur dioxide is in widespread use as an alternative to benzoic acid and has broad antimicrobial and anityeast activity in a pH range of 2.0–4.0. It also confers anitbrowning and anti-oxidant effects. Generally, sulfur dioxide is extremely reactive chemically and is unacceptable for use with orange flavors and fruit, cans, thiamin (vitamin B_1) and some natural colors.

When the preservative system is selected by the beverage formulator it is essential that consideration be given to the effects one or more preservatives will have on the flavoring components so that adequate compensation may be made. This will require sufficient storage testing to reveal longer term oxidative effects as well as immediate reactivity.

6.6.9 ***Acidulants and Acidity Regulators***

We have already seen how the effects of the sweetness/acid ratio can be modified by buffering the acid system in the "American" style ginger ale formulation. The beverage formulator is able further to modify product taste by the selection and use of different acidulants to different effects, e.g., the potentiation of apple taste by the use of malic acid. Certain flavor types react best with certain acidulants, e.g., American cream soda with tartaric acid; it should be remembered that titratable acidity and pH are not the only indicators of the performance of an acidulant. Different acids have different taste effects as an integral part of their character and these can be further modified by the use of buffers.

The best known example of the use of alternative acidulants to citric acid in beverages is cola drinks. Full use is made of the body and sweetness of the sugar content to offset the "punch" achieved by the use of phosphoric acid. An examination of the cola beverage formula in Table 6–20 shows how the use of a cola nut extract alone is inadequate when used alone for flavoring a beverage.The flavor is "filled out" by the formulator with either a single cola flavor (as in Table 6–3) or the individual citrus components and spices of that flavor carefully balanced with the other principle components of the beverage (see Table 6–21).

Table 6–20 Cola carbonated beverage: materials and quantities for 1000 liters finished product

Material	*Quantity*	*Units*
Granulated sugar	94.50	kg
Sodium saccharin dihydrate	0.0210	kg
Sodium benzoate	0.1770	kg
Cola extract	6.0000	kg
Bitter principle	0.0200	liters
Cinnamon flavor	0.00075	liters
Lactic acid (edible grade)	0.1870	liters
Cola flavor	0.1500	liters
Citrus flavor	0.0480	liters
Water to 1000 liters	~937.00	liters
Carbon dioxide to specification		
Total	1000.00	liters

Directions/special notes

Add the sugar as a suitable syrup; dissolve the sodium benzoate in water before adding to the batch; after thoroughly mixing, dilute the cola extract with an equal volume of water and add to the batch; the essence and citrus flavor should be blended together before addition, and added to the syrup just before making up to the final volume; make up into a suitable syrup and mix thoroughly; allow the syrup to stand for at least 2 h before commencing canning, bottling

Analytical characteristics

Refractometric Brix (% w/w at 20° C)	9.5 ± 0.5
Acidity as phosphoric (% w/v)	1.106 ± 0.005

Calculated sweetness

Due to sugar (% w/v)	0.945
Due to saccharin (dihydrate) (% w/v)	0.105
	1.050

Table 6–21 Cola essence (for use in conjunction with cola extract)

45.00 kg	terpeneless lemon oil
38.70 kg	terpeneless orange oil
15.00 kg	terpeneless lime oil
1.00 g	cinnamon flavor
0.30 g	nutmeg flavor
100.00 kg	cola essence to be used with cola extract for flavoring beverages

6.7 SUMMARY

A comprehensive review of all beverages and their compositions as they relate to flavorings is not possible in a single chapter. What it has been possible to achieve is a broad overview of beverage categories and the types of flavorings typically in use with methods of extraction, solubilization and concentration of beverage flavorings. In-depth studies on the use of ginger in beverage formulation and flavoring composition have enabled the reader to gain an insight into the complex and interactive role of the flavorist and applications technologist. An overview of the complexities of beverage formulation has been included with detailed information on the diversity of the choice of sweetness/acid ratio and its interactivity with some perception of flavor. This is one of the most important ingredient areas for the beverage formulator and flavorist to consider when formulating beverages. The interactivity of the flavorist and beverage formulator when the composition of beverages is in discussion is seen to be of the utmost importance.

REFERENCES

K. Bauer and D. Gerbe, in *Common Fragrance and Flavour Materials—Preparation, Properties and Uses,* VCH, Cambridge (1985) 117.

J. Bergstrom and E. Hultman, *J. Am. Med. Assoc* 221 (1972) 999.

H.B. Heath, *Source Book of Flavours,* AVI, Westport, CT (1982).

National Academy of Sciences—National Research Council, Water deprivation and performance of athletes, *Nutr. Rev.* 32 (1974) 314.

L. Schroeter, in *Sulphur Dioxide Applications in Food, Beverages and Pharmaceuticals,* Pergaman Press, Oxford (1966).

The Soft Drinks Regulations 1964—Composition and Labelling, S.I. No. 760, H.M.S.O., London (1971).

E.W. Taylor, *The Examination of Waters and Water Supplies,* J.A. Churchill, London (1958).

D.K. Tressler and M.A. Joslyn, *Fruit and Vegetable Juice Processing Technology* (2nd ed.), AVI, Westport, CT (1982).

FURTHER READING

D. Hicks, in *Non-Carbonated Fruit Juices and Fruit Beverages* (ed. Hicks), Blackie, Glasgow, 1990.

Chapter 7

The Flavoring of Confectionery and Bakery Products

D.V. Lawrence and D.G. Ashwood

7.1 INTRODUCTION TO CONFECTIONERY FLAVORINGS

The principal ingredient in all confectionery is sugar (sucrose), which in its refined form has little flavor apart from its inherent sweetness. Raw (unrefined) sugar has it own particular flavor, which will be dealt with later in the chapter (see Section 7.3). Other important carbohydrates used in confectionery are corn syrup, invert sugar and dextrose, which are added mainly to control or prevent crystallization. The texture of the confection may be altered by their use, and this property is used by confectioners to manufacture many varied products.

Other ingredients such as gums, pectin, gelatine, starch, milk, butter, other fats and cocoa do most to give special textures, although it must not be forgotten that air and water probably have the greatest effect in confectionery. Other ingredients which also play a part include liquorice, honey, nuts, coconut, raw sugar (molasses), malt, extract, dried fruit, fruit, and fruit juices. These ingredients are added usually for their flavoring properties, or for their contribution to the eating quality, mouthfeel or nutritional value of a confection. Some products owe their total appeal to these added ingredients. The flavor industry also provides extracts, concentrates and flavorings to suit requirements for all these confectionery types.

Temperature and cooking (or heating) times also play an important role in determining final taste and texture because they have a significant effect on flavor and flavor development.

Figure 7–1 shows temperature bands for producing various confectionery types. The apparently large range is normal, and takes into account recipe differences and texture required. Lower boiling temperatures enable crystallization to occur and a variation even as small as 0.5° C can make a significant difference to the texture of most types of confectionery.

The principal types of (sugar) confectionery are as follows:

high boilings	cream and lozenge paste
fat boilings	compressed tablets
toffees	jellies and gums
fudge	chewing gum
fondants	panned work
candy	chocolate
paste work	

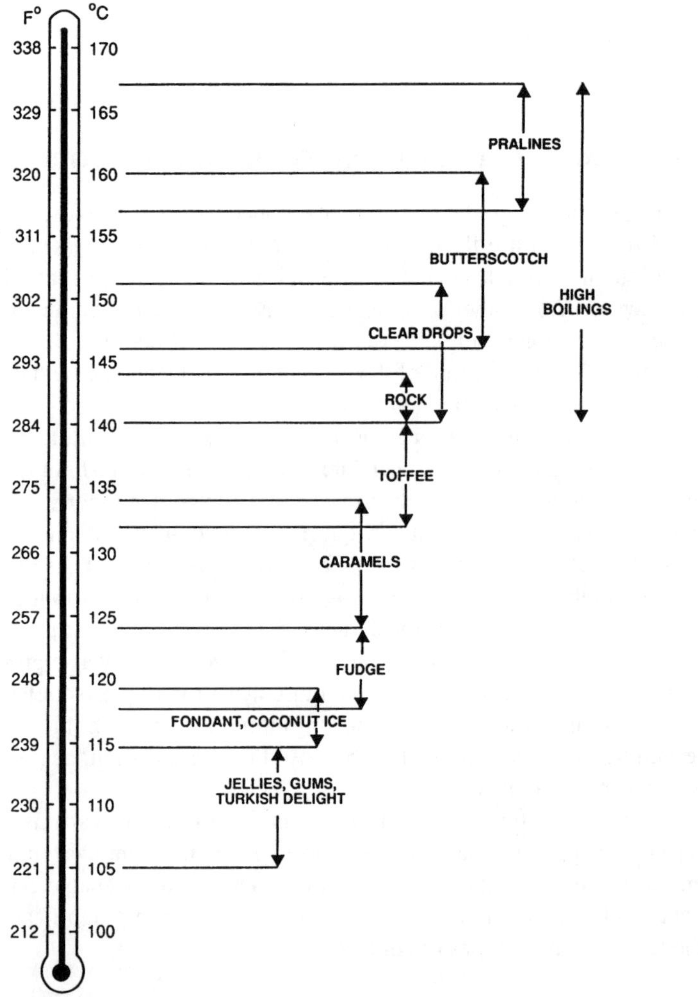

Figure 7–1 Temperature bands.

Typical composition and procedures for the various types are outlined but no account has been taken of water, because it is used mainly to dissolve sugar or other ingredients or to disperse gums. It is then removed by boiling or drying. The role of flavorings is also discussed.

7.2 BASIC CONFECTIONERY TYPES, RECIPES, INHERENT FLAVORS

7.2.1 *High Boilings (Hard Candy)*

Candy is a collective U.S. name for sugar confectionery, whereas in the United Kingdom it describes a special crystallized type.

Typical composition:

Sugar	50–70%
Corn syrup 42DE	30–50%
Acid (citric, tartaric, lactic)	0–2.0%
Flavoring	q.s. about 0.1%
Coloring (synthetic)	q.s. about 0.01%

If natural colorings are used, generally many times more is required.

Method of manufacture: Sugar is dissolved in water and corn syrup added. The mixture is boiled to the required temperature, for example 147° C for a 60/40 sugar/corn syrup mix, and cooled. Acid, flavoring and coloring are then added and the resultant material molded by various means to make the finished confection. In large scale production, liquid corn syrup is metered into sugar solution and cooked in a microfilm cooker (so-called because a thin film of syrup is heated and brought to the required solids content under reduced pressure, in the shortest possible time). Apart from being energy efficient, no browning occurs and therefore little or no cooked flavor is apparent. This syrup is fed into a mixing chamber where calculated quantities of flavoring, coloring and acid are added by means of dosing pumps. The ingredients are then mixed and formed into a ribbon for cooling. The mass is finally transferred onto sizing rollers, prior to spinning to a rope and molding. On a small scale the batch is boiled in a pan, which may also have a facility to remove final amounts of water by vacuum. At the temperature required, the product is transferred to a cooling table (confectioners "slab," which has the facility to have hot or cold water passed through it), where the batch is cooled. When the correct temperature is reached, as determined by the viscosity of the mass rather than any other factor, the flavoring, coloring and acid are added, folded in, and the confection finished as before, or passed through "drop" rollers. "Drop" here means the shape of the sweet (e.g., pear drops) (Figure 7–2).

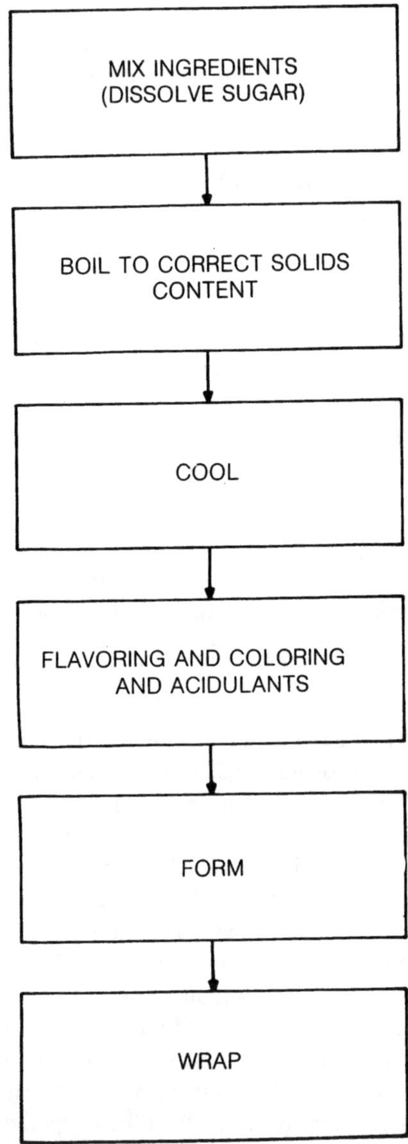

Figure 7–2 Processes used in boiled sweet production.

Another type of high boiling (candy) which should be mentioned is the "deposited" type, where the cooked, flavored, acidified and colored syrup is held at high temperature in special hoppers and deposited into metal molds, prior to cooling and wrapping. The acid used has to be buffered to prevent inversion of sugar, and the coloring and flavoring used must be specially selected to withstand the extra heating necessary. Invert sugar is produced by the addition of acid into the boil, or by long slow cooking. Excessive inversion leads to stickiness, or even a product that will not solidify. Although some confections have invert sugar added because it controls the crystallization of super-saturated sucrose solutions, it is usually added as commercially available material or as golden syrup, treacle or honey. Any additional production of invert sugar during cooking needs very careful control, and buffer salts are added to adjust the pH and consequently the rate of inversion.

Using the same recipe, "pulled sugar work" (seaside rock, figured lollypops, satins) can be made. Their manufacture necessitates "pulling" the previously boiled, flavored, colored and acidified sugar/corn syrup mixture on a machine or over a hook, to incorporate air. When molding (to make fancy shapes or to build up letters) is complete, the batch has to be kept hot for a considerable time, and this inevitably leads to deterioration of the flavor. For this reason heat-stable flavorings or lower quality products without fine top-notes, which would be lost, may be utilized. One way of introducing fine flavors into boiled sugar is to prepare delicately flavored centers. Fruit pulps may be used to produce jams; nut pastes may be prepared, or whole nuts used, as well as all kinds of fillings based on chocolate. They are introduced into the boiled sugar rope by means of a center pump. The high boiled casing then protects and encloses the lower boiled portion, enabling its finer flavor to be retained.

Very many different confections are made using these same basic ingredients, some boiled to the higher temperature range (e.g., satins) whereas those boiled to lower temperatures are allowed to crystallize (e.g., Edinburgh rock).

7.2.2 *Fat Boilings*

Looking at variations on plain boiled sugar as described, the addition of fat is perhaps the most obvious. Traditionally this fat was butter which imparts a smooth mouthfeel, excellent taste, and is self-emulsifying. Butterscotch is made by the addition of 4% or more of butter solids, and the flavor of this product is developed by exposing the raw materials to the high temperature of manufacture.

7.2.2.1 *Butterscotch*

Typical composition:

Sugar	50–60%
Brown sugar	15–20%
Corn syrup	15–20%
Butter	5–15%
Salt	0.5%

Flavor(s):	lemon	
	vanilla	0.1–0.2%
	butter	

Method of manufacture: Sugars and corn syrup are dissolved and boiled together until a temperature of about 145° C is reached. Butter is then added and gently incorporated to preserve as much of its flavor as possible. The batch is then boiled to the final temperature of 145–160° C. Where higher temperatures are preferred, special arrangements for direct heating (gas) may have to be made, because they are often too high for steam heated equipment. The mass is then cooled and flavoring incorporated, before the product is cut and wrapped. Generally lemon, usually in the form of lemon oil is added, because it is said to neutralize the greasy effect of fat. Vanilla flavorings are often used to enhance the character of the product and butter flavorings are popular too, because they increase the overall buttery taste. Commercial flavorings intended for this confection generally contain all these components in carefully balanced amounts.

Variations on butterscotch recipes would be to alter the proportion of white to brown sugar (or the replacement caramelized syrups which are available). If the higher boiling temperatures are used to achieve a special "cooked" flavor and texture, invert sugar has to be either added or made during production, in replacement for all or part of the corn syrup. The inclusion of invert sugar, in one form or another, results in a much less viscous batch, and higher temperatures are required to reach the same solids content. The amount of butter may be varied or replaced totally or partially with other fats. Butter has natural emulsifying agents present, so, if other fats are used, an emulsifier, usually lecithin or glyceryl monostearate (GMS), has to be added in order to ensure proper dispersion of fat through the batch.

7.2.2.2 *Buttermint confectionery*

Incorporation of air, either by means of "pulling," or the addition of "frappé," will result in buttermint types or Mintoes; two typical compositions are as follows:

Typical composition:

Sugar (mix of brown and white)	60%
Corn syrup	30%
Butter or other fat (e.g., hydrogenated palm kernel oil [HPKO])	10%
Salt	0.2–0.5%
Lecithin	q.s.
Peppermint oil	0.1%

Method of manufacture: Sugar syrup and corn syrup are heated to about 130° C, when butter (or other fats with emulsifiers) is added. The mass is reboiled to 138° C, allowed to cool and peppermint oil mixed into the batch which is then pulled to the correct consistency, spun into a rope, formed and wrapped. Frappé is made by beating egg albumen (10%) or gelatin (5%) into previously warmed corn syrup. Both these materials have to be dispersed in minimal amounts of water before addition.

Typical composition:

Sugar (mix of brown and white)	50%
Corn syrup	40%
Fat (HPKO)	4%
Frappé	6%
Salt	0.2–0.5%
Peppermint oil	0.1%
Butter flavor	q.s. about 0.1%
Lecithin	q.s.

Method of manufacture: Sugar and corn syrups are warmed to 140° C. Fat and lecithin are then added and, when dispersed, the frappé and peppermint oil are carefully stirred in. The mass is allowed to cool and finished as usual. The heat of the batch will expand air entrapped in the frappé, and care must be taken to avoid its loss by excessive handling. To avoid inconsistent batches it is necessary to control the temperatures used, as well as manufacture of the frappé.

The quality of peppermint oil used in this type of product is important; an oil without harsh top-notes will enhance the smooth character of the confection. This may be a natural American oil (e.g., from *Mentha piperita*) or an oil from China or Brazil (*Mentha arvensis*) which will have been dementholized at source and subsequently rectified, to remove the harsher top-notes and undesirable residues. Alternatively a mixture of the two may be used. Most essential oil and flavor houses have suitable blends to offer. Butter flavorings may also be added, especially when not all the fat used is butter. These are often made with a vanilla background, which accentuates the smooth character of the confection, together with diacetyl and butyric acid. Both these substances are naturally present in butter and increasing the proportion of them boosts the final flavor considerably and replaces processing losses. It should be noted they are available as both natural and synthetic materials. Butter "esters," i.e., obtained from butter, may also be used. Composite flavorings based on peppermint, vanilla and butter are also available.

As with butterscotch and other confections with "butter" in their description, it should be remembered that statutory requirements may lay down a minimum content of butter solids.

By using only refined sugar and deodorized fat, fruit flavors and acid can be incorporated to make "chews." These are fat boilings prepared without brown sugar and caramelized syrup to achieve a bland product with little background taste that will accept fruit flavor. "Fruit chews" may also contain natural concentrated fruit juice as well as flavoring. Essential oil blends are often used to flavor orange, lemon, lime, mandarin, grapefruit, spearmint and peppermint types, while nature identical or artificial flavors are normally utilized for the rest of the range. Permitted matching colorings are also added.

7.2.3 *Toffees and Caramels*

By the addition of milk solids to confections containing fat, toffees and caramels are produced. It is generally accepted that caramels have a higher proportion of fat and milk solids than toffees, although the difference between toffee and caramel is essentially one of texture and the two types of confection merge into one another without any clear dividing line (*Skuses Complete Confectioner*, 1957). Final boiling temperatures range between 120 and 150° C, the higher the fat and milk content, the lower the cooking temperature becomes, and in consequence the time for flavor development is also reduced. For this reason corn syrup, fat and milk are often pre-mixed and heated separately to allow flavor to develop (see Figure 7–3).

Typical composition:

Sugar (mix of brown and white)	40–70%
Sweetened condensed milk (or reconstituted dried milk)	5–30%
Corn syrup	20–30%
Fat (HPKO and/or butter)	5–15%
Salt	0.5%
Lecithin	q.s. 1% of the fat content
Flavoring	q.s. about 0.1%

Other ingredients often used are flour, eggs and cream, the last two being obligatory (*Butter Regulations*, 1966) if the product is to be named "egg and milk" or "cream" toffee. Malt, treacle, liquorice and chocolate or cocoa can also be added, while nuts and dried fruit can be mixed into the product prior to finishing. Flavorings used are generally vanilla, butter or cream, or mixtures of these, although egg and milk, and other specialized toffee flavors are available.

Method of manufacture: Sugar, with some corn syrup is dissolved in water and heated to about 120° C. Milk, which may be reconstituted dried milk, is heated with fat and the remainder of the corn syrup and mixed to a smooth paste.

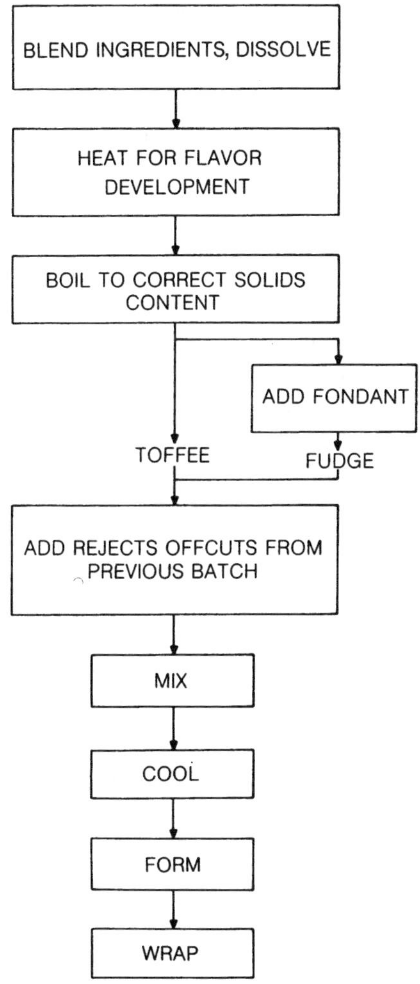

Figure 7–3 Processes used in toffee/caramel/fudge production.

This mixture is then added to the sugar/corn syrup mix and returned to the boil, 120–150° C depending on the composition. At this point extra ingredients may be added, together with any flavoring, and the mass poured onto a cooling table. It is then cut into suitable sized slabs, scored with toffee cutters and presented as such, or passed to forming rollers, after which the "rope" is fed to cut-and-wrap machines. Much of this work is not fully automated, with various mixtures being prepared, blended and fed to a microfilm cooker. Flavorings can then be either added via a mixing chamber or by dosing pump directly into the product. The toffee ribbon produced is then fed to a cooling belt and subsequently to forming rollers and finished as described above.

The characteristic flavor of toffee is produced by the heating of milk solids with corn syrup. The higher the temperature and the longer the heating (and cooling) takes, the more flavored the product becomes. It is therefore most important that the heating processes are carefully controlled to achieve product consistency.

7.2.4 Fudge

It is said that the origin of fudge was "fudged" toffee where sugars were inadvertently allowed to crystallize (*Skuses Complete Confectioner*, 1957). In production, toffee base is made containing a greater proportion of sugar than normal. This is available for crystallization, and fondant is added to seed the process. Production of fondant is described later in Section 7.2.5.

Typical composition:

Sugar (mix of brown and white)	35–45%
Corn syrup	15–25%
Sweetened condensed milk	10–15%
Fat (HPKO and/or butter)	4–10%
Salt	0.2–0.5%
Fondant	15–25%
Flavoring (vanilla)	0.1%

Method of manufacture: A similar initial procedure is used to that for toffee, and the mass boiled to 115–118° C. This is allowed to cool in the pan, fondant is added, and the whole thoroughly mixed. Flavoring is then incorporated, and the mass re-mixed. The fudge is then poured into frames lined with waxed paper and allowed to crystallize. The rate of cooling has to be carefully controlled so that the product does not become too coarse or "marbled" due to uneven formation of crystals. Fudge is usually scored and packed in waxed paper prior to sale.

All the other ingredients mentioned for addition to toffee can be added, together with matching flavorings, to produce special types of fudge, e.g., coffee, walnut, hazelnut, maple and honey. Because the addition of fondant is for the purpose of seeding crystallization of sugar, any offcuts and other fudge waste can be re-used by addition to the batch at the same time as fondant. Because this recycled material has already been crystallized it will start to seed the batch just as well. If recycled material is used it should be remembered that extra flavor will be produced because the scrap has been heated previously. Good control is essential for product consistency.

7.2.5 *Fondant*

This product has already been mentioned in the previous section as an ingredient of fudge. However, it also represents a very large part of the confectionery market in its own right. A good description would be "a controlled crystallization of sucrose in a syrup medium."

Typical composition:

Sugar	60–80%
Corn syrup	20–40%

The ratio of corn syrup to sugar is important because the amount of corn syrup added will have a bearing on the final crystal size of the product. Under the same conditions, i.e., boiling, cooling, beating, the more corn syrup present the smaller the crystal size becomes.

Method of manufacture: The sugar is dissolved in water, taking care that no crystals remain to prematurely seed crystallization in the batch, and corn syrup added. The batch is boiled to 114–118° C, allowed to cool to 45–50° C, then beaten until crystallization occurs. This may be done with a wooden spatula on a confectioners slab, or with a special fondant beating machine. During this operation, latent heat of crystallization is evolved and it is important that the mixing and cooling is continued until this has been dissipated or the crystal size will be increased, resulting in a coarse grained product. The batch is allowed to "rest" to allow complete crystallization. The fondant is then re-melted by heating prior to use, or flavored and colored, then deposited as described later.

Variation of the recipe, boiling temperature, time of beating and the temperature at which beating starts, all play an important part in determining the final crystal size and consequently the texture. In large scale production, syrups are brought together, boiled in a microfilm cooker, cooled through a heat exchanger, then passed to a continuous fondant beater.

Fondant may be stored and re-melted when required for use in other confections such as coconut ice and fudge or, after the addition of flavoring and coloring, as toppings for cakes and centers for chocolate. In order to make these products, or to prepare fondant creams as such, the mass is deposited into shaped rubber mats or prepared starch trays. This is described in Section 7.2.9. As well as being dipped into chocolate, some may be finished by crystallization where they are immersed in a super-saturated sugar solution, which, when dried, leaves a hard crust on the outside of the sweet. Fondant can also be used as an enrobing material to make products like maple brazils.

As variations to the above, honey, maple sugar, chocolate, brown sugars (or the equivalent), coffee, milk solids and fat can be added to give different tastes and textures. Although peppermint is the most popular flavor, many variations such as almond and vanilla, and florals such as rose and violet, can be used in this confection. The addition of a small amount of acid (about 0.1%) will enhance fruit flavors. Concentrated fruit juices may also be added in addition to the flavoring, whereas nuts, dried fruit and similar ingredients can be dispersed through the mass. Apart from added ingredients, fondant does not have any inherent flavor and is extremely useful for flavor testing.

7.2.6 *Candy*

Candy is very similar to fondant, differing only by the size of crystals which are much larger. This is achieved by starting the crystallization process at a much higher temperature (as soon as the batch has been brought to the required temperature) and thereby allowing it to proceed for a longer time.

Typical composition:

Sugar	80%
Corn syrup	20%

Method of manufacture: Sugar is dissolved in water, corn syrup added, and the batch heated to 120° C. It is then allowed to cool slightly, after which the syrup is agitated against the side of the pan with a spatula until it thickens and clouds. Coloring and flavorings are then added, mixed, and the mass poured on to a warm oiled slab to crystallize. It is usually marked as soon as set, and, when cold, broken into pieces and packed.

Candy can be varied with additions in exactly the same way as fondant, a very well-known type being "cough candy." In this product, liquorice and molasses are added together with flavorings based on aniseed oil, clove oil, eucalyptus oil, peppermint oil and menthol. These flavorings are often added for their physiological effect, and many confections of this type do claim to have a beneficial effect. They can be formulated to contain the same amount of these constituents as some medicated syrups.

7.2.7 *Cream and Lozenge Paste*

These products are variations of fondant, but rather than using crystallization techniques, finely ground sugar is dispersed in a gum or other colloidal matrix. Fat may also be added, especially when the product is to be used for biscuit cream.

Typical composition:

Ground sugar	50–95%	
Fine caster sugar	0–25%	
Corn syrup	0–25%	
Gelatin or	1%	or mixtures
Gum arabic or substitute or	4%	
Gum tragacanth	0.5%	
Fat (HPKO)	0–10%	
Flavor	q.s.	

Method of manufacture: Pastes as such can be made by hand by mixing the sugar into gum/gelatin solutions, rolling out into thin sheets and cutting into shapes, and/or sandwiching different colored sheets together and cutting into suitably sized squares or oblongs, then allowing to dry. In production, large trough mixers are generally used. The mass is then passed through sizing rollers, cut into shapes and normally air dried prior to packaging. Lozenges generally contain only ground sugar, gum and flavoring and are "stoved" for 24 h, i.e., placed on trays in warm (30–35° C), well-ventilated rooms to reduce excess moisture to around 3%. Although the high sugar concentration and gum help to bind the flavorings to the mass, there is still a great deal of flavor loss, which is usually compensated by adding extra amounts of flavoring and careful choice of the flavorings used.

This type of confection includes "allsorts," "hundreds and thousands" and similar products. Where fruit flavors are used, acid should be added. Many types of mints and medicated lozenges are made in this way, including "cachous" which are typically flavored with "floral ottos," such as rose and violet. These were originally used as mouth fresheners.

7.2.8 *Compressed Tablets*

This is a different method of producing lozenges, in which a very high compression is used to give a really smooth texture. This can be varied depending on the particle size of the sugar used.

Typical composition:

Freshly ground sugar	95–98%
Gum	1–3%
Stearic acid	q.s. approx. 1%
Peppermint oil	9.25–0.75%
Water	q.s. to disperse gum and "wet" batch

Method of manufacture: Gum is mixed dry with sugar and water then added. Alternatively part of the gum may be added dry, and some as a mucilage (a dispersion of gum in water, usually prepared by overnight soaking). This mass is then granulated (reduced to small free-flowing particles) and dried. Stearic acid, which is added to act as a lubricant for the tableting machine dies, may be predissolved in isopropyl alcohol, together with peppermint oil. This mixture is then added and the whole batch re-mixed and tableted. For fruit flavors, citric acid should be added to the mix prior to granulation.

If effervescent tablets are required, sodium bicarbonate is added prior to drying, followed by citric acid in the form of fine crystals after removal of water. If liquid flavorings are used they should not contain any water which would activate the bicarbonate/acid mix. Dry flavorings are thus often used in these products. Being encapsulated, dry flavorings have the additional advantage of being nonvolatile while dry and therefore have a longer shelf-life. Dextrose compresses well without the need of added gums, so no granulation is required and totally dry mixes can be made. Although dextrose is not so sweet as sucrose, it makes a very acceptable tablet and dry flavorings are therefore essential.

7.2.9 *Jellies and Gums*

Gums were originally made with gum arabic, and though it is still used, many substitutes are now available which may partially or totally replace it. Examples of these alternatives are gelatin, agar, pectin, starch and modified starch. Different choices and mixtures of these ingredients will give textures ranging from hard gums, pastilies, and soft jellies to Turkish Delight and marshmallow. The latter product is made by beating to incorporate air to achieve the desired effect. The flavorist should remember that because gum arabic, gelatin and starch require a large amount of water for dispersion, it is not possible to evaporate the excess by boiling in the usual way, because the mass becomes too viscous. Excess water is therefore removed by heating the sweets in starch molds, sometimes for several days, in order to reach the sugar concentration required to make a satisfactory product. This process, called "stoving," is carried out by placing the starch filled trays containing the confection in rooms kept at elevated temperatures (e.g., 35–50° C) with a flow of dry air. This gives the typical tougher external texture, which is so well liked but it also subjects the flavoring to prolonged heating and oxidation. For this reason only very cheap essential oils are used for the citrus flavors, because the finer top-notes in high quality oils would be lost or degraded. Other fruit flavorings need to be strong and able to withstand the high temperatures required. On the other hand, pectin based jellies are not subject to this sort of treatment, and because they are boiled and cast at relatively low temperatures (approx. 106° C), they are perhaps one of the best mediums for very fine flavors (Figure 7–4).

Typical composition:

Sugar	40–60%
Corn syrup	40–60%
Citric acid	0–1.5%
Gelling agent	see below
Flavor	q.s.

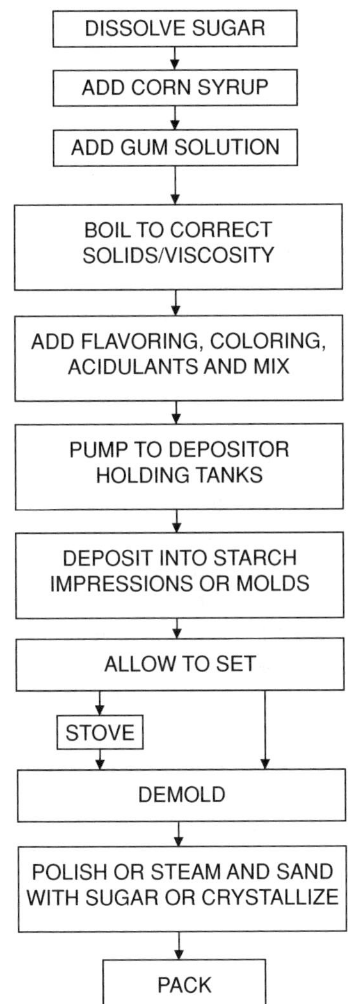

Figure 7–4 Processes used in jelly/gum production.

Alternative gelling agents are used at the following typical levels:

Gelatin	3–10%
Agar	2–3%
Pectin "100"	1.5–2%
Starch (or modified starch)	5–10%
Gum arabic	40–60%

The figures after pectin denote the grade, and 100 means that 1 part pectin will gel 100 parts sugar syrup. Because starch requires prolonged heating and so much water to disperse, pre-gelatinized starches are available, which are in effect pre-cooked or chemically modified. They are used to advantage in gums either wholly or partly in place of gum arabic. They are also used in Turkish Delight to reduce cooking times.

Method of manufacture: Typically, the chosen gelling agent is dispersed in water (sometimes overnight soaking is required) and heated to boiling point. Sugar and corn syrup are added and the batch re-heated to boiling point. The final temperature will depend on the type of jelly, but 106–107° C would be typical for most products. Where the mass becomes too viscous to cast at temperatures lower than this, "stoving" is required to remove excess water. Pectin based products rely on low pH to set, so acid is added just prior to pouring into molds. If the acid levels, which are required for flavoring as well as completing the set, are too high, buffer salts are added for control (a pH of 3.3 is the usual target). Flavorings and colorings are then added and the syrup deposited into rubber mats or impressions in starch. After setting, sweets so produced are de-molded and are oiled, cast into sugar or crystallized to prevent them from adhering.

In large scale production, this whole operation revolves around a starch molding plant. Trays are filled with starch, which receives impressions of the sweets to be made from rotary dies. These are then conveyed under depositing heads where product is metered into the impressions. It is usual to have hot syrup pumped to tanks above the depositor, where each tank has the requisite flavoring, coloring and acid added. By these means, up to six different flavors and colors can normally be handled at the same time. After depositing, the filled trays are dusted with starch and put aside to set, or taken to the drying rooms. When the correct water content has been reached, trays are returned to the plant where the confections are de-molded and sent for finishing. A percentage of the starch is automatically removed after each cycle for drying, and returned for refilling into the trays.

7.2.10 *Chewing Gum*

This product was originally based on natural gum chicle, which is similar to gutta-percha and is the dried Latex of *Sapodilla* trees. Supplies of this were insufficient to meet the demands of chewing gum manufacturers, and, in consequence, substitutes have been developed. These are said to be a blend of resins, oils and rubber, the manufacture and use of which have been covered by patents, mostly of U.S. origin (*Skuses Complete Confectioner*, 1957). Chewing gum is a mixture of gum base and corn syrup, which is thickened with fine ground sugar.

Typical composition:

Corn syrup	40–45%
Ground sugar	40–50%
Gum base	10–15%
Stearates or beeswax	1–3%
Flavor (oil blend)	0.25–0.5%

Method of manufacture: Corn syrup is heated to 124° C and the shredded gum base added and mixed until an even distribution of gum is evident. Ground sugar is placed in another mixer and the gum/syrup blend mixed while still warm when flavoring is added. The product is very sticky and liberal applications of dusting powder, usually consisting of ground sugar and starch, are required to handle the mass. Chewing gum is finished in the same way as described for cream and lozenge paste, i.e., rolled out into thin sheets, marked, cooled and broken. Commercially, very large mixers are used for this product, all with the facility to be heated and cooled, and finishing is carried out by sizing rollers, cutters, cooling tunnels and specialized wrapping machines.

It should be noted that the flavoring plays an extra role in this confection. Not only does it provide the taste but it adds to the pliability of the gum. For this reason essential oil blends or specially formulated flavorings are used. Peppermint and spearmint oils are particularly suitable, as are flavorings with high ester contents (e.g., pear and banana). On no account should these flavorings contain any water, or solvents such as propylene glycol and glycerine. Additional water or humectant would upset the carefully formulated ratio, causing stickiness and result in handling problems.

7.2.11 *Panned Work*

These products are so-called because manufacture is carried out in large revolving pans. There are two types.

7.2.11.1 *Hard panning*

The pans are heated and have a facility to have air blown into them. All kinds of centers are used for this confection—nuts (specially almonds), various seeds, gum, chocolate, boiled sugar, even granulated sugar (used for non-pareils).

The centers are dried, if necessary, by rolling in heated pans, or on trays in drying rooms and, if necessary, again sealed with gum solution. This is done by wetting the centers with gum solution and drying. This action prevents any moisture or oil migrating through the layers of sugar which will be deposited on the centers, while the product is in storage.

Centers are placed in the revolving pans and concentrated sugar syrup added, heat and air are applied and moisture driven off. In this way the sugar forms a fine layer on the center, which can be built up by further applications of syrup. After each syruping, the goods are allowed to fully dry (dusting). Flavorings are added to the pan in small amounts throughout the process, and are absorbed into the coating. Colorings are usually added by dissolving them into the syrups; often only the final coats are colored, although by using several different colors throughout the process, a sweet which changes color as it is eaten can be made. Some confectioners use suspensions in syrup of coloring lakes for this work (permitted synthetic colorings are normally available as the sodium salt which is water soluble, although aluminium salts are made which are insoluble and called lakes). These adhere readily to the coating, and being insoluble in water do not color the mouth when the sweet is eaten. The goods are then finished by coating with edible shellac, and/or polished with wax. Both operations are normally carried out in pans which are retained specially for this purpose (*Skuses Complete Confectioner*, 1957).

7.2.11.2 *Soft panning*

The process of soft panning is mainly carried out under U.K. ambient temperature conditions. Centers which have been prepared previously, and may be almost any type of confectionery, including jellies, gums, high boiled products and compressed work, are placed in the pan. Syrups containing corn syrup and sugar in varying proportions (the ratio depends on the type of center) are used to wet the centers, and fine caster sugar is added. The goods are rolled and further sugar added until the centers are dry. This process of syruping and drying is continued until the confections reach the size required. Finer sugar is used for the final coats to give a smoother finish. Flavorings may be added to the syrups to give an overall effect, although often the centers are highly dosed to give flavor impact when the sweet is chewed, and only the final coats have flavoring added. Quite often this flavoring will be vanilla. Coloring is also added to the final coats and similar solutions or dispersions are used as for hard panning. Flavoring is added, and the goods are glazed and polished in the same way as described for hard panned work.

It is likely that flavoring permeates through the sweets on storage, giving an equalizing effect.

7.2.12 *Chocolate*

Considerably simplifying the process, chocolate is made by intimately mixing previously fermented and roasted ground cocoa beans with finely ground sugar, using cocoa butter or other fats as a diluent. Milk chocolate also has milk solids or "milk crumb" added.

The flavor is dependent on several factors, not the least being the source of the beans. Although originating in South America, cocoa is now grown in many tropical regions, one of the most important being West Africa. The beans are grown in pods containing 20–40 almost white colored seeds surrounded by pulpy flesh. In order to remove this pulp the seeds are carefully fermented for several days. In this process, heat is generated, the germ or embryo is killed, and the initial flavor and brownish color developed. The beans are then separated from the remains of the pulp and dried in various ways, including drying in the sun (*Skuses Complete Confectioner*, 1957). Suitable mixtures from different sources of the fermented beans are made by the user. These take account of the flavor and the market value. Any extraneous matter is removed and roasting takes place in a revolving drum normally heated by hot air at temperatures between 130 and 150° C where the special roasted flavor is developed. After cooling, the germs and husks are removed and the beans broken into small pieces (now called "nibs"). These nibs are then ground to produce "neat work." This material is ideal for addition to products which require a chocolate flavor. It is rich in cocoa butter; some of which may be removed by expression, leaving cocoa. This in turn may be converted to "soluble" cocoa if treated with alkali. Extra cocoa butter is required to increase the fluidity of chocolate, although many substitutes, based on hydrogenated oils are available. These do not have the natural flavor of cocoa butter. The mixture of "neat work," fat, and fine sugar is refined to make plain chocolate, or with milk solids added, milk chocolate (Figure 7–5).

Typical compositions of plain chocolate:

Ground cocoa nib	35–50%
Ground sugar	32–48%
Cocoa butter	5–15%
Salt	0.2–0.5%
Flavor (vanillin)	0.01–0.02%
Lecithin	q.s.

Other fats can be used to replace all or part of the cocoa butter.

Typical composition of milk chocolate:

Ground cocoa nib	10–40%
Ground sugar	20–40%
Cocoa butter	5–20%
Milk powder	10–25%
Salt	0.2–0.5%
Flavor (vanillin)	0.01–0.02%
Lecithin	q.s.

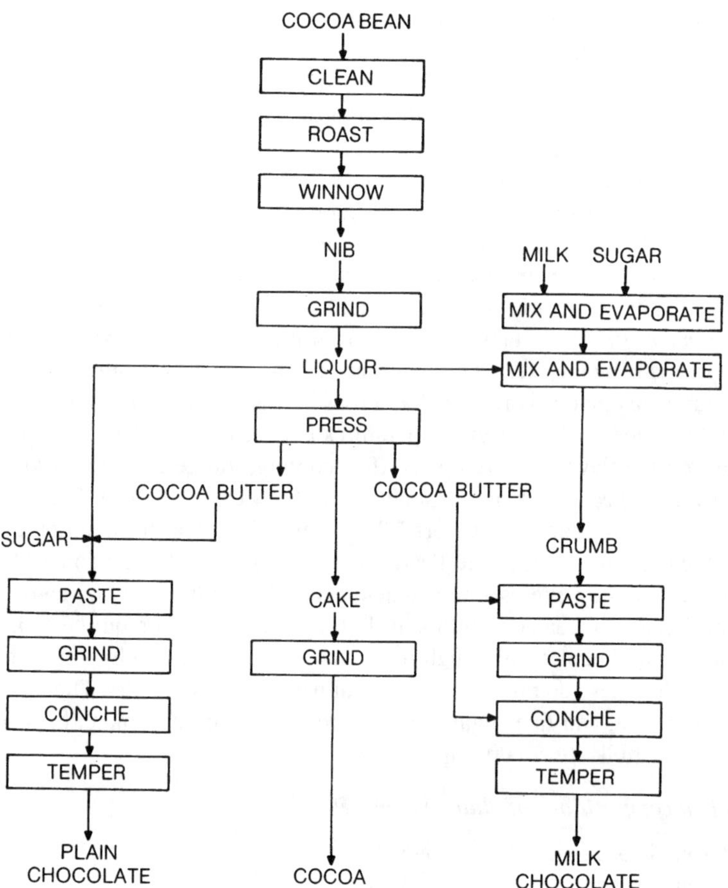

Figure 7–5 Factors affecting flavor in chocolate production (by kind permission of Cadbury Ltd.).

Other fats can be used to replace all or part of the cocoa butter. Lecithin is added to increase the fluidity of the mass. Sometimes milk crumb is used instead of milk powder, cocoa and sugar. A typical composition of milk crumb is as follows:

Sugar	56%
Cocoa mass	7%
Full cream milk solids	37%

Here the sugar and the cocoa mass is dissolved or dispersed in milk before drying and by this means a caramelized, cooked milk taste is imparted, which is characteristic of some brands. Using this material, milk chocolate can be made by merely adding fat, salt and flavoring.

All these operations play an important part in the ultimate flavor. The flavor is also modified by the further treatment of the chocolate mixture (melanger and conche), where the particle size is reduced over a period of several hours and the flavor is modified by driving off more volatile off-flavors.

Although vanilla beans can be incorporated into the chocolate mass, their cost now is prohibitive, and either vanillin or ethyl vanillin are more commonly used. Apart from orange, where oil blends are used, very few other flavors are added to chocolate itself, additional tastes being added as flavored chips or centers. It should also be noted that no water whatsoever is used in chocolate and its presence will completely spoil the setting properties, so flavorings used must be soluble in fat.

Chocolate is re-heated, melted and "tempered" in order to enrobe fondants and other suitable confections to be used as centers.

7.3 FLAVORS FROM INGEDIENTS

Many of these are self-explanatory such as the addition of honey in products like nougat, and liquorice in the many confections containing this ingredient. What may not be so well known in these days of continuous production methods is that, although various grades of brown sugar are readily available (demerara, forths, primrose, dark pieces), they are also sold in liquid form. As well as this, mixtures of the foregoing with refined sugar, in any agreed ratio, can also be obtained. Once the ratio to suit the confection has been agreed, liquid sugar can be delivered by tanker to storage tanks, ready to be pumped to the cookers. This eliminates the weighing and dissolving part of the operation. Most of these liquid sugars are sold at 67° Brix although the high invert products go as high as 81° Brix. The sugar refiners also sell various grades of treacle which is used to advantage in toffee and liquorice confectionery.

It is the effect of heat on these ingredients which is responsible for many of the well-known flavors of products like butterscotch and caramel. If heating is reduced, for example, by the use of vacuum and/or microfilm cooking, stronger flavored syrups are available.

7.4 FLAVORS DEVELOPED DURING PROCESSING

As indicated earlier in the chapter, an underlying flavor of most confections is derived from the heating of sugars and other ingredients. These ingredients undergo the well-known Maillard and Strecker reactions which are well documented (Ames, 1986). Products of these reactions are generally found in most confectionery processes. Using traditional gas or solid-fuel fired pans, very high surface temperatures are common and higher finishing temperatures (up to 170° C) give rise to special flavors, e.g., praline, butterscotch. Steam heating methods, vacuum boiling and microfilm cookers do not show these effects. The sugar industry provides products which produce sought after flavors by heat treatment. In fact by judicious use of these syrups, extremely good effects can be obtained.

In toffee manufacture it is common practice to heat milk and corn syrup together under very rigidly controlled conditions in order to obtain the desired flavor. Some butterscotch, while being produced in steam heated equipment, undergoes a secondary gas heating to increase the temperature, ensuring the correct flavor is developed, before the product is cooled and finished.

Despite this there are some advantages to be had by cooking with steam and vacuum techniques. This is apart from any cost factors. Because of the lower temperatures used, it is possible to produce fruit flavored boiled sweets and clear mints without any hint of the caramelization, which may detract from the desired flavor.

7.5 SELECTION OF FLAVORINGS

Apart from obvious limitations where some flavors are not compatible with other ingredients, e.g., water extracts in chocolate, it is relatively simple to flavor most types of confectionery. "Essential oils" are used widely and are very suitable. They are natural, available on the open market, and from various sources. An exception is in some jelly products, where absolute clarity is required, and a more soluble flavoring would be used. It should be remembered that all flavorings are best added at the lowest temperature possible, to prevent undue volatilization and degradation. It follows then that if the nature of the confection decrees that flavorings have to be added either at high temperature, or the products need to be held in heated conditions for any length of time, there will be loss of flavor. In the main, finer top-notes are lost or degraded so the flavor houses will recommend flavors or oils which have additional top-notes added and are selected for their overall suitability.

The main types of "essential oils" in common use for sugar confectionery are as follows:

Aniseed	Lemon
Clove	Lime
Cinnamon	Orange
Cubeb	Peppermint
Eucalyptus	Rose
Ginger	Spearmint
Grapefruit	Tangerine (mandarin)

As well as these, "oleoresins" can be utilized either alone or in conjunction with essential oils. They are almost without exception miscible together and suitable blends can be compounded. The main types used are:

Bay	Nutmeg
Capsicum	Rosemary
Cinnamon	Thyme
Ginger	Vanilla
Marjoram	

All these have the advantages of being "natural."

Other single chemical substances, either produced by physical extraction or by synthesis, are important in confectionery flavoring, and these include:

Camphor	Eucalyptol
Diacetyl	Maltol
Ethyl maltol	Menthol
Ethyl vanillin	Vanillin

Where compounded flavors are used, higher boiling solvents are preferred in order to reduce volatilization to a minimum. Propylene glycol, diacetin and triacetin are thus preferred to other more volatile solvents.

Some typical formulations to demonstrate a few of the more common flavors generally used for confectionery are as follows (all the following are made in "parts by weight").

Almond flavor:

Benzaldehyde	10.000
Vanillin	0.010
Cinnamon bark oil	0.005
Ethyl alcohol	40.000
Propylene glycol	ad 100.000

Butter flavor:

Vanillin	5.000
Ethyl vanillin	2.500
Butyric acid	5.000
Diacetyl	1.500
Citral	0.002
Propylene glycol	ad 100.000

Butterscotch flavor:

Vanillin	4.000
Maltol	1.000
Maple lactone crystals	0.500
Dihydrocoumarin	0.750
Diacetyl	0.500
Butyric acid	5.000
Lemon flavor	10.000
Propylene glycol	ad 100.000

Caramel flavor:

Vanillin	5.000
Ethyl vanillin	1.500
Dihydrocoumarin	0.500
Gamma-nonalactone	0.100
Caramel color	1.000
Water	15.000
Propylene glycol	ad 100.000

Coconut flavor:

Vanillin	42.000
Ethyl vanillin	1.000
Gamma-nonalactone	5.000
Dihydrocoumarin	1.000
Benzaldehyde	0.010
Water	15.000
Propylene glycol	ad 100.000

Creamy milk flavor:

Vanillin	2.500
Ethyl vanillin	3.500
Maple lactone	0.250
Ethyl maltol	0.350
Gamma-undecalactone	0.200
Gamma-nonalactone	1.000
Delta-decalactone	0.250
Acetyl methylcarbinol	0.300
Diacetyl	0.700
Butyric acid	0.500
Propylene glycol	ad 100.000

Ginger flavor:

Oleoresin capsicum	
4% capsacine	0.200
Oleoresin ginger	1.000
Ginger oil	1.500
Lemon oil	0.250
Ethyl alcohol	ad 100.000

Raspberry flavor:

Raspberry ketone	5.000
Maltol	0.250
Ionone alpha	0.050
Ionone beta	0.045
Ethyl butyrate	1.500
Isoamyl acetate	0.400
Isobutyl acetate	2.000
Propylene glycol	ad 100.000

Rose flavor:

Rose geranium oil	2.000
Citronellol	0.500
Geraniol	0.750
Phenyl ethyl alcohol	0.250
Ethyl alcohol	45.000
Propylene glycol	ad 100.000

Strawberry flavor:

Ethyl maltol	3.000
Ethyl butyrate	2.500
Ethyl valerate	0.150
Isoamyl acetate	0.010
Phenyl ethyl acetate	0.020
Benzyl acetate	0.400
Gamma-decalactone	0.250
Anisaldehyde	0.300
Orange oil	0.100
Isoamyl butyrate	0.040
cis-3-Hexenol	0.100
Ethyl alcohol	10.000
Propylene glycol	ad 100.000

Vanilla flavor:

Maltol	0.250
Dihydrocoumarin	0.500
Vanillin	8.000
Ethyl vanillin	2.000
Heliotropin	0.020
Cinnamon bark oil	0.005
Water	20.000
Propylene glycol	ad 100.000

"Winter" flavor:

Aniseed oil	10.000
Peppermint oil	72.000
Menthol	5.000
Eucalyptol	10.000
Clove oil	2.500
Camphor	1.000
	100.000

All the above flavorings are designed to be used at the rate of 0.1–0.2% in the finished product.

7.6 INGREDIENTS OF BAKERY PRODUCTS

Bakery products are, in general, based on three major ingredients and a number of minor, but nevertheless extremely important, components.

7.6.1 *Flour*

Flour is in most cases the major ingredient and is usually derived from wheat. It may be whole wheat as in the case of wholemeal flour, or part of the wheat berry as in white flour. It can be of different grades depending on its protein content: high protein (11%+) for bread making, and low (approx. 9%) for cakes and biscuits. In addition, flour can be treated in various ways to increase the amount of damaged starch cells, a factor which in turn increases its water holding power. It is also possible to treat flour with oxidizing agents (typically ascorbic acid) either to increase the apparent strength of the protein fraction, or, as is the case in the Chorleywood Breadmaking Process, to reduce the time required for fermentation when used in conjunction with mechanical development. It is also a requirement in U.K. law (*The Bread and Flour Regulations*, 1998) that white flour is nutritionally supplemented with calcium and iron plus vitamins. None of these processes has effects likely to cause major flavor problems. Anyone interested in understanding more about flour and its properties should read *Modern Cereal Chemistry* (Kent-Jones & Amos, 1967) or some of the other major works on the subject.

There are of course other cereals that can be used—rye, oats and maize being the most common. Rye is of particular interest in that it was normal in continental practice to produce a "rye sour" by a long period of fermentation, which gives the product a particular flavor and helps to improve keeping properties of the bread. It is possible to reproduce this effect and its shelf-life improvement by the use of flavorings and chemical additions. By this means, the risk, inherent in long fermentation periods, of the culture of undesirable microbiological organisms is avoided. The practice of sour dough systems is now little used as it is being replaced by flavorings or specially prepared dried sours made under strictly controlled conditions.

7.6.2 *Sugars*

Sugars are the second major ingredient to be found in most bakery formulae. There are many different sugars that can be used.

(a) Sucrose is the most common, both in granular form or as a ground powder; it is also available in its partly refined stage as brown sugars, the best known of which is probably demerara.
(b) Molasses or its partly refined stage golden syrup is another sweetening ingredient manufactured as a byproduct of sugar manufacture; it is used not only for its sweetening property but as a flavoring in many products.
(c) Dextrose is produced by the acid or enzymic conversion of starch derived from maize or as a byproduct of protein extraction from wheat flour. It is typically available either as a component of liquid glucose syrup or as a powder (dextrose monohydrate). The liquid syrup is supplied with different levels of conversion of starch to sugars, which is measured as dextrose equivalent (DE). The syrup is sold within a range of solids (72–84%) depending on its intended end-use.
(d) Invert sugar is a product of acid or enzyme treatment; in this case the substrate is sucrose. It has many properties similar to glucose and is often used in formulae for its ability to act as a humectant. We must also consider honey in this group; it is used as a flavoring ingredient as well as a sweetener. There are many honey types all of which have different flavors, characterized by the flowers visited by the bee at the time the honey was produced.
(e) Other bulk sweeteners, fructose and polyols such as sorbitol, are used to make products suitable for diabetics because they do not require the human digestive system to provide insulin.

7.6.3 *Fats*

Fats are the third major ingredient. They are derived from animal and vegetable sources and have to undergo several purification steps before being suitable for bakery use. The processes of filtration, color and flavor removal, fractionation and hydrogenation allow the manufacturer to produce tailor-made fats for the particular application. The very special flavor characteristics of butter that are changed during the baking process must also be considered. These are the target of much research in the flavor industry, the results of which have produced some excellent flavorings which can add a special note to many baked products.

7.6.4 *Liquids*

Liquids are one of the minor but very important ingredients in bakery formulations, and are normally added in one or more of many forms including egg, milk and water. The prime function of liquid is to bind all the various additions of the formulae, holding them together in the early stages of the baking process. Later, as the temperature rises, a secondary function of the protein fraction of egg coagulates to produce structure. Free liquid then enters the starch grain of the flour and allows gelatinization to take place, again adding to the structure of the product. The retention of moisture in the finished item is important in its taste sensation when eaten. Lack of moisture can for example make a cake unacceptable, whereas too much of it can make a biscuit equally unpleasant to eat.

7.6.5 *Gases*

Gases, producing the effect of aeration, are another minor but important ingredient.

Mechanical aeration, although not strictly an added ingredient, can be produced by beating or whisking, and here egg has a very important role in baked products in that it can hold air in its protein structure. Fat of the correct type will entrap air when beaten. Aeration can also be produced by chemical and biochemical components.

Chemical aeration is possible using ammonium carbonate which on heating decomposes to produce ammonia, carbon dioxide and water. Unfortunately, ammonia tends to re-dissolve into any available water in the product and is therefore not acceptable in high moisture products such as cake. It does form a very useful aerating ingredient for biscuits and low moisture items. Sodium bicarbonate upon heating will release some carbon dioxide, however, the reaction can be made to produce more carbon dioxide when used in conjunction with a variety of acids. Although chemical residues remain after both reactions, they are considered acceptable tastes in powder aerated products. The common acids used are:

Tartaric acid
Cream of tartar
Monosodium orthophosophate
Acid sodium pyrophosphate
Calcium hydrogen phosphate
Glucono-delta-lactone

Glucono-delta-lactone is preferred in that it produces the minimum aftertaste.

Biochemical aeration by the use of yeast, usually the specially cultivated variety (*Saccharomyces cerevisiae*), is the prime source of biochemical aeration although it is possible to take advantage of the yeast spores floating in the air or found on the surface of fruits. There are obvious risks in using these so-called wild yeasts as they can be a very unreliable source of aeration and so are little used.

Yeast breaks down the available carbohydrates by the use of enzymes in a fermentation process which produces carbon dioxide gas and a large range of other organic chemicals including ethyl alcohol, pyruvic acid and acetaldehyde, some of which can then go on to further reactions. Many of the chemicals produced have flavor and it is this complex combination which gives bread its unique taste.

7.6.6 *Other (Minor) Ingredients*

7.6.6.1 *Salt*

Salt is an important consitutent of most bakery items and often classified as a flavoring. Its function is that of a flavor enhancer as most products without salt have a flat unappetizing taste. It is usually added at the rate of 0.2–1.0% of the finished product depending on the degree of sweetness of the formulation.

7.6.6.2 *Color*

This is often the first clue to what flavors to expect from a product, and as such can be an important addition to any bakery formulae. The present trend toward natural materials limits the range of colors available and many bakery products are now manufactured without color addition, requiring added flavoring to be capable of instant recognition.

7.6.6.3 *Flavoring*

Some flavors, such as orange, lemon, strawberry and vanilla, are acceptable world-wide whereas others are peculiar to more specific markets. With the increase of international travel, the range of flavoring suitable for bakery products continues to grow in all countries.

7.7 BAKERY PRODUCTS

Figure 7–6 shows the composition of the major groups of baked items; the x-axis showing increasing sugar content, the y-axis the fat content. It is not intended that this should be a precise guide as to how to formulate a recipe for any particular item, but it serves to show the relationships among the various broad groups of products. These broad product categories are now examined individually.

7.7.1 *Bread*

Bread is the basic fermented item consisting mainly of flour with small amounts of fat and sugar. It does, however, contain yeast, which, by breaking down the carbohydrate content of flour, produces carbon dioxide gas and a range of complex substances which have natural flavoring effects (see Section 7.6.5). These flavors become pronounced once they are subjected to the heat of the baking process (see Section 7.7.7). Many attempts have been made to simulate the odor and flavor of baked bread, and some close approximations have been achieved. However, an all-purpose bread flavoring remains one of the flavorists' goals.

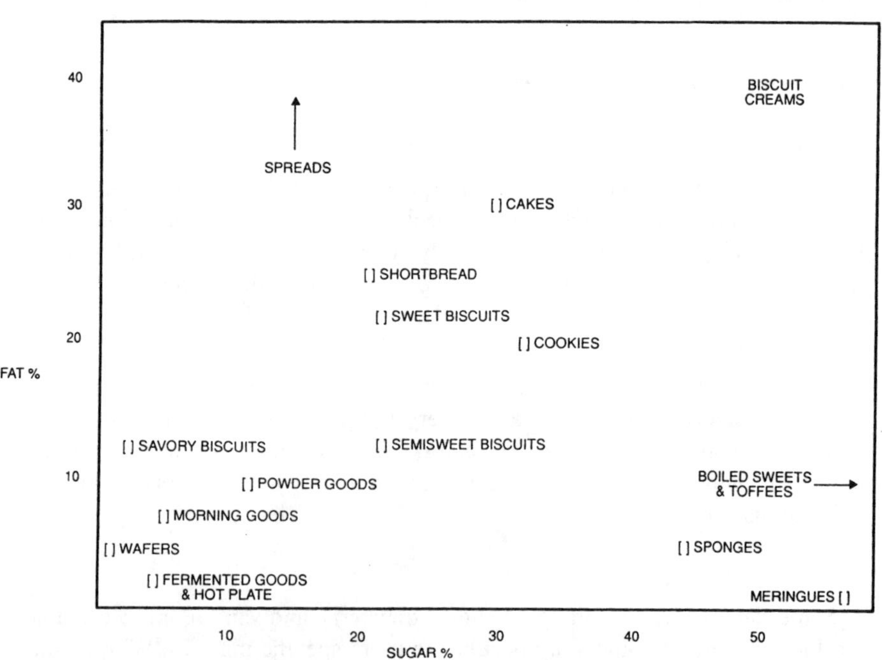

Figure 7–6 The relationship of fat and sugar in baked products.

7.7.2 *Hot Plate Goods*

To give a complete picture in this area of low fat and sugar, we must consider products such as scotch pancakes, muffins and pikelets, all of which are baked on a hot plate. This often requires that the article is turned over or cooked on both sides to complete the baking process. These products can be yeast or powder aerated. Wafers are also in this group, and consist of a thin batter baked between two hot plates in a very short time. These are conditions that very few added flavorings can withstand, and, therefore, many wafers are bland, tasting only of the flour and possibly milk that has been part of the original batter. They rely on being layered together as sandwiches with biscuit cream or toffee and can be chocolate covered. Vanilla, fruit flavors and nougat are some of the popular flavorings used. Wafers can also be served with ice cream to provide the flavoring element.

7.7.3 *Morning Goods*

These products are so-called because they are normally made by the baker first thing in the morning following the production of bread. They contain higher levels of fat and sugar than fermented and hot plate goods and are often enriched by the addition of dried fruit and sometimes by the addition of spices such as cinnamon and cardamom. Citrus flavorings are also used although both citrus oils and spices can, and do, slow down the activity of yeast. This can be compensated for by higher levels of yeast although, in the worst cases, yeast activity can be stopped altogether resulting in a dense, unacceptable product. The choice of any flavoring addition must therefore be made very carefully.

7.7.4 *Powder Goods*

These are products that are aerated by the use of baking powder rather than yeast. This category includes scones and raspberry buns, so-called because they contain raspberry jam. These products usually have the characteristic chemical note produced by the aerating agents; they are rarely flavored with more than vanillin or vanilla flavoring.

7.7.5 *Biscuits*

These products can be subdivided into groups dependent on their fat and sugar content.

Savory biscuits fall very close to morning goods in terms of fat and low sugar content. Typical flavoring used in such products are cheese and cheese plus products (e.g., pizza); also the use of meat flavoring, spices and additions such as monosodium glutamate with its recognized flavor enhancing ability.

Crackers and laminated pastry (e.g., puff pastry) are made from dough, low in fat and sugar, that has fat laminated into it to produce alternating leaves of dough and fat. During the baking process, steam generated from the dough forces the leaves apart to produce a multi-layer structure. These products are rarely flavored and are used in the main as carriers for toppings and fillings.

Semisweet biscuits can be misleading description in that some of the effect produced by fat is achieved using a chemical reducing agent, typically sulfur dioxide added as sodium meta-bisulfite. This has the effect of chemically breaking linkages in the protein matrix that hold the structure rigid and by doing so allow the dough to become extensive and machineable into thin sheets from which biscuits can be made. These chemical conditions make the survival of any added flavoring difficult as they too can be attacked by the reducing agent.

In this product area, the effect of fat adsorption of certain flavor notes starts to become noticeable. This is particularly so in biscuits as they are usually long shelf-life products (8 months shelf-life for a biscuit is normal). During this time, flour tastes begin to reappear and the fat starts to oxidize. This is where a good flavoring can help by suppressing these taints of aging to an acceptable level.

Sweet biscuits are the major group of biscuits with typical fat content of 17–22% and sugars around 30%. Crisper types can have sugar levels much higher than 30% and ginger nuts are a good example of this. With this higher level of fats and sugars, the baking temperature tends to be lower (e.g., 180° C), and the product can be enhanced by flavoring additions that have quite subtle notes. Vanilla and butter flavoring are used as a basis for many of these product variations.

Shortbread and pastry have higher levels of fat without a corresponding increase in sugar, and in pastry used for savory fillings no sugar is added. Shortbread and some pastry is made with butter so the use of flavoring, if used at all, is limited to vanilla types so as not to detract from the baked butter notes.

English cookies are so called to distinguish them from the generic use of the word which covers all types of biscuits in America. English cookies have higher levels of fat and sugar than sweet biscuits, and may contain ingredients such as egg, which makes them almost identical in composition to cake, the only difference being lower moisture content (usually below 10%). They can be flavored more easily than other biscuit types because the baking temperature can be lowered and the higher moisture content retained in the finished article. Many products in this group contain high levels of added ingredients such as chocolate chips, nuts, vine fruits and citrus peel. Part of their sweetening may come from fructose in order to maintain moistness in the finished product.

As the fat and sugar percentage has increased, the technology of manufacture has also had to change. The increases in these ingredients has changed the dough from one capable of being stretched and rolled into a thin sheet, to one that can be rolled into a sheet but not thinly, or alternatively, molded into a biscuit shape, through to an almost batter-like consistency that requires to be extruded through a nozzle and cut off in an appropriate size piece.

7.7.6 *Cakes*

Cakes are high in fat and sugar and have finished product moisture content above 10%. As a result they have a shorter shelf-life than biscuits. They invariably contain egg protein as part of their structure and so have this characteristic as a background note to any added flavoring. The baking temperature has again been lowered, and, with retention of moisture, flavoring with more pronounced top-notes can be used successfully. It is possible to go even higher in fat and sugar content into the so-called high ratio cake area, whereby, with the use of specialized fats and flours, very moist sweet cakes can be produced. Like other cake products these can be successfully flavored if allowance is made for their very sweet taste.

Sponges and meringues are products which do not typically contain fat or oil, although in some sponges small amounts of fat are used. A meringue is egg white and sugar beaten together to produce a stable foam which can be piped out to a suitable shape. This is then cooked at a low temperature which coagulates the protein of the whites and dries out the meringue. They are difficult to flavor because of the long baking time, their low finished water content, and the interaction of the flavoring with the protein content of the product. Sponges on the other hand contain flour and are baked at a higher temperature with the idea of leaving in some moisture. Their high protein content, resulting from the egg, makes them flexible when first baked and they are capable of being rolled into products like Swiss rolls after the application of jam and/or cream. The higher moisture content makes flavoring possible, with chocolate being a particular favorite.

7.7.7 *Baking Process*

It will be obvious that some of the above ingredients have flavors of their own that can be enhanced or masked depending on the end result required. To add to this medley of flavors, most bakery goods are subject to a heating process, baking, boiling in water or frying in hot oil which can again add, modify or destroy flavors. These process stages can be very short at high temperatures or long at low temperatures, each giving particular problems in achieving the desired effect.

The process of heating flour-based products brings about complex changes which are still the subject of debate. The increase in temperature of the mass first causes any gas trapped in the structure to expand. This gas can be entrapped air during the mixing process or gas produced chemically as referred to earlier, and at high temperatures, steam from the added water. As the temperature of the mass rises, a point at which the starch starts to gel is reached, and the structure which up to that point is soft and changing, becomes firm and rigid.

The final stage is the onset of browning and the development of the baked flavor notes. This used to be considered as purely the effect of sugar caramelization; more recent evidence confirms the predominance of the "Maillard reaction" in the coloring and flavoring of baked products. It is now accepted that many different flavoring chemicals can be produced naturally during the baking process by this reaction. It first produces aroma compounds before the development of color. As the reaction proceeds, low molecular weight color and flavor compounds are made, followed in the final stages of the reaction by the production of nitrogenous polymers and copolymers. Condensation of the free amino groups with reducing sugars produces the reaction and it has been shown that the amino acid component has the greatest influence on the resulting flavor compounds produced. The polymerization and dehydration reactions involved produce complex compounds, which, depending on the temperature, moisture and pH, can be extremely varied. Furfural, hydroxymethylfurfural, pyrazines, and short chain aldehydes have all been identified as being produced and research into this area continues (Figure 7–7).

The baking process can be particularly hard on added flavoring. The effect of high temperatures and low moistures in biscuits, for instance, amounts almost to steam distillation of the flavoring with the loss of many of the highly volatile substances and degradation of some of the less volatile components.

This can be seen most readily in the use of citrus oils in biscuits; the clean notes of the oil become musty and unpleasant. The use of so-called folded oils (terpeneless), which have been concentrated by removing the terpenes, can be used to advantage in this application.

7.8 BAKERY FILLINGS

The various fillings and toppings used with bakery products is an area where flavoring can play an important role.

7.8.1 *Jams and Jellies*

The use of conventional jams and the more specialized agar and pectin jellies has always been an accepted way in which bakery products can be given a flavor after the baking process. Bakery jams often derive their body constituent from sources such as plum or apple. Often, the effect of the named fruit will be achieved by the addition of flavoring to produce strawberry, raspberry and other flavored jams. Acid conditions exist in most jams and this is particularly true in pectin jellies where it is an essential constituent to bring about the set of the jelly. This acid background will influence the types of flavoring that can be considered, as some flavoring ingredients can be chemically changed in the presence of organic acids.

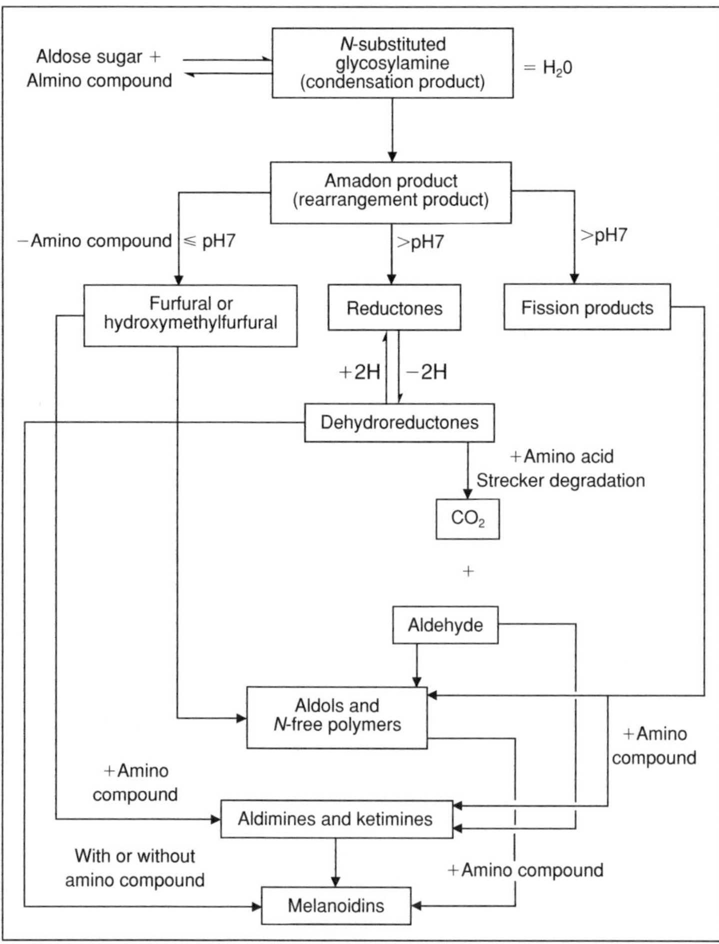

Figure 7–7 An outline of the Maillard reaction. Redrawn by permission of *Food Manufacture,* November (1989).

7.8.2 *Marshmallow*

The original marshmallow was made from an extract of the root of a shrubby herb, *Althoea officinalis*, which produces a gum with a distinct flavor and also the property that when mixed with water, it can be aerated to produce a stable foam. Today, marshmallow is manufactured using egg albumen, milk albumen, and gelatin, or a mixture of these in combination with sugar and added flavoring. The nature of the flavoring used is important in that the foam stability of the materials described is destroyed if small amounts of fat or fat-like substances are introduced. It is therefore necessary to use flavoring free from fat and oil. For citrus flavors the use of water extracts from the oil, although weaker in flavor, can produce satisfactory results.

7.8.3 *Creams*

Real dairy cream is not usually flavored with anything more than vanilla, and until recently even that was strictly not allowed by code of practice. This restriction was agreed by the flour, confectionery and dairy industry to avoid any confusion between dairy cream and imitation creams. Imitation creams are an important area for flavoring additions. They are oil in water emulsions with added emulsifiers, stabilizers and sugar and they have little flavor of their own. Dairy flavorings such as butter, cream and vanilla are required to make them acceptable as a filling and they have advantages in use because they are less likely to support microbial contamination than creams of dairy origin.

7.8.4 *Biscuit Creams*

Biscuit creams are fat and sugar mixtures made without added water, which would soften the biscuit and make it unacceptable. The use of creams is a convenient way of adding flavor to a biscuit which if added to the biscuit base would be unlikely to survive the baking process. In this way it is possible to produce lemon, strawberry, coffee and many more different flavored biscuits. However, fat can selectively adsorb flavoring components and can, in extreme cases, totally mask a flavoring during the shelf-life of a biscuit. Figure 7–8 shows the sort of changes a bakery flavoring undergoes from its initial balanced position through mixing, baking, fat adsorption and during the product's life. By recognizing and compensating for these changes, the goal of a highly acceptable product can be achieved.

7.8.5 *Icings*

Icing on the top of bakery goods is yet another way to add flavoring after the baking process. The result of rapidly crystallizing sugar and glucose mixtures from saturated solutions produces fondant, a bland material which on warming, flavoring and coloring, can be poured over a product. Variations on fondant can be made by the addition of fat, gelatin and other stabilizers and aerating agents to give icings that remain soft or set hard.

7.9 SUMMARY OF FLAVORING CHARACTERISTICS

Some ideas of the complexity of flavoring bakery items and the important characteristics of bakery flavoring are summarized below.

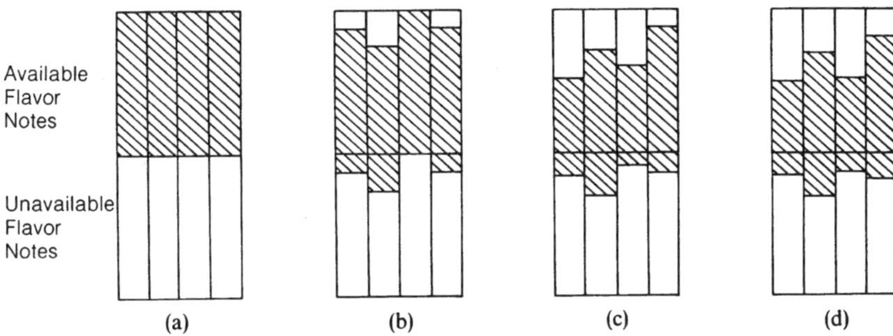

Figure 7–8 Pictorial representation of the changes that take place in a bakery flavoring during processing and product storage. Bars indicate flavor strength: (a) initial balance; (b) balance once included in formulae; (c) balance after baking; and (d) balance after storage.

(a) It must be capable of withstanding heat processes varying between 100 and 300°C.
(b) It must be capable of retaining its character in variations of pH between 6 and 8.
(c) It must not be subject to flavor distortion in high fat conditions or must be so formulated to allow for this condition.
(d) It must retain the required flavor impact in low moisture and long storage conditions, such as those occurring in biscuit manufacture or the high moisture conditions of cakes.
(e) It must be compatible with the medium in which it is used (e.g., fat and oil free) when used in meringues and mallows.
(f) it must be cost effective.

REFERENCES

J. Ames, Maillard reaction, *Food Manufacture,* November (1986) 6.

The Bread and Flour Regulations, Statutory Instrument 1963, No. 1435, H.M.S.O., London (1963).

Butter Regulations, Statutory Instrument 1966, No. 1074, H.M.S.O., London (1966).

Food Labelling Regulations, Statutory Instrument 1984, No. 1305, H.M.S.O., London (1984).

Kent-Jones and Amos, in *Modern Cereal Chemistry,* Northern Publishing Co., London (1967).

Skuses Complete Confectioner, 13th ed., W.J. Bush & Co., Ltd., London (1957).

FURTHER READING

A.R. Daniel, in *Up to Date Confectionery,* Maclaren & Sons, London (1947).

Flavor Chemistry, Advances in Chemistry, Series 56, American Chemical Society (1969).

D.J.R. Manley, in *Technology of Biscuits and Cookies,* Ellis Horwood, Chichester (1983).

S.A. Matz, in *Cookie and Cracker Technology,* AVI, Westport, CT (1968).

R.J. Taylor, in *Food Additives,* John Wiley, New York (1980).

P. Wade, in *Biscuits Cookies and Crackers,* Vol. 1, Elsevier Applied Science, London (1988).

Chapter 8

Savory Flavors for Snacks and Crisps

D.C.F. Church

8.1 INTRODUCTION

This chapter will cover the area of flavors and seasonings that are used on savory snacks. These include crisps, extruded maize snacks, fried snacks, nuts, snack biscuits, tortillas and similar products.

8.2 HISTORY OF SAVORY FLAVORS FOR SNACKS AND CRISPS (POTATO CHIPS)

One of the first flavorings used on crisps was in the 1950s when barbecue flavoring was developed for crisps in the U.S.A. This was followed by sour cream and onion flavor because of the popularity in the same country of "dips."

One of the first flavors in Europe some 30–35 years ago was cheese and onion, which originated in Ireland. Cheese and onion flavored crisps originally made by the two companies in Ireland, Tayto and Murphys, are still the largest volume of flavored crisps in Ireland and are being produced to the same basic recipe. In the U.K. cheese and onion flavored crisps were being made by Meredith Drew (part of United Biscuits), in the early 1960s. At about that time a flavor called "vinegar powder" was being developed for crisps and this later became salt and vinegar flavor. The key ingredient used in this product was sodium diacetate.

The advent of flavored crisps was also related to the development, in the mid 1960s, of packaging film that was both moisture proof and flavor retentive. This was arguably the major breakthrough that enabled the snack food market to develop.

In the United Kingdom, Smiths, who had the monopoly on crisps with their little blue salt twists, which had been made since the early 1920s, suddenly found themselves with a major rival, Golden Wonder. This new company really hit the market very hard with its flavor range. It established a flavor pattern little changed today for crisps. The five main flavors are salt and vinegar, cheese and onion, smoky bacon, prawn cocktail and beef/chicken.

8.3 SNACKS

In the 1950s in the U.S.A., snacks called "corn curls" or "cheese curls" were becoming increasingly popular. These are produced by high pressure extrusion of maize flour and the main flavoring added was cheese flavor. Without flavoring, extruded maize snacks are unappealing, unlike crisps. Cheese is still today the most popular flavor world-wide for extruded snacks. The idea of the American corn curl was imported into the U.K. in the middle to late 1960s with "Chipitos," but another brand captured the market with cheese flavor, "Wotsits," which still sells well over 20 years later.

Also in the mid-1960s another form of snack was marketed, fried pellets based upon potato granules which, when fried in fat at a high temperature for a short time, expand and produce a man-made crisp. These are then seasoned in the same way as potato crisps, i.e., by sprinkling on the flavor. Fried snacks also tended to have their own flavors developed for them.

Since the early days of the flavored snack market, nearly 30 years ago, many new flavors have been developed. Literally dozens have come and gone. What really has happened is the market has moved into sectors. This is of course very important to the flavor houses, because they have to know for which snack or potato chip or savory biscuit the specific flavor type is required.

8.3.1 *Savory Biscuits*

Until recently, flavor variations on savory biscuits have been limited to cheese types. Cheese biscuits are still major lines and there has been only limited success with other flavor variations. There is great scope for developments in this area and bacon or pizza flavor biscuits are popular new products.

8.3.2 *Market Separation*

The savory snack market divides into several product areas. Extruded corn curls/maize puffs have developed mainly as children's snacks, where a shape can be given to the extruded corn. Many different names have been given to these over the years, "Fangs," "Monsters," "Jaws," "Outer Spacers," etc. The flavors then become very important in the success of such products.

In the late 1960s marketing departments realized that maize snacks had a major advantage. They were very cheap to produce, because maize is relatively inexpensive, could be produced as fun packages and with interesting shapes, color, texture, name and flavor. As such they would sell in large quantities, mainly to children.

The savory snack market in the U.K. alone in 1994 is worth around £1.5 billion (sterling) with a very large number of new adult flavors. These include spicy types and "natural" flavors, where the flavors are free flowing powders and are usually a blend of many ingredients. The most popular flavors for various product groups are shown below.

8.3.3 *Potato Crisps/Chips*

Normal application is by dusting of a powder flavor at around 6–8% on freshly dried and still warm chips. The U.K. top ten flavors are:

salt and vinegar
cheese and onion
smoky bacon
spring onion
prawn cocktail
roast chicken
Worcester sauce
tomato ketchup
beef
Marmite

8.3.4 *Extruded Maize Snacks*

The normal application method is by mixing a powder flavoring with vegetable oil to form a slurry (2 parts oil, 1 part flavoring) and spraying on to the warm, dried maize extruded snack to give a final dosage of around 12% flavoring and 24% vegetable oil (higher levels are usual for cheese flavors). The U.K. top ten flavors are:

cheese
fried onion
pickled onion
spicy sauce type
tomato sauce type
salt and vinegar
bacon
sausage type
burger
cheese and onion

8.3.5 *Fried Snacks*

Application is by dusting in a similar way to crisps at around 6–8% on freshly fried snack pellets. Top flavors are:

cheese
barbecue
prawn cocktail
salt and vinegar
cheese and onion
bacon
spicy types

8.3.6 *Nuts*

Flavoring of nuts in the U.K. is virtually limited to peanuts. The main flavor type is "dried roast," which is a paprika, monosodium glutamate (MSG), and salt mix dusted onto nuts that have been previously wetted by a starch/gum solution and then slowly roasted. Strong flavors can also be dusted onto freshly roasted and oiled (fried) peanuts, such as spicy flavors: chili, barbecue and bacon types. Flavors can be used on peanuts which have been given a crisp coating. Paprika is the main flavor used and such products are very popular in Europe.

8.3.7 *Tortilla Snacks*

This is a rapidly growing adult snack area and these products are produced from a specially prepared maize base (Masa). Adult flavors are mainly used such as spicy chili, salsa, barbecue and "cool ranch."

8.3.8 *Snack Biscuits*

This is an area dominated by baked-in cheese flavors. The predominant ingredient is cheese powder with yeasts and flavorings also being used. This is a growth area with new dust-on type seasonings being used including cheese, spicy flavors and bacon flavors.

8.4 BASIC RECIPES FOR CRISPS AND SNACK SAVORY FLAVORS

8.4.1 *Salt and Vinegar Flavor*

	Approx. %
Salt	35.0
Sodium diacetate	20.0

Citric/malic acid	3.0
Lactose/whey powder	36.0
Monosodium glutamate	5.0
Anti-caking agent	1.0
	100.0

8.4.2 *Cheese and Onion Flavor*

	Approx. %
Onion extracts	0.1
Cheese flavorings/extracts	1.0
Cheese powder	10.0
Salt	25.0
Yeast/yeast extracts	12.5
Monosodium glutamate	5.0
Pepper	0.5
Rusk/wheat flour	25.0
Whey powder	20.0
Anti-caking agent	0.9
	100.0

8.4.3 *Smoky Bacon Flavor*

	Approx. %
Paprika extract	0.2
Smoke flavorings	1.0
Yeast powder	7.5
Smoked yeast powder	5.0
Hydrolyzed vegetable protein	3.0
Monosodium glutamate	7.5
Rusk	25.0
Salt	25.0
Dextrose	25.8
	100.0

8.4.4 *Beefy Barbecue Flavor*

	Approx. %
Onion	0.01
Paprika extract	0.09
Caramel color	3.0

Hydrolyzed vegetable protein	20.0
Yeast powder	4.0
Monosodium glutamate	5.0
Salt	19.4
Onion powder	25.0
Rusk	9.5
Yeast extract	8.0
Anti-caking agent	1.0
Dextrose	5.0
	100.0

8.4.5 *Cheese Flavor (for Corn Curls)*

	Approx. %
Cheese flavors (Nat/N.I.)	1.5
Paprika color	0.5
Lactic acid	1.0
Cheese powder	40.0
Monosodium glutamate	3.0
Rusk	10.0
Salt	15.0
Whey powder	29.0
	100.0

8.4.6 *Paprika Flavor*

	Approx. %
Paprika extract	1.0
Monosodium glutamate	5.0
Ground paprika	30.0
Dextrose	5.0
Sugar	6.5
Rusk	20.0
Salt	20.0
Hydrolyzed vegetable protein	2.5
Yeast extract	2.5
Ground pepper	1.5
Onion powder	5.0
Anti-caking agent	1.0
	100.0

8.5 INGREDIENTS FOR SAVORY FLAVORS

8.5.1 *Production Methods*

The operation normally used to produce savory flavors is by batch blending, where accurate mixing is essential to obtain standardization of color, flavor and particle size. Generally slurry flavors for extruded maize snack need to be of finer particle size than those used for potato snack and crisp dustings. A flow chart (Figure 8–1) gives the outline of a typical mixing process.

8.5.2 *Quality Systems and Methods*

As the majority of savory snack and chip seasonings are added to the snack product without further heat processing or sterilization, many manufacturers now insist that all seasonings and flavorings supplied meet many strict quality control requirements and, in particular, microbiological standards. Raw materials such as spices, onion powders and some dairy products can be a source of bacteriological contamination and all U.K. savory flavor manufacturers now only use raw materials that have been subjected to some form of heat processing unless they are produced with low bacteriological counts. Ingredients in this latter category would include salt, acids and sugar. Most companies also now employ hazard analysis and critical control point (HACCP) methods in order to maintain a good quality control relating to microbiological standards, foreign bodies and high quality ingredients. Table 8–1 is a summary of a typical HACCP system.

8.6 MAJOR RAW MATERIALS AND INGREDIENTS USED IN POWDER SAVORY FLAVOR BLENDS

8.6.1 *Acids and Acidity Regulators*

The main acid used is citric acid powder (E330). Citric acid is available in a natural form as spray dried lemon powder. Also widely used are tartaric acid (E334), lactic acid (E270), which is a liquid acid absorbed onto a base and mainly used in cheese flavors, and malic acid (E296). The main acidity regulator used is sodium diacetate (E262), which, in addition, is the major ingredient and flavor used for salt and vinegar and pickled onion flavors. Its use is also common at a lower level in tomato ketchup and other sweet and sour flavors.

8.6.2 *Anti-Caking Agents*

These are normally added in small quantities (maximum 2% in the flavoring) in order to maintain a free flowing powder for application onto crisps or snacks.

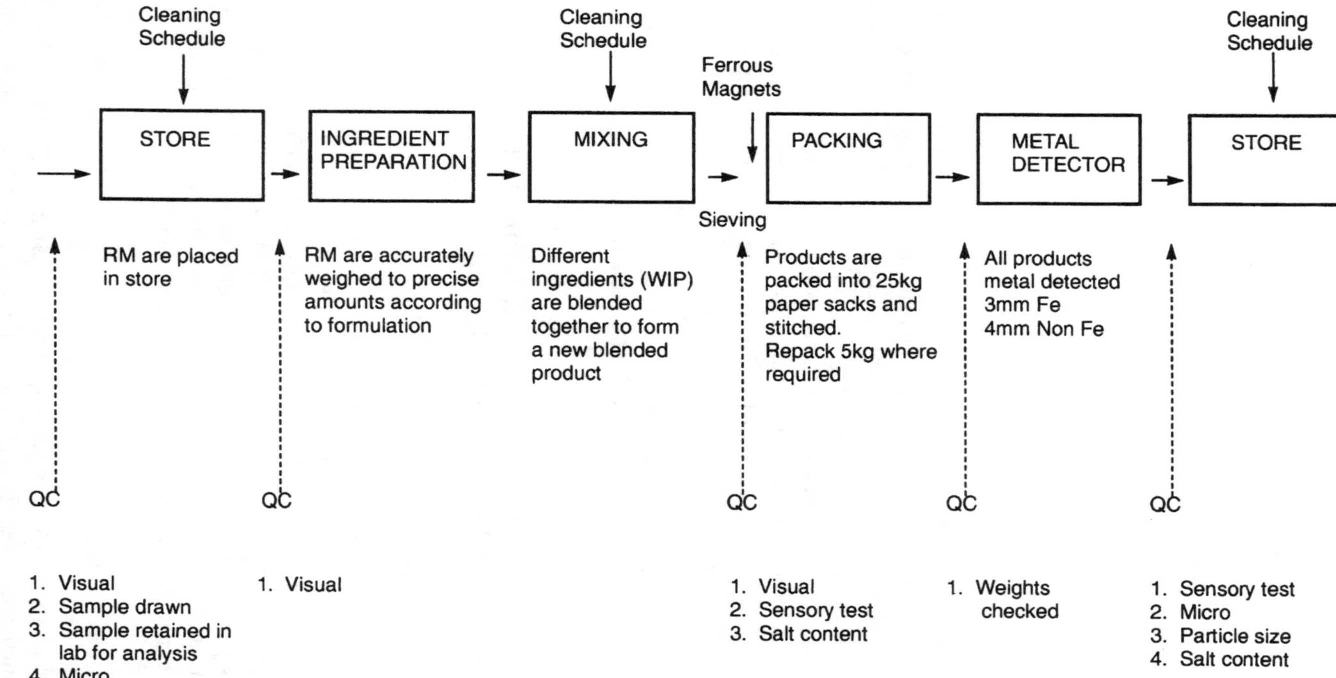

Figure 8–1 Powder savory flavor manufacture. Flow chart—mixing process. RM = raw materials, WIP = work in progress.

Table 8–1 Hazard analysis critical control point summary

CCP No/ process step	*Hazards*	*Control measure*	*Target levels and tolerances*	*Monitoring procedure*	*Corrective action*
Raw materials received	Foreign bodies	(for 1–3) (a) QC	100% No damaged goods	As per QC raw material manual	(a) Reject load
	Microbiological levels unacceptable	(b) Stores/goods control	100% Correct goods		(b) Inform supplier
	Incorrect raw materials	(c) Documentary checks	100% Correct physical		(c) Monitor future supplies
	Damaged bags or pallets	(d) Physical examination	Micro within spec.		
		(e) Microbiological testing including yeasts and molds			
		(f) QC positive release system			
Raw materials stores	Pests	Pest control contract. Insectocuters and moth trap. Fly screens and suitable doors.	Nil	Contractor check	Inform contractor and action as required
	Use of rejected material	Isolation of unsuitable materials	N/A		Advise QC of out of date stock
	Incorrect rotation	Operator control			
	Damaged bags or pallets				

continues

Table 8–1 continued

CCP No/ process step	*Hazards*	*Control measure*	*Target levels and tolerances*	*Monitoring procedure*	*Corrective action*
Batch weigh off	Foreign bodies	Visual inspection	Nil	QC/Auditing (continuous)	Taken up with management
	Bacterial contamination from personnel	Staff hygiene training. Hand sanitizer on entry into factory			
	Equipment	Cleaning schedule		Swabbing of cleaning effectiveness	
	Incorrect ingredients or amounts	Operator training and QC checking		Traceability on formulae	
Blending	Foreign bodies	Sieving/magnets/ metal detector	Nil	Metal detection records Sieve/inspection of finds	See metal detector procedure Sieve finds investigate and inform supplier
	Bacterial contamination from personnel	Staff hygiene training			
	Machines	Cleaning schedule		Swabbing of cleaning effectiveness	
	Incorrect use of raw materials or amounts	Operator control		100% Organoleptic testing. Analytical tests on final products to consumer specs.	

Packaging	Foreign bodies in packaging	Control of suppliers manual preparation of packaging	Nil		Investigation and correction
Sealed finished product	High micro levels storage	Micro testing	See customer specs	Micro testing. Bacterial test on final products to consumer specs	Reject/retest destroy if serious contamination
	Storage	Warehouse supervision			Inform appropriate personnel
Transport and delivery to customers	Damage taint contamination of packaging	Checks of lorries	Nil	Despatch operator	Advise management
Maintenance and cleaning	Failure of hygiene Incorrect use of cleaning chemicals Failure of equipment	Cleaning schedules staff training Use as specification in cleaning manual Routine maintenance		Swabbing to determine cleansing effectiveness Cleaning schedule Keep records	

The main agents used are silica (SiO_2/E551) and other silicates such as magnesium silicate. Sodium aluminum silicate is also used but its use is currently under review.

8.6.3 *Colors*

In the U.K. natural colors are mainly used and these are generally extracts of spices and similar products. Coloring extracts used include:

Paprika (Capsanthin, E160c)
Annatto (Bixin, E160b)
Turmeric (Cumin, E100)
Carmine (E120)
Beta-carotene (E160a)
Beetroot (Betanin, E162)

In general, "natural" colors are not as stable to light as are the artificial counterparts and may be more difficult to incorporate into powder seasonings. Artificial colors are still used in many countries and are incorporated into snack flavors. These include sunset yellow (E110) and tartrazine (E102), both of which are commonly used to color cheese snacks. Also used is carmoisine (E122) which is widely used to color tomato ketchup flavorings.

8.6.4 *Carriers and "Fillers"*

These bulk powders may be considered as fillers. However, they normally have a function such as keeping good free flow qualities after a lengthy period, which is most important when some of the ingredients are hygroscopic as in tomato ketchup flavors. The main carriers used are wheat flour and its processed form, rusk, corn starch, soya flour, lactose, dextrose and maltodextrin. Whey powder is also widely used as a carrier mainly in cheese flavor mixes.

8.6.5 *Dairy Powders*

This group of ingredients is mainly used in cheese and cheese and onion flavors but some, such as whey powder, can occur in other products. Cheese powders are available in many different types, mainly cheddar tasting varieties, although stilton and sharp American style cheese powders are gaining popularity. Most cheese powders are available as uncolored or naturally colored.

Sour cream powder is also being more widely used in "cool ranch" type flavors and also sour cream and onion. Yogurt powder similarly is beginning to be used more widely in spicy type flavors. Whey powder is mainly used as a carrier, particularly in cheese flavors.

8.6.6 *Fat Powders*

Spray dried hydogenated fats can be added to some dairy type mixes such as cheese or cheese and onion to enhance mouthfeel.

8.6.7 *Flavor Enhancers*

Flavor enhancers are mainly added to bring out a more savory taste, particularly in products such as bacon or meat flavorings, although small levels of around 5% are common in most crisp and snack flavorings. The main flavor enhancer used is monosodium glutamate (E621) in a fine crystal or powder form. Used to a lesser extent are the ribonucleotides (E627, E621, E635) which are nominally added in combination with monosodium glutamate.

8.6.8 *Flavorings*

A very wide variety of natural and nature identical flavor blends are used as "top-notes," to boost or modify natural raw materials. In this category are onion and cheese flavorings. A complex mixture of flavorings may be added to produce flavors such as prawn cocktail and smoky bacon. Usually a combination of liquid flavors and sprayed dried products are used to give both an initial impact and lasting flavor. Enzyme modified cheese (EMC) extracts and powders are used to boost cheese-based seasonings. Mixtures of spice oils and extracts are used in adult chili, tortilla type products. Thermal reaction flavors are often used to boost beefy and other meat flavors (see Chapter 9).

8.6.9 *Herbs and Spices*

Over thirty common herbs and spices may be used in powder form in various snack flavors, particularly in products such as pizza flavor, spicy chili, curry flavors, etc. Because of possible microbiological problems, most herbs and spices are heat treated prior to blending, although irradiated products are used in European countries such as France and Holland. Herbs and spices are also available in the concentrated form of essential oil and extract, which gives a wide potential of natural flavor ingredients.

8.6.10 *Hydrolyzed Vegetable Proteins*

Hydrolyzed vegetable proteins (HVP) powders are widely used, particularly in beef, barbecue, chicken and other meat flavors. Their use is quite common in many products to give a savory taste. They are usually produced by acid hydrolysis of a protein source, such as soya protein. There is some concern over chloropropanol residues resulting from this chemical processing (see Chapter 9).

8.6.11 *Salt*

Salt is the most common ingredient in the majority of savory flavors; the total salt content of snack seasoning being typically between 20–30%. Normally, fine crystal pure dried vacuum salt is used, although milled salt and sea salt are also used in certain recipes. Substitution of sodium chloride by potassium chloride has been tried with limited success to produce a "lower salt" seasoning.

8.6.12 *Sweeteners*

The main sweetener used is still sugar in products such as tomato ketchup flavors. Dextrose is also very widely used, partly as a sweetener and also as a carrier, for example, in bacon flavors. Intense sweeteners such as saccharin and aspartame are used in some products including prawn cocktail, tomato ketchup and sweet and sour flavors. Saccharin in particular also has a slight flavor enhancing effect. Dosage level is normally around 0.1% maximum.

8.6.13 *Vegetable Powders*

The most commonly used vegetable powder is, of course, onion powder. A wide variety of excellent qualities are now available, the main origins being the United States, France, Egypt and Eastern Europe. Similarly, garlic powders are widely available. Tomato powder, particularly from Spain, is used broadly in many formulations specially in tomato ketchup flavors and barbecue flavors. Other vegetable powders include carrot powder, celery root powder and red pepper powder.

8.6.14 *Vitamins*

In the U.K. an increasing amount of children's snack products are being prepared with added vitamins. These are normally added to the flavoring mix to produce around 15% of the recommended daily allowance (RDA) of the vitamin in the finished snack product. Provided this level is achieved in a single serving, a "claim" for the vitamin may be made under the U.K. law.

8.6.15 *Yeast and Yeast Extracts*

A wide variety of powder products are available for use in savory flavors. Brewers yeast is normally not suitable unless it has been debittered. Similarly, bakers yeast may need further treatment before use as a savory flavor ingredient. They are, of course, inactive products and yeast powders such as torula yeast are largely used in bacon flavors and cheese and onion flavors. Further heat processing produces a roasted taste as in autolyzed yeast powder to which it gives a very pleasant savory character. Powdered yeast extracts are widely used particularly where a "natural" flavor enhancing effect is required. Autolyzed yeast and yeast extract are commonly used to give rounder cheese type flavor and are regularly used in biscuit dough in the manufacture of savory cheese biscuits.

8.6.16 *Pre-Extrusion Flavors*

Due to the lack of "impact," flavors are not widely used pre-extrusion. However, they may be useful in pelleted snack products. Most commonly used ingredients which survive extrusion and subsequent frying are spices, onion powders and flavor, yeast extracts and strong flavorings such as smoke. This flavoring is receiving growing attention and may be used to good effect when combined with mild dusting flavoring applied after frying.

8.7 NEW DEVELOPMENTS AND TRENDS

Although there is little evidence of change in the main varieties of potato crisps and children's snack flavors in the U.K. market, the adult sector continues to receive a lot of attention and there is a wide diversion of flavors used. The trend is toward spicy and Far Eastern flavors such as hot curry and tandoori flavors as well as spicy Mexican flavors such as chili, jalapeno and salsa which are mainly used on tortilla type snacks. "American" flavors, such as sour cream and onion, western barbecue and "cool ranch" are now widely accepted.

New snack bases are constantly being developed, some of which are a combination of biscuit and snack technology. These often require stronger dusting flavorings of which examples are T-bone steak and chili sauce. Fried pasta snacks have also been developed, to which Italian style flavors such as pizza and bolognaise are often added. Savory popcorn is also becoming more acceptable and the usual flavors here are sharp cheese or barbecue.

The development of flavor application equipment for snack manufacture has also been receiving attention and electrostatic application systems are now common.

Low fat crisps and "light" snacks, with reduced oil content, are increasingly being made available to the consumer. Subtle flavors are often needed on these products.

8.8 CONCLUSIONS

Although many countries now have an established pattern for use of flavors on their snack products, the market is still growing world-wide and with the availability of hundreds of ingredients it will still be possible for flavorists to develop new snack ideas for many years.

FURTHER READING

R. Gordon Booth, *Snack Food,* available through Snack Food International magazine.

Directive on Colours for Use in Foodstuffs 89/107/EEc.

Labelling Directive 79/112/EEC.

S.A. Matz, *Snack Food Technology,* available through Snack Food International Magazine.

Snack Food International Magazine, 12 Blackstock Mews, London N4 2BT, England.

The Snack Magazine (official journal of the European Snack Association), Creative Business Communications, 4 Russell Street, Leek, Staffs, ST13 5JF England.

Chapter 9

Thermal Process Flavorings

Charles H. Manley, Belayet H. Choudhury and Peter Mazeiko

9.1 INTRODUCTION

The history of flavor development and creation had its beginning with the use of natural extracts and synthetic aromatic chemicals produced to match the flavor notes of those found in nature. This chapter reviews many of these historic beginnings and the materials which have become the flavorist's palette. Those materials first used were extracts of natural herbs, spices and fruits, like those used in the fragrance trade. These materials could be skillfully compounded into high impact flavors. Although many natural materials have very low amounts of volatile components which give rise to their aromatic impact they are readily available and could be isolated at useful levels. With the advent of the science of chemistry and, in particular, the discipline of organic and analytical chemistry, many of the "secrets" of the natural extract were discovered and duplicated. Today more than 2,000 synthetic organic chemicals are used by the flavorists to create high quality flavors.

Humans' first interest in flavors was to add flavor quality to foods by the use of these extracts or flavor substances. Confectionery products could be produced or enhanced by the use of these materials and so too could other fine foods and beverages. However, as we all know, many of the foods we enjoy are cooked. The flavor of the material we eat is a function of the cooking or processing given it. Raw meat, except in rare instances, is not consumed as such. That is also true of such materials as the coffee bean. The flavor we appreciate is developed by heating the food to a certain temperature for a certain amount of time. Cooked meats have always been a craving for humans, yet one which is not always available. Some 150 years ago a meat flavor substitute was created and two men have been noted for their contribution in developing this product. They were Liebig of Holland and Maggi of Switzerland and the product of their efforts was an acid hydrolysate of vegetable protein. Both men were involved in the development of commercial businesses based on their discovery. Today the Nestlé Company still manufactures hydrolyzed plant proteins (HPP) under the Maggi name whereas the Liebig name has disappeared, but the commercial tradition lives on under the Unilever banner.

These first flavors made under thermal processing conditions are used around the world by food manufacturers (May, 1995). They are referred to as HVP or HPP (hydrolyzed vegetable protein or plant protein) or HAP (hydrolyzed animal protein). It is estimated that nearly 9,000 metric tons of these materials are produced annually in the United States with a similar volume in Europe (IHPC). Until the late 1950s these were the only "processed flavors" available. Because of the heat treatment given these materials during processing one may consider them the first "thermal process flavorings." As you read this chapter you will start to understand how these materials based on the acid hydrolysis of vegetable proteins, such as corn, wheat, soy or other grains, could generate aroma and taste components that are similar to meat flavor and taste.

The term "thermal process flavorings" was coined by the flavor industry to categorize flavors that are generated by a cooking or heat process. There are a number of groups, including regulatory bodies, around the world that have adopted this term for defining flavors made by heating various materials to create flavor profiles which include meat types, coffee, cocoa and even nut flavors. This chapter will discuss the chemistry, flavor creation, processing and application of the group of flavors known as "thermal processed flavors" and the flavor compounds made from them.

9.2 HISTORY

Although the first useful HVP product was developed because of the Napoleonic War's needs for a meat extract–like substitute, it was the World War II years that gave rise to the development of more sophisticated meat flavors. The British government embarked on the ground nuts scheme to produce ground nut oil. One result of this project was to stimulate industry to find ways of utilizing the protein by-product from the ground nut oil production; this led to research into thermally generated flavors from proteins and amino acids (Boyer, 1953).

Research into the flavor of beef was started by Unilever in 1951 at their Colworth Laboratories in Bedford, England. At the same time, other laboratories around the world were applying science to the understanding of meat aroma.

The original work of Crocker (1984) established the relationship of components in meat and their development of characteristic flavors during thermal processing or cooking. It was found that the water-soluble fraction of meat gave rise to the typical cooked or roasted notes of meat flavor. The species specific character of cooked meat was found to be related to the nature of the fat found in the meat. A fuller discussion of this will be given later in this chapter.

The early work at Unilever has been reviewed by May (1995). They approached the research from two directions. One was to isolate and identify the aroma arising from cooked beef and the other was to isolate and characterize the flavor precursors, which give rise to the basic beef flavor character. Their conclusion indicated that many of the volatile compounds were either sulfur containing or carboxyl compounds and that the low molecular weight water-soluble material from beef was the major precursor of cooked beef aroma. These components were composed of a significant number of amino acids or small peptides and/or amino sugar and reducing sugars. They found that some of the amino acids and sugars, particularly cysteine and ribose, were reduced or disappeared. These findings and the development of processed flavors based on heating precursors of amino acids and sugar led to the first patent on this type of flavor being issued to Unilever (May & Akroyd, 1960).

9.3 THE MAILLARD REACTION

The basic reaction between amino acids and sugars had been studied by Maillard in France in the early 1900s. A significant number of research publications and reviews have been written about this reaction mechanism which now carries Maillard's name. One of the early discussions of the relationship between the Maillard reaction and food flavor was published by Hodge (1953). In model systems, he showed how the early reaction between the carbonyl group of reducing sugar and the free group of an amino acid (or peptide/protein) gives a Schiff's base. This, via cyclization and the Amadori rearrangement, gives the 1-amino-1-deoxy-2-ketose. A general outline of the types of reactions involved in the Maillard reaction is given in Figure 9–1.

There are many fine review articles on the organic mechanisms of the Maillard reaction which will enhance your basic understanding of that chemistry. This chapter will focus on the results of such chemistry and its contribution to the aroma and taste of cooked foods.

The types of reactions, precursors and types of changes that are generated from their thermal reactions are shown in Table 9–1. During the nearly 50 years since the first focused research on meat flavors, thousands of research articles have been published and nearly as many compounds have been identified. Advances in separation methods (gas chromatography and liquid chromatography), characterization methods (mass spectroscopy) and computer technology have turned a once difficult scientific project into an afternoon's routine analysis.

9.4 AROMATIC COMPOUNDS FROM PRECURSORS

Let us first review the significance of the major components and their precursors. This will give a basic understanding of the chemistry behind creating a thermal process flavor.

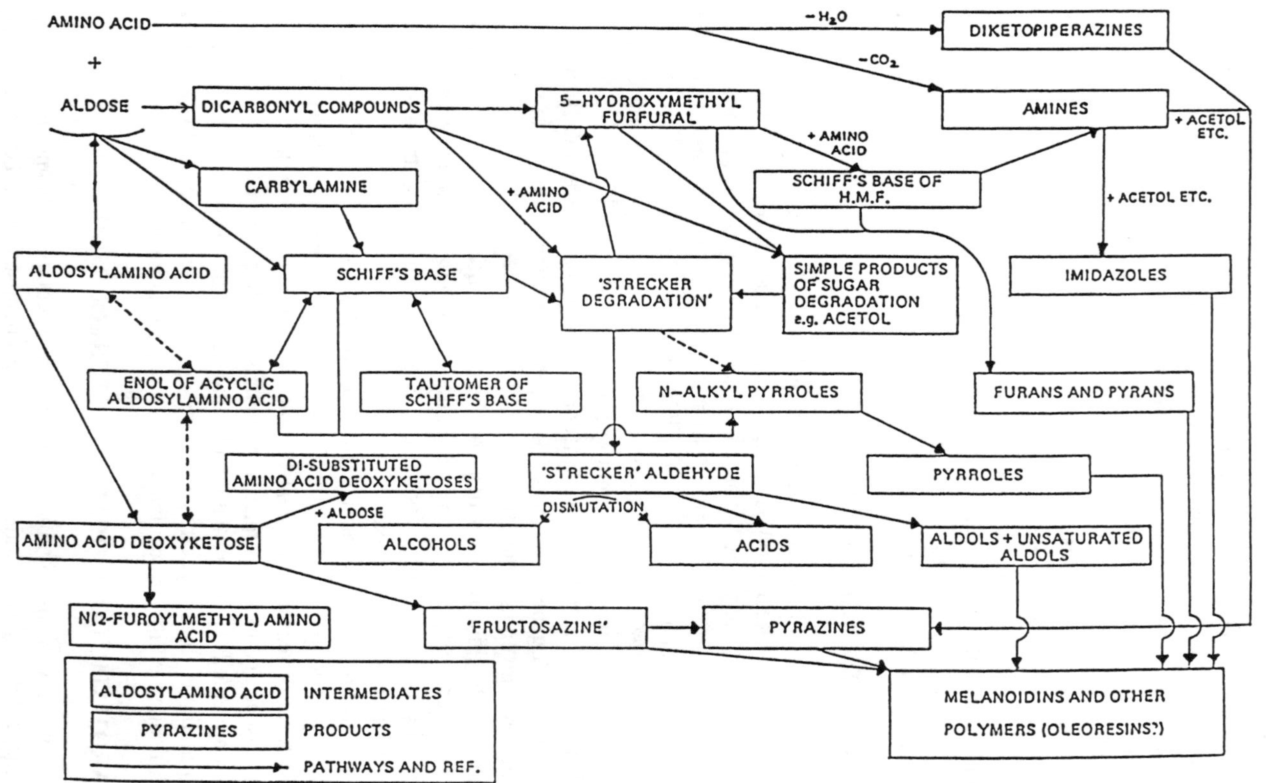

Figure 9–1 Maillard reaction pathways. *Source:* Reprinted from T.F. Stewart, *Scientific and Technical Survey*, No. 61, © 1969. Reproduced by permission of the Director Leatherhead Food R.A.

Table 9–1 Reactions Involved in Process Flavors

Ingredient		*Reaction Type*		*Changes*
Proteins		Maillard		Aroma creation
Amino acids		Oxidation		Color development
Peptides		Strecker degradation		Taste changes
Fats	→	Polymerization	→	Contaminant
Fatty acids		Sulfur replacement		Off-flavor formation
Sugars		Hydrolysis		Taste enhancement
Carbohydrates				Texture changes
Nucleotides				
Vitamins				

9.4.1 *Pyrazines*

Pyrazines have been identified as being created through the Maillard reactions that are initiated by carbohydrate cleavage. In the Maillard reaction the main cleavage pathway is retroaldolization. Pyrazines are one of the major groups of volatile chemicals produced from the Maillard reaction. These aromatic components are well characterized as the compounds that directly contribute to roasted type of flavor notes (Maga & Sizer, 1973). Some pyrazines have an extremely low odor threshold. For example, the odor threshold of 2-isobutyl-3-methoxypyrazine, a bell pepper type note, has been found to be 0.002 ppb in water solution (Shibamoto, 1986).

Pyrazines have been identified in many cooked foods, and Table 9–2 indicates some of those found. Many model systems have been studied to determine the chemical pathways which give rise to each type of pyrazine, and a simple outline of the mechanism of formation is also shown in Figure 9–2.

9.4.2 *Thiazoles, Thiazolines and Thiazolidines*

The thiazoles have been shown to develop from thermal degradation of thiamine (vitamin B_1). Thiamine gives three thiazole derivatives during heat-degradation (Dvirvedi et al., 1971) and the general mechanism for their development is shown in Figure 9–3. Thiazoles are considered one of the main constituents that give the meaty flavor character to meats. They are extensively used to produce imitation meat flavors. Table 9–3 indicates some of the thiazoles and thiazolines found in foods.

A number of thiazoles have been identified in a reaction mixture of D-glucose, ammonia and hydrogen sulfide (Shibamoto & Russell, 1977). The major products of the system were 2-ethyl-3-hydroxy-6-methylpyridine and 2-ethyl-3,6-dimethylpyrazine.

Table 9–2 Commonly Found Pyrazines

2-Methylpyrazine
2,3-Dimethylpyrazine
2,5-Dimethylpyrazine
2,6-Dimethylpyrazine
2-Methyl-3-ethylpyrazine
2-Methyl-5-ethylpyrazine
Trimethylpyrazine
2,5-Dimethyl-3-ethylpyrazine
2,3 Dimethyl-6-ethylpyrazine
Tetramethylpyrazine
2,5,6-Trimethyl-3-ethylpyrazine

9.4.3 *Thiophenes*

The thiophenes are widely distributed in vegetables such as onion and are also found in cooked meats (Boelens et al., 1971). Model Maillard systems that consist of a sugar and a sulfur-containing amino acid produce significant numbers of thiophenes (Ho et al., 1983; Zhang et al., 1988).

Their odor thresholds are very low and the flavor character very pronounced. These components are not very stable and it may be concluded that their importance in process flavors may be of little significance.

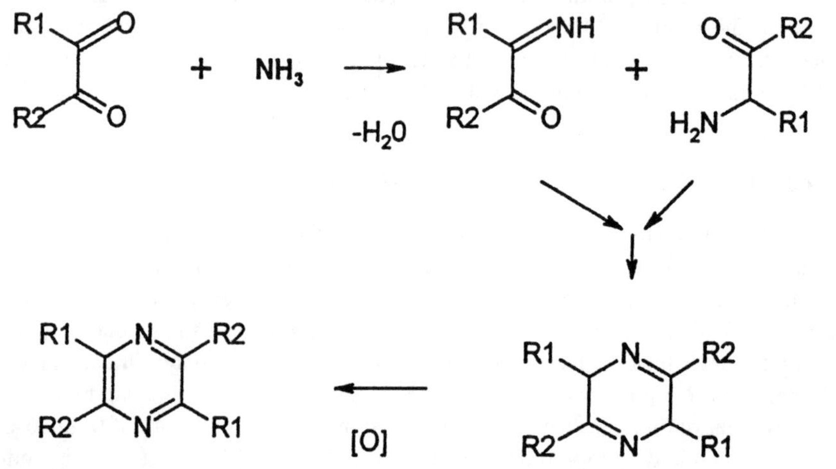

Figure 9–2 Mechanism for pyrazine formation.

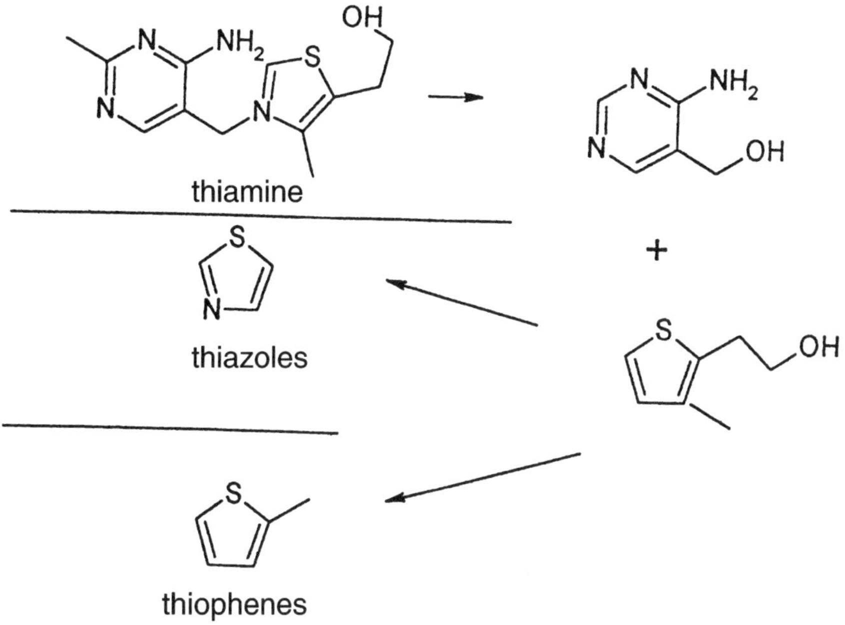

Figure 9–3 Mechanism for thiazole formation.

9.4.4 *Furans and Furanones*

The furans are the most abundant volatile products of the Maillard reaction. They account for the burnt sugar/caramel odor of heated carbohydrates. Many of the sweet type notes of process flavors and cooked foods contain degradation products of sugars. Those of major significance are shown in Table 9–4 and include maltol, isomaltol, 2-hydroxy-5-methyl-3(2H)-furanone (furaneol), 2,5-dimethyl-4-hydroxy-3(2H)-furanone (Maggi lactone) and 2-hydroxy-3-methyl-3-methyl-2-cyclopentene-1-one. The generalized pathway of formation is shown in Figure 9–4.

The furanones undergo further reaction in the presence of a sulfur donor to form thiophenone and thiophene. These components are reported to possess odor character of cooked meat and have been found in many cooked foods.

The furanone components are considered major contributors to the "hydrolysate or Maggi" note of many hydrolyzed vegetable proteins, perhaps one of the reasons that HVP is a popular ingredient in meat flavors.

Table 9–3 Thiazoles and Thiazolines

2-Methylthiazole
4-Methylthiazole
2-Propylthiazole
2-Methyl-4-ethylthiazole
Trimethylthiazole
2-Acetylthiazole
2-Methyl-4-propyl-5-ethylthiazole
2-Isopropyl-4,5-dimethylthiazole
2-Propyl-4-ethyl-5-methylthiazole
2-Butyl-4,5-dimethylthiazole
2-Butyl-4-methyl-5-ethylthiazole
2-Pentyl-5-methylthiazole
Benzothiazole
2-Acetyl-2-thiazole

9.4.5 *Pyrroles*

Pyrroles are heterocyclic compounds that have not received too much attention as flavor components in cooked foods. They have been created from Maillard reaction by the pathway shown in Figure 9–5. A model system consisting of rhamnose and ammonia was shown to produce eight pyrroles (Shibamoto & Bernhard, 1978).

The pyrroles have been characterized as having sweet and corn-like aroma. Some, like 2-acetylpyrrole, have a caramel-like aroma similar to that of the furanones. A number of pyrroles are approved as safe for use as flavor components (Table 9–5).

Table 9–4 Furans

Furfuryl alcohol
2-Ethyl furan
5-Methyl-2-furfural
Furfural
2-Acetyl-5-methyl furan
5-Methyl-2-propionyl furan
Furfuryl ethylethane
2-Propronyl furan
4-(2-furyl)-3-buten-2-one
2-methyl-3-(2-furyl)-2-propanal
2,2′-Difurfuryl ether
2-Furfuryl methyl ether
2-*n*-Butyl furan
3-Phenyl furan

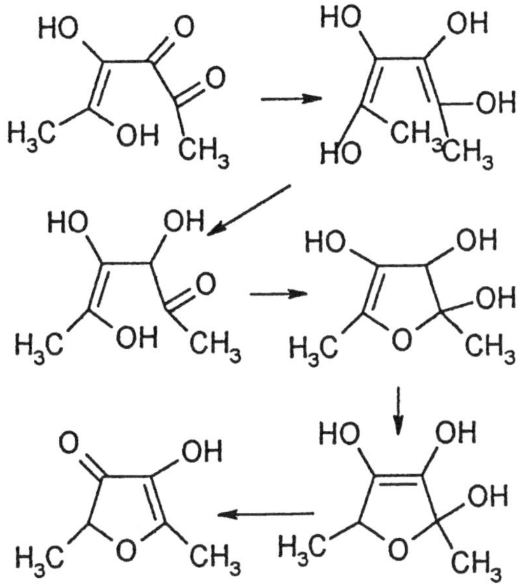

Figure 9–4 Mechanism for furanone formation.

9.4.6 *Pyridines*

Many akylpyridines have been isolated in a number of cooked foods including roasted lamb fat (Buttery et al., 1997). These components appear to give the species character aroma due to the fat interactions on thermal treatment. Ammonia, coming from glycine, can react with nonanal arising from beef fat, producing 2-butylpyridine in a beef fat/glycine model thermal reaction (Ohnishi & Shibamoto, 1984).

Pyridines have a unique aroma that can be considered both a flavor contributor and an off flavor. Concentration plays an important role in the perception of the aroma character of these materials.

9.4.7 *Amino Acids*

The role that small water soluble precursors play in the development of meat flavor was established very early by such workers as Hornstein and Crowe (1960), Batzer et al. (1962), Wasserman and Gray (1962), Pipen et al. (1969) and many others. It was concluded that the water soluble relatively low molecular weight materials such as amino acids, peptides and carbohydrates played an important role as precursors of meat flavor. It was particularly recognized that the amino acid cysteine was one of the most important precursors in beef aroma.

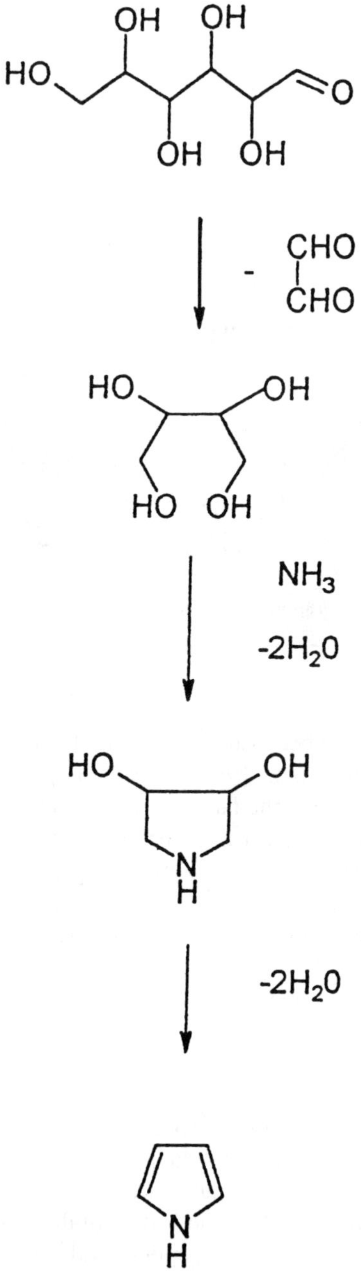

Figure 9–5 Mechanism for pyrrole formation.

Table 9–5 Commonly Found Pyrroles

2-Butylpyrrole
2-Acetylpyrrole
3-Methyl-4-ethylpyrrole
n-Methyl-2-formylpyrrole
Indole
n-Methylpyrrole

May (1974) and Heath (1970) confirmed that belief and showed that cysteine contributed sulfur to reactions as an important route to the creation of the many sulfur containing aromatic chemicals listed in other parts of this chapter. May (1974) showed that cysteine and ribose (also a small water soluble chemical found in meat) completely disappeared on cooking. The composition of amino acids and reducing sugars in beef, pork, lamb or vegetable hydrolysates were found to be very similar. As we noted, one of the first patents to be granted on process flavors to Unilever was in 1960 and it reported the production of a meat-like flavor produced by the heating of an aqueous solution containing free amino acids, a sulfur compound and a monosaccharide. Many other patents granted since then have been based on the optimization of the various reactants such as the nitrogen source (Morton et al., 1960; Scheide, 1973; Chhuy & Day, 1974; van Delft & Giancino, 1978; Maizena, 1978; Huth & Schum, 1983; Tandy, 1985; Parker & Pawlott, 1986; Lieke et al., 1988), the sulfur donor (Giacino, 1965; van Pottelsberghe, 1971; Heyland, 1975; Mosher, 1978) and the amount of fat (Kyowa Hakko Kogyo, 1967; Soeters, 1970; Rothe et al., 1976; van den Ouweland & Swaine, 1980; Schrodter et al., 1980, 1986; Lee et al., 1984; Solms, 1969; Nishimura & Kato, 1988; Schallenberger et al., 1969; Tallent, 1979; Schliemann et al., 1987).

One of the reasons that protein hydrolysates became very useful as meat flavor substitute flavorings is the fact they have amino acid profiles that are similar to meat and they are reasonably cheap compared to meat extracts. During the manufacture of protein hydrolysates, strong mineral acid (usually hydrochloric) is used to catalyze the lysis of the protein to free amino acid or small peptides. There are off flavor notes formed during the reaction and they are usually removed by an activated charcoal treatment. The amino acid profiles of various protein hydrolysates from different protein sources are shown in Table 9–6. Other sources of amino acids can be autolyzed yeast extracts, meat extracts and pure amino acids extracted or sourced from fermentation or synthesis.

The simple compounding and reaction of amino acids, however, does not assure one that a meat flavor can be reproduced. This emphasizes the complexity of this area of flavor chemistry. Although science has shown us the bases of meat flavors, the subtleties are still an art relying on the flavor chemist to build the profile needed from his or her knowledge of both the art and the science.

Table 9–6 Amino Acid Profiles of Various Protein Hydrolysates

Amino acid	*Lean beef*	*Wheat gluten*	*Soya*	*Yeast (autolysate)*
Aspartic acid	7.3	3.8	11.7	10.1
Threonine	4.4	2.6	3.6	0.5
Serine	3.7	5.2	4.9	4.5
Proline	4.3	10.0	5.1	5.1
Glutamic acid	15.9	38.2	18.5	11.6
Glycine	4.0	3.7	4.0	5.4
Alanine	5.6	2.9	4.1	7.3
Valine	4.3	2.9	5.2	6.0
Isoleucine	4.3	1.6	4.6	4.6
Leucine	7.3	2.8	7.7	6.9
Tyrosine	2.5	0.2	3.4	2.9
Phenylalanine	3.0	3.3	5.0	3.9
Lysine	7.5	1.9	5.8	7.1
Histidine	2.4	2.0	2.4	2.2
Arginine	6.0	3.4	7.2	7.1
Methionine	2.1	1.1	1.2	1.6

Another aspect of flavor is contributed by certain amino acids. For nearly a century the taste enhancing effects of these amino acids, known as umami, have been known, studied and used commercially (see Chapter 10). The properties of the monosodium salt of L-glutamic acid found in high amounts in most proteins, but particularly high in vegetable proteins, was found to have the high value as a flavor enhancer. Since the early 1950s this salt of an amino acid has been extensively used by the food industry around the world. Monosodium glutamate (or MSG as it is commonly known) because of its extensive use, some may believe overuse, has become the focus of public and regulatory concern and discussions. The U.S. Food and Drug Administration has made several reviews regarding the safety of the use of MSG and has yet to take any action on its use. The FDA's policy is that it is safe for its intended use, however, because of public concern, MSG must appear on all food labels where it is used and the flavor labeling exemption that the FDA has for flavors is not allowed for flavor enhancers and specifically MSG. Under both U.S. Department of Agriculture and FDA label requirements MSG must appear on the product's ingredient declaration. This extends, as we stated, to the bulk label on a compounded flavor. This is true of regulations throughout Europe.

The flavor enhancing effect known as umami has been found to play an important role in the overall development of a high quality savory flavor. There is a synergistic effect between the base (soup, gravy or sauce) and the added MSG to enhance the overall flavor and taste impression of the final product. However, MSG alone does not contribute to making a meat flavor.

9.4.8 *Nucleotides*

The nucleotides are another material found commonly in foodstuffs. As with MSG these materials have been found to have an umami effect. They are not only more powerful than MSG; they also have a synergistic effect with MSG, so that smaller amounts of both of the materials can be used to establish the same enhancing effects. The two major nucleotides are disodium-5′-inosinate (IMP, as inosine-5′-monophosphoric acid) and disodium-5′-guanylate (GMP, as guanosine-5′-monophosphoric acid). They have been commercially available since the late 1950s as a 50/50 mixture prepared by fermentation process.

Not only are the nucleotides good flavor enhancers, they are also involved in the actual formation of meat flavor. Van den Ouweland and Peer (1975), another Unilever team, made significant additions to the science of meat chemistry when they demonstrated that a number of meat-like mercapto-substituted furan and thiophene derivatives were developed when hydrogen sulfide was reacted with 4-hydroxy-5-methyl-3(2H)-furanone. They proposed the reaction mechanism where the dihydrofuranone was derived from ribose-5-phosphates for the ribonucleotides. There is evidence that the furanone could also come from the heat decomposition of a sugar (Hicks et al., 1974).

9.4.9 *Aldehydes and Ketones*

Although aldehydes and ketones are found in all roasted products, they do not play a major role in meat flavors but do make a significant contribution to other roasted notes like coffee or chocolate (Holscher & Steinhart; Eichner et al.). The Strecker degradation of sugars with amino acids gives rise to α-aminoketones. Strecker aldehydes of various chemical structures are formed from transamination and decarboxylation reactions to such compounds as 2-methylpropanal, 2-phenylacetaldehyde, 2- and 3-methylbutanal depending on the precursor amino acid. Table 9–7 shows some of the aldehydes and ketones found in meat volatiles.

To understand the usefulness of the components we have discussed we need to look at the research literature on cooked foods themselves.

9.5 AROMA COMPONENTS FOUND IN COOKED FOODS

To create flavors it is useful to know the significance of each volatile component in the organization impression of the food. This section will identify the components of various cooked foods and their role in flavor. We cannot discuss all the aroma components identified in the various thermally prepared foods. Although the components discussed are reported as significant, reacting a mixture of only those components will not create a complete flavor. The art of flavor creation rules again.

Table 9–7 Aldehydes and Ketones

3-Hexanone
4-Methyl-2-pentanone
Hexanal
2-Hexenal
3-Methyl-4-heptanone
2-Heptanone
3-Heptenal
(E)-2-Octenal
2-Nonenal
4-Ethyl benzaldehyde
Decanal
2,4-Nonadienal
2-Undecanone
Undecanal
2,4-Decadienal
5-Tridecanone
2-Dodecenal
Dodecenal
2,4-Undecadienal
Tetradecanal
2,4-Decadienal
Benzaldehyde
Nonanal
Octanal
Heptanal

9.5.1 *Beef Flavor*

A number of major contributors of beef flavor have been reported by Lawrie (1982):

- Pyrrolo-[1,2-a]-pyrazine
- 4-Acetyl-2-methyl pyrimidine
- 4-Hydroxy-5-methyl-3-(2H)-furanone
- 2-Alkylthiophene
- 3,5-Dimethyl-1,2,4-trithiolane

Other volatile compounds of interest that have been found in meat are listed in Table 9–8 (May and Ashurst, 1991). Many of those chemicals are on the flavor chemist's palette to use (i.e., they have some regulatory approval, FDA/FEMA GRAS in the U.S.A. or the Council of Europe List). However, some of these ma-

Table 9–8 Distribution of Various Chemical Classes Found in the Volatiles of Meat

Class of Compound	*Beef*	*Chicken*	*Lamb*	*Pork*	*Bacon*
Alcohols and phenols	64	32	14	33	10
Aldehydes	66	73	41	35	29
Carbonylic acids	20	9	46	5	20
Esters	33	7	5	20	9
Ethers	11	4	—	6	—
Furans	40	13	6	29	5
Hydrocarbons	123	71	26	45	4
Ketones	59	31	23	38	12
Lactones	33	2	14	2	—
Misc. nitrogen compounds	6	5	2	6	2
Misc. sulfur compounds	90	25	10	20	30
Oxazole and oxazolines	10	4	—	4	—
Pyrazines	48	21	15	36	—
Pyridines	10	10	16	5	—
Thiazoles and thiazolines	17	18	5	5	—
Thiophens	37	8	2	11	3

terials are not available to make a flavor, but the flavor chemist has found ways to substitute other chemicals which give similar notes, or these components are generated in the process flavor.

9.5.2 *Chicken Flavor*

The aroma of cooked chicken is certainly very different from that of beef. As we mentioned, the fat portion of the meat being cooked will play the deciding factor as to the final profile. That is so with chicken where the decomposition products of the chicken fat will make a significant contribution. The character of cooked chicken is largely due to *cis*-4-decenal, *trans*-2-*cis*-5-undecadienal and *trans*-2-*cis*-4-*trans*-5-tridecatrienal (Shahidi et al., 1986). These chemicals are all used in flavor formulation. Although the unsaturated aldehydes contribute a significant effect, the other components are of flavor value. Many of these chemicals have been available for some time, but the use of a process flavor still can contribute desirable notes.

9.5.3 *Pork Flavor*

The aroma of pork is very different from beef or chicken. The major character is that of sulfury notes. The breakdown of thiamine has been shown to create a significant portion of the aroma (Giancino, 1968, 1970). Compounds such as

3-mercaptopropanol, 3-acetyl-3-mercaptopropanol and 4-methyl-5-vinylthiazole are characteristic of pork flavor.

9.5.4 *Bacon Flavor*

Most of the early work in this area involved analysis of related product types with the emphasis being on hams, sausages and smoke type flavors. Recently work has been done to analyze volatile fractions of fried bacon. These studies have yielded the identification of more than 135 compounds.

Overall the hydrocarbons, alcohols and carbonyl compounds provided the largest groups of compounds found, although many of these compounds have no bacon or meaty flavor or aroma. Compounds found such as 2-hydroxy-3-methyl-2-cyclopenten-1-one (known to flavorists as maple lactone) and acetyl propionyl (2,3-Pentane diione) provide more of the sweet brown and buttery character of a bacon note (Lustre & Issenberg, 1970). Phenols, such as phenol, guaiacol and 4-methyl guaiacol, typically found in wood smoke have been found in the volatiles of cured bacon.

Twenty-two pyrazines were found in fried bacon, including 2,6-dimethyl pyrazine, trimethyl pyrazine and 5,6,7,8-tetrahydroquinoxaline; twelve furans, including 2-pentylfuran; and three thiazoles, two oxazoles and six pyrroles including 2-acetyl pyrrole. These compounds have been identified in other meat volatiles and are typical of pork products.

9.5.5 *Roasted Nuts and Seeds*

Roasted nuts and seeds derive their flavor character from the reaction of the precursors and the high temperature treatment given to the products.

Over 221 volatiles have been identified in extracts obtained by steam distillation of roasted sesame oil. The overall flavor character of the roasted seeds, which is often described as roasty and nutty, is strongly dependent on the roasting conditions. Flavor studies have shown that components such as 2-furfurylthiol, 2-methoxyphenol, 2-phenylethylthiol and 4-hydroxy-2,5-dimethyl-3(2H)-furanone are the most important odor components (Ho et al., 1983; Nakamura et al., 1989; Shu et al., 1985).

Roasted peanuts have been well studied with over 279 volatile compounds identified. The major groups of compounds found have been pyrazines, sulfides, furans, oxazoles, aliphatic hydrocarbons, pyrroles and pyridines.

9.5.6 *Coffee Flavor*

One of the most complex thermally generated flavors is that generated from roasted coffee beans. The creation of the aroma profile is due to the fermentation process given the beans and the final roasting conditions. Over 1,000 compounds have been identified as volatile flavor components from the fermentation/roasting (Maarse & Visscher, 1989; Flament, 1989). Research indicates that 60 to 80 of the compounds contribute to the character of roasted coffee. Significant amounts of pyrazines, aldehydes, furans, acids and thiol compounds are found in the volatiles. Of the more than 80 pyrazines formed during roasting, 2-methoxy-3-isopropyl pyrazine and 2-methoxy-3-isobutyl pyrazine, which possess strong green type vegetable-like odors, are present in green unroasted coffee and contribute to the final roasted coffee aroma impression (Vitzthum et al., 1976). These and other pyrazines which are formed by the Maillard reaction between amino acids and reducing sugars result in the production of 14% of the overall volatile content of roasted coffee flavor (Silwar, 1982).

Methional is essential for roasted coffee aroma. Methional has a cooked potato like aroma, but it can undergo degradation to form the more volatile methanethiol (Gasser & Grosch, 1988). Low concentrations of this compound produce the pleasant aroma arising from freshly roasted or ground coffee. The "skunky notes" derived from the sulfur containing compounds are very reminiscent of fresh roasted coffee.

Three furanones are also considered to contribute a significant amount of aroma character to roasted coffee. They are 4,5-dimethyl-3-hydroxy-2(5H)-furanone, 5-ethyl-4-methyl-3-hydroxy-2(5H)-furanone and 2,5dimethyl-4-hydroxy-3(2H)-furanone (Blank et al., 1992). These compounds have caramel-like aroma and are common to many kinds of thermally treated foods (Baltes, 1979).

9.5.7 *Cocoa/Chocolate Flavor*

Cocoa aroma is the result of fermentation to liberate amino acids and sugars and the subsequent roasting process. We find basically the same types of compounds in the volatiles of cocoa that are found in coffee aroma. Fermentation and roasting conditions (time and temperature) play an important part in the character of the aroma. Pyrazines, again, contribute a major character note. Roasting in the temperature range of 120 to 135° C for 15 minutes produces the highest amount of pyrazines (Ziegleder, 1982). Cocoa and chocolate, like coffee, are very complex aromas with many components contributing to the overall aroma and quality of the flavor.

One of the interesting aspects of the cocoa bean is that neither cystine or cysteine has been reported to occur, and the only other amino acid containing sulfur, methionine, is present at lower concentrations than the other amino acids occurring in the fermented bean. Cocoa would have a very different aroma profile if more sulfur containing amino acids were present.

A cocoa substitute known as St. John's Bread has been produced commercially from roasted carob (locust bean) pods. It is a byproduct of the production of locust bean gum. The deseeded pods are removed, dried (roasted) and broken or ground as a cocoa type material. The aroma profile can be similar to a lower grade of cocoa powder.

9.5.8 *Hydrolysate Flavor*

We have noted that HVPs have been in use commercially as a meat flavor substitute and as bases for process flavors. The hydrolysis process is shown in Figure 9–6. The amino acids liberated in the hydrolysis and the subsequent thermal processing give rise to volatile compounds found in other cooked or roasted products. Summaries of some of the more important volatiles are found in Table 9–9 (Manley et al., 1981).

9.5.9 *Caramel, Molasses and Maple Flavor*

All these flavors have a common flavor profile developed by the thermal degradation and rearrangement of the carbohydrates (the thermal degradation of sugar). Once again we find that furanone compounds such as 2,5-dimethyl-4-hydroxy-3(2H)-furanone and 2-hydroxy-3-methyl-2-cyclopenten-1-one are the major aroma components (Healey & Carnevale, 1984; Sugisawa, 1967; Underwood et al., 1961).

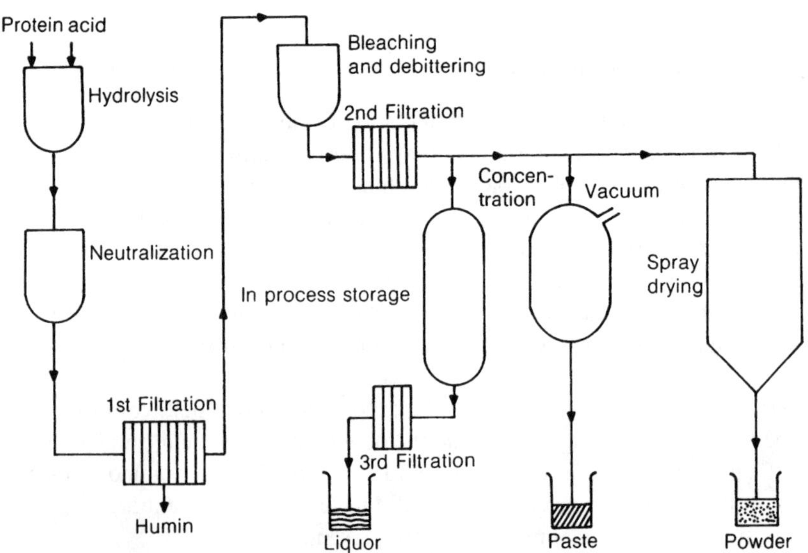

Figure 9–6 Protein hydrolysis process.

Table 9–9 Volatile Components Found in Hydrolysates

Pyrazines	Sulfur Containing
Pyrazine	Methyl mercaptan
Methyl-	Dimethyl sulfide
2,5-Dimethyl-	Ethyl methyl sulfide
Ethyl-	Diethyl sulfide
2,6-Dimethyl	2-Methyl thiophene
2,3-Dimethyl	Dimethyl trisulfide
Isopropyl-	Furfuryl methyl sulfide
2-Ethyl-6-methyl-	Benzyl methyl disulfide
2-Ethyl-5-methyl-	3-Methyl thio-1-propanol
Trimethyl-	
2-Ethyl-3-methyl-	**Furanones**
2-Vinyl-	3-Hydroxy-4-methyl-5-2(5H)-furanone
2-Vinyl-6-methyl-	
2-Ethyl-3,6-dimethyl-	
2-Ethyl-3,5-dimethyl-	
2-Ethyl-2,5-dimethyl-	**Phenols**
2-Isobutyl-6-methyl-	Guaiacol
2-Isobutyl-5-methyl-	4-Ethyl guaiacol
2(2′-Furyl)-5(6)-methyl-	*p*-Cresol
Tetratmethyl-	*m*-Cresol
6,7-Dihydro-5H-cyclopenta-	
Acids	**Aldehydes**
Lactic	5-Methyl furfural
Succinic	Benzaldehyde
Acetic	
Formic	
Levulinic	
Pyroglutamic	

9.5.10 *Bread Flavor*

A discussion of thermally generated flavor would not be complete without a few comments on the aroma generated on the baking of bread. Nearly 300 volatile compounds have been identified in baked bread aroma. The largest number of compounds are the bases [CFR Title 21, 102.22(a)] followed by aldehydes and ketones (Mannie, 1997). The most potent of the cracker-like odor components was reported to be 2-acetyl-1-pyrroline (Schieberle & Grosch, 1989).

9.6 COMPONENTS USED TO CREATE A PROCESS FLAVOR

The flavorist has a wide range of materials to compound flavor with. The mainstay of his or her creative development focus is on the major "building blocks" which have significant flavor profiles of the cooked product being created. These materials can be of the named food such as meat extracts or cocoa powder, or they could be the reaction mixtures or blend of aromatic chemicals (GRAS flavor materials). The criteria for the flavorist's choices are based on the flavor profile desired by the customer, the application that the flavor will be used in (a very important part of the flavor development activity) and the cost parameters suggested by the customer. Our brief discussion of the materials used will give some insight into the creative task of the flavorist.

9.6.1 *Meat Extract*

As one might have surmised, meat extract is a major ingredient in many process flavors. The precursor amino acids that are responsible for the development of meat flavor are contained in meat extract and therefore it is a fine base material to react with the addition of other amino acids, reducing sugars, hydrolysates, autolysates and fats. Because of the expense of meat extracts and the fact that process flavors are being used in many cases to mimic meat extracts, it would not be economical to use too much meat extract in the creation of a process flavor.

9.6.2 *Hydrolysates*

Hydrolysates can act as a major substrate for the creation of flavor as they contain many of the amino acids in the free form needed to produce meat or roasted aroma when heated. HVP offers the flavor manufacturer an economical source of amino acid for creating process flavors.

Some food companies, however, have requested flavors without HVPs due to certain contaminants formed during the processing (this issue will be discussed under the section on safety). Flavors where HVPs are excluded have been deemed "clean-label" flavors in the United States. Labeling regulations in the U.S.A. allow for the bundling of certain flavor ingredients under the labeling term natural flavor, flavor or artificial flavor or a combination term such as natural flavor *with other natural flavors* (called in the U.S.A. a WONF flavor). Under these labeling terms certain ingredients need not be disclosed. Although the FDA considers HVP a natural flavor, the presence of these materials in a flavor or a food product must be disclosed on the label. The nature of the protein(s) used in the hydrolysis must also be disclosed. Therefore, HVP based on the hydrolysis of corn and soy protein must be declared on the flavor label as containing hydrolyzed corn and soy protein. "Clean-label" flavor would not contain any HVP or other materials, such as MSG, which would have to appear on the flavor label or on the label of the final food product.

9.6.3 *Yeast Products*

Yeast materials such as autolyzed yeast extracts also play an important role in process flavors. Typically bakers yeast grown for flavor properties (primary yeast) or de-bittered spent brewers yeast (secondary yeast) is used for flavor. The yeast by itself can have a nice nutty, savory or bitter character which is useful to the flavorist in compounding a process flavor. If the yeast is ruptured to allow the native endogenous (autolysis) or added enzymes to break down the proteins to peptides and free amino acid and the water soluble components of the yeast are removed, the resulting material, termed autolyzed yeast extract (AYE) or yeast extract, is very useful as a reactant in creating process flavors.

AYE and other yeast materials made by using exogenous enzymes (commercially available enzymes) have been found to have good flavor enhancing properties due to the amino acids and certain ribonucleotides naturally occurring in them. This makes these yeast products great precursors for meat flavor and as materials that can be blended into a flavor for flavor enhancing effects.

9.6.4 *Amino Acids and Peptides*

As we have indicated, many amino acids are used as precursors to developing strong aromas. The major amino acids used are the sulfur-containing ones such as cysteine, its dimer cystine, and methionine. Glutathione (L-glutamyl-L-cysteinyl-glycine) is a great precursor for meat flavor, but it is costly for general use. The patent literature is full of examples of sulfur containing amino acids as precursors of strong meat flavors. The reaction of 4-hydroxy-5-methyl-3-(2H)-furanone with cysteine creates its thio analogue which has a very significant meat flavor profile (van den Ouweland & Peer, 1975).

Various other amino acids are used in process flavors with reducing sugars to create aromatic compounds from the Maillard reaction route discussed before. Meat extracts, HVPs and AYE are excellent sources of free amino acids and peptides for use as precursors for flavor.

9.6.5 *Sugars and Other Carbohydrates*

A number of reducing sugars are typically used in reaction mixtures. Sugars such as ribose, xylose, arabinose, glucose, fructose, lactose and sucrose are utilized in many process flavors. The sugar is needed in the Maillard reaction as part of the Strecker degradation reaction that was noted earlier. There are a number of carbohydrates that are economical sources of certain sugars and therefore useful in flavors. These materials include dextrins, gum arabic, pectins and alginates.

9.6.6 *Aromatic Compounds*

There are nearly 2,000 permitted aromatic flavor chemicals, many of which are useful to the flavorist as he or she compounds the final flavor. We have discussed the generation of meat process flavors and described the aromatic chemicals formed during the thermal treatment. These same components have been synthesized and approved as safe for use as flavoring components. As many as 30 to 50 may be used in a flavor compound. Table 9–10 reports the major aromatic chemicals that a flavorist would have on the shelf to create the final flavor.

With certain exceptions (see the International Organisation of the Flavour Industry's guidelines for process flavors in Appendix 9-A) synthetic aromatic

Table 9–10 Aromatic Chemicals Used by Flavorists To Top-Note Process Flavors

Volatile Chemicals	*FEMA #*	*CE*	*CAS#*
Pyrazines			
2,3 Dimethyl Pyrazine	3271		5910-89-4
2,3,5 Trimethyl Pyrazine	3244	735	14667-55-1
2 Acetyl Pyrizine	3126	2286	22047-25-2
2-Mercapto Methyl Pyrizine	3299		59021-02-2
Thiazoles			
Thiazole	3615		288-47-1
4-Methyl-5-Thiazoleethanol (Sulfurol)	3204		137-00-8
2-Acetyl Thiazole	3328	4041	24295-03-2
2-Isobutyl Thiazole	3134		18640-74-9
Aldehydes			
T-2 Decenal	2366	2009	3913-71-1
3-Methylthio Propionaldehyde (Methional)	2747	125	3268-49-3
Isovaleraldehyde (3-Methylbutyraldehyde)	2692	94	590-86-3
Isobutyraldehyde	2220	92	78-84-2
2,4 Decadienal	3135	2120	25152-84-5
2-Furfuraldehyde (Furfural)	2489	2014	98-01-1
Ketones			
Diacetyl	2370	752	431-03-8
Acetyl Methyl Carbinol (Acetoin)	2008	749	513-86-0
Trithioacetone	3475	2334	828-26-2
P-Mentha 8 Thiol-3-One	3177	38	462-22-5

continues

Table 9–10 continued

Volatile Chemicals	*FEMA #*	*CE*	*CAS#*
Alcohols			
1-Octene 3-OL	2805	72	3391-86-4
Furanones			
4-Hydroxy-2,5-Dimethyl-3(2H)-Furanone	3174	536	3658-77-3
5-Ethyl-3-Hydroxy-4-Methyl-2(5H)-Furanone	3153		698-10-2
Pyridine			
Pyridine	2966	604	110-86-1
2-Acetyl Pyridine	3251	2315	1122-62-9
Lactones			
Gamma Decalactone	2360	2230	706-14-9
Gamma Dodecalactone	2400	2240	2305-05-7
Gamma Octalactone	2796	2274	698-76-0
Delta Dodecalactone	2401	624	713-95-1
Delta Decalactone	2361	621	705-86-2
Gamma Nona Lactone	3356	2194	3301-94-8
Acids			
Oleic Acid	2815	13	112-80-1
Isovaleric Acid	3102	8	503-74-2
2-Mercaptopropionic Acid (Thiolactic Acid)	3180	1179	79-42-5
4 Methyl Octanoic Acid	3575		54947-74-9
Esters			
Butyl Butyryl Lactate	2190	2107	7492-70-8
Phenols			
Phenol	3223		108-95-2
Guaicol	2532	173	90-05-1
Isoeugenol	2648		97-54-1
Sulfur			
Dimethyl Sulfide	2746	483	75-18-3
Dimethyl Disulfide	3536		624-92-0
2,3-Butanethiol	3477	725	4532-64-3
2-Methyl-3-Furanthiol	3188		28588-74-1
Mercaptans			
Methyl Mercaptan	2716	475	74-93-1
Furfuryl Mercaptan	2493	2202	98-02-2
Benzyl Mercaptan (a-Tolune Thiol)	2147	477	100-53-8
2,5 Dimethyl Furan-3-Thiol	3451		55-764-23-3
2 Methyl Furan-3-Thiol	3188		28588-74-1

chemicals are not allowed to be used in the process flavor prior to thermal treatment because the industry's guideline for preparation of these types of flavors does not allow it. These components are generally very volatile and would be lost during the thermal treatment. They may therefore be added to the process flavor after reaction to establish a strong characterizing top-note.

9.6.7 *Other Materials*

Certain other food materials are allowed to be added to the reaction mixture prior to thermal processing. The list of materials can be found in the IOFI guideline (Appendix 9-A). An important ingredient is thiamine (vitamin B_1). Because of its chemistry it can generate strong meat type aromas as noted in our prior discussions.

9.7 PROCESS TECHNIQUES

Process flavors can be developed by numerous techniques and are subjected to various pH levels, water activity, buffering, temperatures and processing times. Pyrazines, for example, are formed at pH values above 5.0 whereas the brown color formation in the Maillard reaction is found at pH values above 7.0. Furfurals and some sulfur compound production are favored at lower pH conditions.

Commercial methods of processing are very variable allowing the manufacturer the ability to produce aroma profiles and to control the end point of the flavor being developed. A further consideration in the manufacturing of the flavor is, of course, economics.

9.7.1 *Liquid Reactions (Kettle with Water, Oil or Both)*

Liquid reactions are carried out in reaction vessels which can be made of stainless steel or are glass lined. The production of HVP requires glass lined reactors due to the very corrosive nature of the reaction mixture (acid, base and high salt levels). Heating may be done electrically, but more typically it is by high pressure steam. The reaction mixture is agitated by a mixer(s) with or without scraped surface blades. A conventional atmospheric kettle will employ a reflux condenser and some air cleaning devices to return aroma or remove it from the reaction mixture. Aroma production during the reaction can be very intense. Pressure kettles are used to allow for better control of reaction conditions above refluxing temperature. Typical reaction temperatures are well below 150° C, with a typical temperature in the 100–120° C range. Very high temperatures present problems of control of the reaction, economics of processing and high initial capital costs. As we shall discuss later, temperature of greater than 150° C may create constituents of safety concern. In many ways this process resembles the manufacture of a soup product where only key ingredients that produce an intense aroma profile are used.

A major amount of process flavors are prepared in the liquid form and either directly dried by one of the methods to be discussed later or may have other flavor materials added to the mixture prior to dehydration or packaging.

9.7.2 *Roller Dryer (Lower Moisture Reactions)*

The roller or drum dryer allows the components to be reacted and dehydrated in one continuous process. The control of the process is governed by the temperature of the drum, the time spent on the drum and the nature of the components (moisture content for example) in the reaction mixture. Just as with cooking, the raw materials will play a major role in the resulting flavor profile. A major drawback to this process is the significant loss of volatile components that are flashed from the mixture while on the hot surface of the drum.

Many food materials may be used in this process; for example, yeast can be dried to give very nice toasted savory notes useful in many flavor applications.

9.7.3 *Paste Reactions (Higher Temperature and High Solids)*

Another typical method deals with the reaction of high solids mixtures. Both the roller dryer and kettle methods accommodate reaction mixtures which are fairly fluid. The reaction of high solids paste requires the use of special equipment capable of operating at high temperature and moving very viscous materials. The Lodiger mixer, Z-mixer or other types of paste reactors can handle these types of products. The equipment typically has rotating plows and high speed choppers in the side of the mixer. They are capable of operating at very high temperatures if engineered properly. One clear advantage of this process is the ability to work with high levels of fat in the reaction mixture. Although this process may lead to the loss of some volatile components, the use of high fat levels may effect the retention of important ones. It is an excellent method for the development of strong roasted, sautéed and/or fried notes during the reaction.

A major drawback of this process is the initial capital investment for the equipment and supporting heat generation. Due to the corrosive nature (high salts and amino acids) of some of the reaction mixtures, special consideration must be given to the design of the equipment. For example, normal bearings and gaskets cannot tolerate the type of reactants typically used to create a process flavor.

9.7.4 *Extrusion*

The extruder offers yet another way to produce a flavor by a continuous method. The extruder is designed to allow materials to be added separately, mixed and thermally reacted in one continuous process. High solids mixtures can be accommodated with great control of the time/temperature given to the reaction mixture. This process is known to help preserve the flavor components (*Food Formul.,* 1997). Some volatile top-notes are lost, because there is a significant difference in pressure when the reaction mixture reaches the end of the heating barrel of the extruder.

A few flavor companies are using the extrusion chamber as a reaction vessel for the development of heat-stable flavors. Heat, moisture and pressure inside the extruder accelerate flavor development of these "process" or "reaction" flavors. Factors such as time, temperature and pH, as previously noted, influence the outcome of the final flavor. Flavor precursors such as reducing sugars and amino acids can react inside the extruder through Maillard reactions to form finished flavors. Because the material already has undergone extensive heat processing, it is less likely to be affected by subsequent heat treatment and therefore is more heat-stable.

When the extrusion process is used to create a snack product it is not uncommon for flavors added to the mixture to disappear or change their character during the extrusion process. In general, extrusion temperatures range from about 150 to 235° C. Very few traditional flavors are able to survive these temperatures. Certain process flavors have been developed to allow for increased flavor retention in the extruded food product. Flavor perception of extruded food products differs depending on the base composition. Extruded foods are largely starch or protein based materials, with small amounts of additives. As starch and protein are subjected to heat, pressure and shear in the extruder, they undergo various molecular changes. For example, starches undergo hydrogen bond breakage and gelatinization and/or dextrinization. Proteins undergo denaturation, association and coagulation (Mannie, 1997). As a general procedure in formulating food bases for extrusion, the nonvolatile components are added at the beginning. For example, precursor ingredients that react to form a flavor can be mixed into the main base. These ingredients will then help build up and supplement the flavor in the finished product's flavor profile.

9.7.5 *Spray Drying*

Once the reaction is completed, it may be necessary to dehydrate the mixture for commercial sale. The dehydration step is an important one to ensure that the materials are in a useful and stable form. The processed flavor is mixed with a suitable carrier such as starch, modified food starch, gum or combinations of these materials. Although the process flavor is exposed to higher temperature again, the spray drier exposure is usually short and gentle enough not to create any further aroma components. The resulting flavor is free flowing and typically not too hygroscopic. Some process flavors high in HVP or salt tend to be difficult to dry because of the hygroscopicity of those materials.

The initial cost of a dryer is considerable and it requires a major working area in a plant. These negatives are overcome by the efficiency of the operation and control of the dehydration process.

9.7.6 *Tray Drying*

The tray drying method offers a different method of dehydrating the process flavor. Because of the hygroscopic nature of some process flavors, this method can be very useful. The product is placed in trays and the trays are placed on heated shelves (plates) in a vacuum chamber. The pressure in the chamber is reduced and the temperature on the plates increased to drive off the moisture. The resulting "cake" is ground to a specific particle size and packaged. Typically the products are very hygroscopic and need to be handled in an air conditioned room. Manufacturers of HVP have found this process useful in creating HVP with high quality flavor character. The process can drive off some unwanted flavor notes and allow for some further development of positive flavor character (Hoch, 1997).

The process does have the disadvantage of being a batch process, unlike the spray drier method. It also requires further processing (grinding and sizing) of the material prior to packaging.

9.8 FINAL FLAVOR COMPOUNDS

We have explored the major methods of producing process flavors. Let us define the "process flavor" as that mixture that has been exposed to high temperature in order to generate a characteristic flavor profile. In many cases the flavor company does not sell the process flavor directly, but uses it internally as a building block to create the final flavor which is sold commercially. The flavorist will select the various components from the groups that we have discussed. He or she will evaluate various creations with the help of applications specialists and perhaps a staff chef to determine if the final flavor works in the food matrix and process, has the shelf-life desired by the customer and fits the economic considerations of the request. When the work is done the formulation will become the finished commercial flavor.

We offer examples of process flavor formulation that have been reported in the literature.

9.8.1 *Beef Flavor Formulation*

α-Ketobutyrate	0.25 g
Disodium inosinate	1.30 g
Monosodium glutamate	9.60 g
Sodium chloride	10.00 g
HVP	43.30 g

Twenty grams of this formulation is heated at 98° C in 684 ml of water for 15 minutes. The resulting product has roasted beef notes. Suggested use level is 0.4% in a beef soup application.

9.8.2 *Roast Beef Flavor Formulation*

HVP	22.00 g
Sodium guanylate	0.64 g
Malic acid	0.38 g
L-Methionine	0.31 g
Xylose	0.57 g
Water	76.10 g

The mixture is refluxed at 100° C for 2.5 hours to develop a strong roast beef like flavor and aroma. Suggested use level is 0.5 to 2.0% in a finished food product.

9.8.3 *Chicken Flavor Formulation*

Water	60.0 ml
L-Cysteine HCl	13.0 g
Glycine HCl	6.7 g
Dextrose	10.8 g
L-Arabinose	8.0 g
Sodium hydroxide (50%)	10.0 ml

The mixture is heated to 90–95° C for 2 hours and the final mixture is then adjusted to a pH of 6.8 with 20 ml of the sodium hydroxide. The suggested use is 0.2 to 2.0% in finished foods for a roasted chicken note.

9.8.4 *Pork Flavor Formulation*

L-Cysteine HCl	100 g
D-Xylose	100 g
Wheat gluten hydrolysate liquid	1000 g

The mixture is heated at reflux temperature for 60 minutes.

9.8.5 *Bacon Flavor Formulation*

L-Cysteine HCl	5.0 g
Thiamine HCl	5.0 g
Liquid soy hydrolysate	1000.0 g
Glucose or xylose	1.0 g
Smoke flavor	1.0 g
Bacon fat	20.0 g

Cook at reflux temperatures for 90 minutes.

9.8.6 *Lamb Flavor Formulation*

L-Cysteine HCl	110 g
Protein hydrolysate liquid	1500 g
D-Xylose	140 g
Oleic acid	100 g

The mixture is heated under reflux condition for 2 hours.

9.8.7 *Chocolate Flavor Formulation*

L-Leucine	1.0 g
L-Tyrosine	3.0 g
L-Serine	3.0 g
L-Valine	2.0 g
Tannic acid	0.5 g
Fructose	1.0 g
Cocoa extract	5.0 g
Glycerine	10.0 g
Propylene glycol	24.5 g
Water	50.0 g

The above mixture is heated at reflux temperature for 30 minutes.

9.9 APPLICATIONS OF THERMAL FLAVORS

Process flavors are designed to be used in a wide variety of food products; for example, processed meats, soups, sauces, gravies, dressings, baker goods, snack products and entrées. Growth of the food service industry has also increased the need for savory bases on process flavor building blocks. The ease of preparation, quality and stability of the flavor are important features of a successful flavor use in any of these applications. These types of flavors also offer a great opportunity to control the development of flavor during the thermal processing of the food

(Kevin, 1997). Precursor materials in a process flavor will allow for further development of flavor during the heat treatment of the product prior to its distribution. A process flavor can also add grill notes to products that have never been grilled or roasted notes to food that has never been exposed to heat. Process flavors can enhance, reinforce or duplicate the real flavor of food products. Table 9–11 summarizes the use and typical usage level of various process flavors in food application. The type of flavors may be characterized as follows:

Process Flavors (PF)—We define these flavors as the main type of products discussed in this chapter.
Compounded Flavors (CF)—These are process flavors enhanced by the addition of natural or artificial top-notes and other materials such as flavor enhancers (MSG, etc.).
Precursor Systems (PS)—These are mixtures of materials which when subjected to thermal treatment will liberate flavor.
Savory Blends (SB)—These are mixtures of process flavors and spices or herbs.

Table 9–11 Typical Usage Level of Various Types of Process Flavors in Common Applications

	% Flavors used in application			
Applications	*PF*	*CF*	*PS*	*SB*
Soups—dry	X	0.1%–0.5%	X	2.0%–5.0%
Soups—canned	0.5%–2.0%	0.1%–0.5%	0.5%–1.0%	2.0%–5.0%
Sauces—dry	X	0.1%–0.5%	X	1.0%–3.0%
Sauces—prepared	1.0%	0.1%–1.0%	0.2%–0.5%	2.0%–4.0%
Gravies—dry	X	0.1%–1.0%	X	0.5%–1.0%
Gravies—prepared	1.0%	0.1%–1.0%	0.2%–0.5%	0.2%–0.5%
Snack products	X	0.25%	X	4.0%–8.0%
Meat analogues	1.0%–5.0%	0.2%–1.0%	0.5%	0.5%–1.0%
Reformed meat products	1.0%–5.0%	0.1%–1.0%	0.2%–0.5%	0.5%–2.0%
Beverages	X	0.05%–0.25%	X	0.01%–0.05%
Bakery goods	1.0%–2.0%	0.1%–2.0%	0.25%–1.0%	0.05%–0.50%

PF, Process Flavors; **CF,** Compounded Flavors; **PS,** Precursor Systems; **SB,** Savory Blends with Spices and Herbs.

9.9.1 *Soups*

Soups represent a major application for process flavors. The homemaker's or chef's way of creating a fine soup is one of the basic ways for the creation of a process flavor. Placing meat, meat drippings, fat, seasonings, sugar, salt and other food materials into a pot and heating for a period of time is the essence of what we have been discussing in this chapter. The flavors formed from the cooking process are stable as they have been created by a heat process. Process flavors may be used in the creation of a liquid soup product to enhance the savory character of the final product. Combinations of herbs, spices, flavor top-notes and enhancers are typically used by the manufacturer to produce high quality soup products.

Precursor systems may be added prior to heating so that flavor is developed during the heating or cooking. Supplementation with process flavors may also provide the missing flavors due to processing changes or ingredients left out of the commercial recipe. Sautéed, roasted or grilled notes from process flavors will add quality to the taste of the finished product even when the processing or recipe does not allow it.

Dry soup mixes rely solely on process flavors to build the flavor, taste and quality of the final product when reconstituted. These instant soups are not generally produced by dehydrating a real soup. They are fabricated from dry ingredients and therefore are not flavorful on their own without the addition of flavors. Dry soup mixes account for a significant use and volume of the process flavors being produced (Kevin, 1997).

9.9.2 *Sauces and Gravies*

These products are similar in their need for process flavors as the soup product, except that their usage is considerably higher. After all sauces and gravies are concentrated products which should have high flavor profiles and impact. The flavor of most sauces and gravies that have meat flavor character are solely created by the use of process flavors.

9.9.3 *Snack Foods*

Savory snacks (these are salt based or for the most part non-sweet) rely heavily on the use of seasoning and spice blend, but many of the unique snacks have process flavors incorporated into them. A wide variety of natural and artificial flavors are used. Bacon, beef, chicken and cheese flavors with roasted, grilled and barbecue notes are some of the more popular flavors used in snack products (Church, 1995). Snacks consisting of potato chips, reformed potato snacks, nuts, and extruded and fried snacks usually are flavored by adding flavors, seasonings and enhancers topically. Extruded or expanded snacks may have a process flavor or precursor system added to the feed prior to processing or frying. Details of the process used to make these types of products will be covered in another chapter.

9.9.4 *Other Foods*

A great variety of other foods, such as meat analogues, reformed meat products, beverages and bakery goods benefit from the use of process flavors. Again the character and quality of a meat, savory or brown flavor may be enhanced by these types of flavors (*Food Process.,* 1997).

9.10 REGULATORY ISSUES

9.10.1 *Process Flavors*

The complexity of process flavors has caused confusion in understanding how they are regulated. In the United States these materials come under the Generally Accepted as Safe (GRAS) concept. GRAS substances include any substance which is generally recognized by experts, qualified by scientific training and experience, to evaluate the substance's safety under its conditions of intended use. In 1971 the Code of Federal Regulations (CFR), the legal basis for commercially using substances in food, further defined the GRAS concept as:

> General recognition of safety based on history of common use in food does not require the same quality and quantity of scientific evidence required of a food additive, but shall ordinarily be based on generally available data and information. General recognition of safety through scientific procedures must ordinarily be based on published literature, and require the same quality and quantity of scientific evidence that would be required for approval of a food additive regulation.(CFR, Title 21, 121.3)

The FDA has indicated that the GRAS determination of process flavors lies with the manufacturer. This type of determination is considered as an independent GRAS determination made by individuals outside the FDA. Such a determination is subject to the risk that the FDA may object to it and may challenge its use in food on the grounds that the ingredient is an unapproved food additive. The FDA has noted that process flavors may be considered GRAS for the following reasons (Lin, 1994; Easterday & Manley, 1992):

> The manufacturing of process flavors mimics high-temperature cooking such as barbecuing. The major difference between the flavors in cooked meat and processed flavors are mixtures of selected food ingredients rather than a raw agricultural commodity. Also most process fla-

> vors are produced at temperatures below 150° C, a much lower temperature then barbecuing. When preparing gravy, a chef mixes selected ingredients and cooks at a certain temperature for a specific time. Such a gravy is similar to the way process flavors are created. The use level of process flavor is low, as is true of other flavors.

Although there is no legal FDA definition of a process flavor, the Flavor and Extract Manufacturer's Association worked with the world flavor trade organization, the International Organisation of the Flavour Industry (IOFI), to establish a working definition of a process flavor. The definition from IOFI's code of practice is given in Appendix 9-A.

The United States Department of Agriculture (USDA) has also established guidelines for preparing a composition that may be called a process flavor or reaction flavor (Edwards, 1990). They have indicated that with the following exceptions, and under the conditions described below, ingredients consumed in the reaction may be listed collectively as reaction flavors (because there is definition of "reaction flavors" [process flavor] under the FDA codes, it is assumed that the title of the flavor should be natural flavor, natural flavor with other natural flavors or artificial flavor, whichever applies).

The exceptions are (which must be listed in the flavor ingredient disclosure statement in descending order of predominance):

- All ingredients of animal origin (identified by species and tissue if appropriate), e.g., beef fat, chicken meat extract, gelatin.
- All non-animal proteinaceous substances, e.g., monosodium glutamate, hydrolyzed vegetable protein, autolyzed yeast extracts or yeast.
- Thiamine hydrochloride, salt, and complex carbohydrates.
- Any other ingredient that is not consumed in the reaction.

The conditions of the reaction:

- Reaction contains amino acid(s), reducing sugar(s), and protein substrates.
- Treated with heat of 100° C or greater for a minimum of 15 minutes.

The FDA citing public health (allergenic reactions to certain proteins), cultural and religious rationale in a 1990 final rule directed that flavors must declare proteinaceous ingredients as well as fermented, hydrolyzed, autolyzed and enzyme modified ingredients (*Federal Register,* 1990). The January 6, 1993 FDA regulations amended the labeling requirements for protein hydrolysates and requires the identification of the source from which the protein was derived (CFR Title 21, 102.22(a)).

In Europe the details found in the IOFI guideline are becoming the legal definition for the European Union (EU). The EU has established process flavors as a category of flavors, which must be labeled as such on food product.

9.10.2 *Hydrolyzed Proteins*

The definition of hydrolyzed protein use as a flavor material is found in the *Food Chemical Codex* (1996). The specification for HVP is:

> Acid hydrolysates of proteins are composed primarily of amino acids, small peptides (peptide chains of 5 or fewer amino acids), and salt resulting from the essentially complete hydrolysis of peptide bonds in edible proteinaceous materials catalyzed by heat and/or food-grade acids. Cleavage of peptide bonds typically ranges from a low 85% to essentially 100%. In processing, the protein hydrolysates may be treated with safe and suitable alkaline materials. The edible proteinaceous materials used as raw materials are derived from corn, soy, wheat, yeast, peanut, rice, or other safe and suitable vegetable or plant sources, or from milk. Individual products may be in liquid, paste, powder, or granular form.
> *Functional use in food:* Flavoring agent; flavor enhancer; adjuvant.
> *Requirements:* Calculate all analyses on the dry basis. In a suitable tared container, evaporate liquid and paste samples to dryness on a steam bath, then, as for powdered and granular form, dry to constant weight at 105° C.
> *Assay (total nitrogen; TN):* Not less than 4.0% total nitrogen.
> α-*Amino nitrogen (AN):* Not less than 3.0%.
> α-*Amino nitrogen/total nitrogen (AN/TN) percent ratio:* Not less than 62.0% and not more than 85.0%, when calculated on an ammonia nitrogen-free basis.
> *Ammonia nitrogen (NH_3-N):* Not more than 1.5%.
> *Glutamic acid:* Not more than 20.0% as $C_5H_9NO_4$ and not more than 35.0% of the total amino acids.
> *Heavy metals (as Pb):* Not more than 10 mg/kg.
> *Insoluble matter:* Not more than 0.5%.
> *Lead:* Not more than 5 mg/kg.
> *Potassium:* Not more than 30.0%.
> *Sodium:* Not more than 20.0%.

9.10.3 *Autolyzed Yeast Extract or Yeast Extract*

These materials also are defined in the *Food Chemical Codex.* Their specification is:

> Yeast extract comprises the water-soluble components of the yeast cell, the composition of which is primarily amino acids, peptides, carbohydrates, and salts. Yeast extract is produced through the hydrolysis of peptide bonds by the naturally occurring enzymes present in edible

yeast by the addition of food-grade enzymes. Food-grade salts may be added during processing. Individual products may be in liquid, paste, powder, or granular form.

Functional use in foods: Flavoring agent; flavor enhancer.

Requirements: Calculate all analyses on the dry basis. In a suitable tared container, evaporate liquid and paste samples to dryness on a steam bath, then, as for powdered and granular form, dry to constant weight at 105° C.

Assay (protein): Not less than 42.0% protein.

α-Amino nitrogen/total nitrogen (AN/TN) percent ratio: Not less than 15.0% or more than 55.0%.

Ammonia nitrogen: Not more than 2.0% calculated on a dry, salt-free basis.

Glutamic acid: Not more than 12.0% as $C_5H_9NO_4$ and not more than 28.0% of the total amino acids.

Heavy metals (as Pb): Not more than 10 mg/kg.

Insoluble matter: Not more than 2%.

Lead: Not more than 3 mg/kg.

Mercury: Not more than 3 mg/kg.

Microbial limits:

- *Aerobic Plate Count:* Not more than 50,000 CFU per gram.
- *Coliforms:* Not more than 10 per gram.
- *Yeast and molds:* Not more than 50 CFU per gram.
- *Salmonella:* Negative in 25 g.

Potassium: Not more than 13.0%.

Sodium: Not more than 20.0%.

9.11 THE SAFETY QUESTION

The flavor industry has always had a great record of concern for the safety of flavor materials. Through their trade organization, the Flavor and Extract Manufacturers Association (FEMA), the industry has invested in significant amounts of research to continually monitor the safety of substances used in food flavors.

9.11.1 *Safety of Thermal Process Flavorings*

As we mentioned, process flavorings are *generally accepted as safe* (GRAS) in the United States. Some of the earliest process flavors have been evaluated by the Food and Drug Administration and FEMA. The FDA reviewed short-term and long-term animal feeding studies on beef and chicken flavors (cysteine/HVP/xylose or glucose) and smoked ham flavor (cysteine/HVP/xylose/liquid smoke preparation) and found no toxic effects (SCOGS, 1978).

In the late 1970s a Japanese team discovered several polycyclic heteroaromatic amines (PHAAs) in charred fish and meat (Powrie et al., 1983; Sugimura et al., 1989). These materials were shown to have mutagenic activity by the Ames test. One early study showed the relationship between cooking time/temperature and the development of revertants in the TA1538 *Salmonella* test (Kato & Yamazoe, 1987).

Further studies during the 1980s uncovered the existence of more than 25 PHAAs with various reported degrees of mutagenicity. Jaerstad et al. (1983) have identified the major precursors of the PHAAs as the amino acids creatine and creatinine. Both of these amino acids are found in the muscle proteins of animals. Analytical surveys of various meats have supported the view that meat when cooked at high temperature contains one or more of the PHAAs.

In the late 1980s carcinogenicity studies in mice, rats and primates showed that the PHAAs are potent carcinogens. Because the production of process flavors is similar to the cooking of meat, the flavor industry in the United States has joined with the FDA to study the chemistry of process flavors and review the practice of the manufacture of process flavors. An analytical method for the quantitation of PHAAs has been established to allow for the quantitative review of process flavors. A survey of process conditions indicated that only very small amounts of process flavors are produced at temperatures which would allow for the production of PHAAs. Figure 9–7 shows the major ones found in cooked meat and selected for the analytical study.

IOFI guidelines control the methods of manufacture of process flavors to ensure that toxic, carcinogenic or mutagenic materials are kept to an absolute minimum and to propose analytical criteria for the flavors. The Council of Europe's

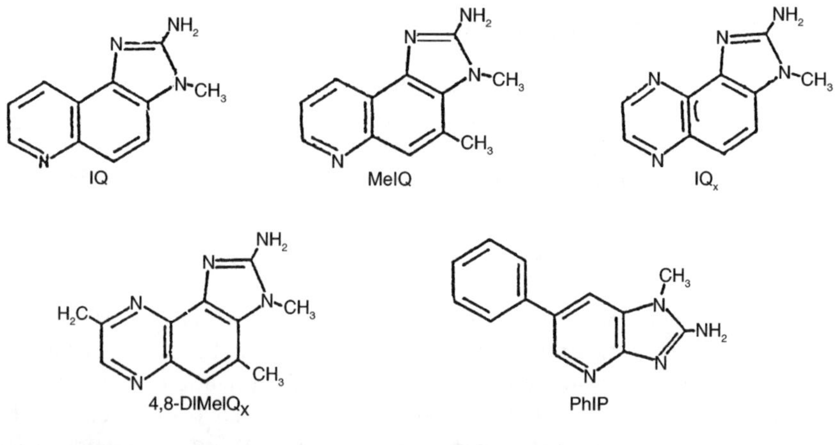

IQ NH_2 N $N-CH_3$

MeIQ NH_2 N $N-CH_3$ CH_3

IQ_x NH_2 N $N-CH_3$

4,8-DIMeIQx H_3C N NH_2 $N-CH_3$ CH_3

PhIP CH_3 N NH_2

Figure 9–7 Heterocyclic aromatic amines found in meat extracts.

expert panel on flavor has indicated that an acceptable level of PHAAs occurring in process flavor would be 50 ppb. These flavors are typically used in some food products at levels of 1% or less. That means that the exposure to consumers is extremely small and compared to exposure from fried, grilled and even cooked meats it is insignificant.

9.11.2 *Hydrolyzed Proteins*

In the late 1970s it was reported that the process manufacturing HVP created some chlorine-containing compounds from the reaction of the hydrochloric acid on the lipids remaining in the proteins used to make the HVP. These materials were reported as monochloropropandiols (MCPs = 2-monochloro-1,3-propandiol and 3-monochloro-1,2-propandiol) and dichloropropanol (DCPs = 2,3-dichloroproan-1-ol and 1,3-dichloropropa-2-ol). The main chlorohydrins found in HVPs that have been of concern to regulatory agencies in the U.S. and Europe have been 1,3-dichloropropan-2-ol (DCP) and 3-monochloro-1,2-propandiol (3MCP). The U.K. Ministry of Agriculture, Fisheries and Food has established a level of 50 ppb and 1 ppm respectively for the maximum allowed levels for commercial products.

In the United States the FDA's Center for Safety and Applied Nutrition (CFSAN) Committee on Cancer Assessment has considered the DCP and 3MCP as genotoxic carcinogens based on reports from several international organizations including the Joint FAO/WHO Expert Committee of Food Additives (JECFA). JECFA concluded that the two chloropropanols are "undesirable contaminants in food" and their levels in HVP "should be reduced to the lowest technologically achievable levels." The United States "Delaney clause" found in the U.S.'s 1958 Food Additives Amendment bans any food additive which is shown to be carcinogenic. These materials are considered contaminants and come under the FDA's constituents policy. Under that policy, if a contaminant or constituent of a food additive (that is not in itself a carcinogen) is carcinogenic, a quantitative risk assessment may be performed to determine whether the contaminant is present in the food at levels of public health concern.

The hydrolysate manufacturers trade organization, International Hydrolyzed Protein Council (IHPC), is working with the *Food Chemical Codex* to establish analytical specifications for DCP and 3MCP (at 1 ppm and 50 ppb respectively) which CFSAN will accept as protective of public health. The industry is taking the initiative to reduce the levels of chloropropanols in all their products to below the specification and, therefore, the FDA may not need to take any regulatory action and the industry can self-regulate at these levels (Dern, 1997).

We have indicated that HVP is used in processed flavors at various levels, but typically at levels well below 100% of the formulation. Therefore, the levels of chloropropanols found in process flavors would be significantly below levels of safety concern. Process flavors are generally not used at the levels of HVP in foods which add a further reduction of the risk from exposure to these compounds.

9.12 CONCLUSIONS

Process flavors are a complex and varied group of flavors. Their creation is very similar to that of flavors generated by some home-cooking operations or commercial food preparation processes. The commercial flavors find use in a wide range of food and beverage applications and represent one of the major creative and application challenges of the flavorist today.

The flavor industry will continue to contribute a great deal of effort to the understanding of how flavor forms when heat treated and to use that knowledge in the creation of new and more useful flavor systems. The industry will also continue to determine the safety of all of the inventory of flavor materials used by the flavorist to create commercial flavors.

REFERENCES

W. Baltes, *Dtsch. Lebensm. Rdsch.* 75 (1979) 2.

O.F. Batzer, A.T. Santoro, W.A. Landmann, *J. Agric. Food Chem.* 10 (1962) 94.

I. Blank, A. Sen, W. Grosch, *14eme Coll. Sci. Int. Café, San Francisco 1991 ASIC, Paris,* 1992, 117.

M. Boelens, P.J. de Valois, H.J. Wobben, A.J. van der Gen, *J. Agric. Food Chem.* 19 (1971) 984.

R.A. Boyer, US Patent 2,682,466 (1954). British Patent 699,692 (1953).

R.G. Buttery, L.C. Ling, R. Teranishi, T.R. Mon, *J. Agric. Food Chem.* 25 (1997) 1227.

L.C. Chhuy, A.E. Day, US Patent 4,081,565 (1974).

D.C.F. Church, in P.R. Ashurst (ed.), *Food Flavorings* (1995) 224.

E.C. Crocker, *Food Res.* 13 (1984) 179.

A. Dern, *Food Chemical News,* July 7, 1997, 16.

B.K. Dvirvedi, R.G. Arnold, L.M. Libbey, *J. Food Chem.* 19 (1971) 1014.

O. Easterday, C.H. Manley, A Report to the 1992 Toxicology Forum; Aspen, CO.

C.R. Edwards, US Department of Agriculture Letter date November 1, 1990.

K. Eichner, K. Schnee, K.M. Heinzler (eds.), *ACS Symp. Ser.* 218.

Federal Register 55-41 (1990) 7289.

I. Flament, *Food Rev. Int.* 5 (1989) 317.

Food Chemical Codex, Fourth Edition, National Academy Press, Washington DC, 1996.

Food Formul. May (1997) 62.

Food Process. November (1997) 75.

U. Gasser, W. Grosch, *Lebensm. Unters. Forsch.* 186 (1988) 489.

C. Giacino, German Patent DE-AS 1,517,052 (1965).

G. Giancino, U.S. Patent 3,394,015 (1968); Patent 3,519,437 (1970).

K. Healey, J. Carnevale, *J. Agric. Food Chem.* 321 (1984) 1363.

H.B. Heath, *Flavour Ind.* 9 (1970) 586.

S. Heyland, German Patent DE-AS 2,546,035 (1975).

K.B. Hicks, D.W. Harris, M.S. Feather, R.N. Loeppky, *J. Agric. Food Chem.* 22 (1974) 724.

C.T. Ho, K.N. Lee, O.Z. Jin, *J. Agric. Food Chem.* 31 (1983) 336.

G.J. Hoch, *Food Process.* April (1997) 76.

J.E. Hodge, *J. Agric. Food Chem.* 1 (1953) 928.

W. Holscher, H. Steinhart, in T.H. Parliament, M.J. Morello, R. McGorrin (eds.), *ACS Symp. Ser. 543* 206.

I. Hornstein, P.E. Crowe, *J. Agric. Food Chem.* 8 (1960) 494.

H. Huth, H. Schum, German Patent 3,206,587 (1983).

International Hydrolyzed Protein Council (IHPC), Washington DC.

M. Jaerstad, A.L. Reutersward, R. Oste, A. Dahlqvist, S. Grivas, K. Olsson, T. Nyhammar, in The Maillard Reaction in Foods and Nutrition (G.R. Waller and M.S. Feather, eds.) *ACS Series* 215, 1983, 507.

R. Kato, Y. Yamazoe, *Jpn. J. Cancer Res.*, 78 (1987) 297.

K. Kevin, *Food Process.* December (1997) 49.

Kyowa Hakko Kogyo Co., Ltd., UK Patent GB-PS 1,115,610 (1967).

R.A. Lawrie, *Food Flav. Ingred. Proc. Pack.* 4 (1982) 11.

E.C. Lee, P.J. van Pottelsberghe, J.S. Tandy, European Patent EP-PS 0,134,428 (1984).

B. Lieke, G. Knrad, O. Deitel, East German Patent DD-WP 272,031 (1988).

L.J. Lin, in Thermally Generated Flavors (T.H. Parliment, M.J. Morello and R.J. McGorrin, eds.) *ACS Symp. Ser. 543*, Washington, DC, 1994.

A.O. Lustre, P.J. Issenberg, *J. Agric. Food Chem.* 18 (1970) 1056.

H. Maarse, C.A. Visscher, in *TNO-Civo Food Analysis Institute,* The Netherlands, 1989, 661.

J.A. Maga, C.E. Sizer, *CRC Crit. Rev. Food Technol.* 4 (1973) 39.

Maizena GmbH, German Patent 2,841,043 (1978).

C.H. Manley, R. Swaine, J. McCann, in *The Quality of Foods and Beverages,* Academic Press, 1981, 61.

E. Mannie, *Prepared Foods* July (1997) 61.

C.G. May, *Food Trade Rev.* 44 (1974) 7.

C.G. May, in PR. Ashurst (ed.), *Food Flavoring,* Blackie, Glasgow, 1995.

C.G. May, in P.R. Ashurst (ed.), *Food Flavoring* (1991) 283.

C.G. May, P. Akroyd, US Patent 2,594,379 (1952); US Patent 2,934,435 (1960).

J.D. Morton, P. Akroyd, C.G. May, US Patent 2,934,437 (1960).

A.J. Mosher, German Patent DE-PS 2,851,908 (1978).

S. Nakamura, O. Nishimura, H. Masuda, S. Mihara, *Agric. Biol. Chem.* 53 (1989) 1891.

T. Nishimura, H. Kato, *Food Res. Int.* 4 (1988) 175.

S. Ohnishi, T.J. Shibamoto, *J. Agric. Food Chem.* 32 (1984) 987.

D.M. Parker, D. Pawlott, European Patent EP-PS 0,223,560 (1986).

E.L. Pipen, E.P. Mecci, M. Nonaka, *J. Food Sci.* 34 (1969) 436.

W.D. Powrie, C.H. Wu, H.F. Stuich, in *Carcinogens and Mutations in the Environment* (H.F. Stich, ed.) CRC Press, Florida (1983) 121.

M. Rothe, M. Specht, E. Bohme, East German Patent DD-WP 143,556 (1976).

R.S. Schallenberger, T.E. Acree, C.Y. Lee, *Nature* 221 (1969) 555.

J. Scheide, German Patent DE-OS 2,335,464 (1973).

P. Schieberle, W. Grosch, in Thermal Generation of Aromas (T.H. Parliment, R.J. McGorrin, and C.H. Ho, eds.), *ACS Symp. Ser.* 409 (1989) 258.

J. Schliemann, G. Wolm, R. Schrodter, H. Ruttloff, *Nahrung* 31 (1987) 47.

R. Schrodter, J. Schliemann, G. Wolm, *Natrung* 30 (1986) 799.

R. Schrodter, G. Wolm, J. Schliemann, *Nahrung* 26 (1980) 625.

Select Committee on GRAS Substance (SCOGS) *Evaluation of the Health Aspects of Protein Hydrolysates as Food Ingredients,* Report, Life Sciences Research Office, Fed. Amer. Soc. For Exptl. Biol., Maryland (1978).

F. Shahidi, L.J. Rubin, L.A. d'Souza, *CRC Crit. Rev. Food Sci. Nutr.* 24 (1986) 141.

T.J. Shibamoto, *Food Sci.* 51 (1986) 1098.

T.J. Shibamoto, G.F. Russell, *J. Agric. Food Chem.* 25 (1977) 109.

T.J. Shibamoto, R.A. Bernhard, *J. Agric. Food Chem.* 26 (1978) 183.

C.K. Shu, B.D. Mookherjee, A.H. Bondarovich, M.L. Hagerdon, *J. Agric. Food Chem.* 33 (1985) 130.

R. Silwar, Ph.D. Thesis, Technical University Berlin, 1982.

C.J. Soeters, US Patent US-PS 3,493,395 (1970).

J. Solms, *J. Agric. Food Chem.* 17 (1969) 686.

T. Sugimura, K. Wakabayashi, M. Nagao, H. Ohgaki, *in Food Toxicology, A Perspective on the Relative Risks* (S.L. Taylor and R.A. Scanlan, eds.) Marcel Dekker, Inc., New York (1989) 31.

H. Sugisawa, *J. Food Sci.* 32 (1967) 381.

W.H. Tallent, *J. Am. Oil Chem. Soc.* 56 (1979) 378.

J.S. Tandy, European Patent EP-PS 0,160,794 (1985).

J.C. Underwood, C.O. Willits, H.G. Lento, *J. Food Sci.* 26 (1961) 288.

United States *Code of Federal Regulation* (CFR), Title 21; 121.3; Title 21; 102.22(a).

A. van Delft, C. Giancino, US Patent 4,076,852 (1978).

G.A.M. van den Ouweland, H.G. Peer, *J. Agric. Food Chem.* 23 (1975) 501.

G.A.M. van den Ouweland, R.L. Swaine, *Perf. Flavor.* 5(6) (1980) 15.

P.J. van Pottelsberghe, German Patent GP-PS 1,318,460 (1971).

O.G. Vitzthum, P. Werkhoff, E. Ablanque, *7 eme Cool. Sci. Int. Café, Hamburg 1975 ASIC,* Paris, 1976, 115.

A.E. Wasserman, N. Gray, *J. Food Sci.* 30 (1962) 801.

Y. Zhang, M. Chien, C.T. Ho, *J. Agric. Food Chem.* 36 (1988) 992.

G. Ziegleder, *Dtsch. Lebensm. Rdsch.* 78 (1982) 77.

Appendix 9–A

International Organisation of the Flavour Industry (IOFI) Guidelines for the Production and Labelling of Process Flavourings

INTRODUCTION

Process flavourings are produced by heating raw materials which are foodstuffs or constituents of foodstuffs in similarity with the cooking of food.

The most practicable way to characterize process flavourings is by their starting materials and processing conditions, since the resulting composition is extremely complex, being analogous to the composition of cooked foods. They are produced every day by the housewife in the kitchen, by the food industry during food processing and by the flavour industry.

The member associations of IOFI have adopted the following guidelines in order to assure the food industry and the ultimate consumer of food of the quality, safety and compliance with legislation of process flavourings.

1. *Scope*

1.1. These Guidelines deal with thermal process flavourings, they do not apply to foods, flavouring extracts, defined flavouring substances or mixtures of flavouring substances and flavour enhancers.

1.2. These Guidelines define those raw materials and process conditions which are similar to the cooking of food and which give process flavourings that are admissible without further evaluation.

2. *Definition*

A thermal process flavouring is a product prepared for its flavouring properties by heating food ingredients and/or ingredients which are permitted for use in foodstuffs or in process flavourings.

3. *Basic Standards of Good Manufacturing Practice*

The chapter 3 of the Code of Practice for the Flavour Industry is also applicable to process flavourings.

4. *Production of Process Flavourings*

Process flavourings shall comply with national legislation and shall also conform to the following:

4.1. Raw materials for process flavourings. Raw materials for process flavourings shall consist of one or more of the following:

4.1.1. A protein nitrogen source:

- protein nitrogen containing foods (meat, poultry, eggs, dairy products, fish, seafood, cereals, vegetable products, fruits, yeasts) and their extracts
- hydrolysis products of the above, autolyzed yeasts, peptides, amino acids and/or their salts

4.1.2. A carbohydrate source:

- foods containing carbohydrates (cereals, vegetable products and fruits) and their extracts
- mono-, di- and polysaccharides (sugars, dextrins, starches and edible gums)

4.1.3. A fat or fatty acid source:

- foods containing fats and oils
- edible fats and oils from animal, marine or vegetable origin
- hydrogenated, transesterified and/or fractionated fats and oils
- hydrolysis products of the above

4.1.4. Materials listed in Table 9–A.1

4.2. Ingredients of process flavourings

4.2.1. Natural flavourings, natural and nature identical flavouring substances and flavour enhancers as defined in the IOFI code of practice for the flavour industry.

4.2.2. Process flavour adjuncts. Suitable carriers, antioxidants, preserving agents, emulsifiers, stabilizers and anticaking agents listed in the lists of flavour adjuncts in Annex II of the IOFI code of practice for the flavour industry.

4.3. Preparation of process flavourings. Process flavourings are prepared by processing together raw materials listed under 4.1. and 4.1.2. with the possible addition of one or more of the materials listed under 4.1.3. and 4.1.4.

4.3.1. The product temperature during processing shall not exceed 180°C.

4.3.2. The processing time shall not exceed ¼ hour at 180°C with correspondingly longer times at lower temperatures.

Table 9–A.1 Materials Used in Procession

Herbs and spices and their extracts
Water
Thiamine and its hydrochloric acid salt
Ascorbic acid
Citric acid
Lactic acid
Fumaric acid
Malic acid
Succinic acid
Tartaric acid
The sodium, potassium, calcium, magnesium and ammonium salts of the above acids
Guanylic acid and inosinic acid, and its sodium, potassium and calcium salts
Inositol
Sodium, potassium and ammonium sulfides, hydrosulfides and polysulfides
Lecithine
Acids, bases and salts as pH regulators:
  Acetic acid, hydrochloric acid, phosphoric acid, sulfuric acid
  Sodium, potassium, calcium and ammonium hydroxide
  The salts of the above acids and bases
Polymethylsiloxane as antifoaming agent (not participating in the process)

4.3.3. The pH during processing shall not exceed 8.
4.3.4. Flavourings, flavouring substances and flavour enhancers (4.2.1) and process flavour adjuncts (4.2.2) shall only be added after processing is completed.
4.4. General requirements for process flavourings.
4.4.1. Process flavourings shall be prepared in accordance with the General Principles of Food Hygiene (CAC/Vol A-Ed. 2 (1985)) recommended by the Codex Alimentarius Commission.
4.4.2. The restrictive list of natural and nature-identical flavouring substances of the IOFI code of practice for the flavour industry applies also to process flavourings.

5. *Labeling*

The labeling of process flavourings shall comply with national legislation.
5.1. Adequate information shall be provided to enable the food manufacturer to observe the legal requirements for his products.
5.2. The name and address of the manufacturer or the distributor of the process flavouring shall be shown on the label.
5.3. Process flavour adjuncts have to be declared only in case they have a technological function in the finished food.

Chapter 10

The Development of Dairy Flavorings

Suzanne White and Geoff White

10.1 INTRODUCTION

The phone rings. It's a customer wanting a sample of "Cheese Flavoring"; but which one, of several hundred? Over the last decade things have progressed, and customers are more aware of the wide range of dairy profiles available. They are also generally aware that end application is important in selecting the right flavoring. The "everything must be natural" stance has eased and is used more selectively. Some customers are happy for us to add MSG; others are not. We have survived the BSE scare well, at least until the emergence of latest CJD concerns about cysteine from human hair! Current specters have to be allergens (especially peanuts) and genetic modification. These issues should be simple to address for dairy flavorings; but don't believe it!

Sadly, it is often still the case that flavor requirements are not tangibly defined, end product and processing are not (fully) revealed, and the legal and marketing implications of the target market are not clear. This makes the job of the flavorist trying to satisfy customers' flavor needs much more difficult. It is very important that the flavorist engages the customer in a dialogue about the flavor needs. Only then can the flavorist use broad experience to help satisfy those needs.

In this chapter, we shall attempt to categorize the basic areas of knowledge that the experienced dairy flavorist draws upon to understand and fulfil the flavor needs of today's demanding food development technologist. It has been written mainly for the flavorist venturing into dairy flavor types for the first time, but should prove of interest to all interested in other flavorists' views. It is not highly technical, but gives the basics, which help to build up a "feel" for dairy flavors.

The breadth of coverage of different types of real cheeses, butters and other dairy products may surprise you, but it shouldn't. Such an understanding is vital to the process of defining the flavor target; the target must be adequately defined if you are to have any chance of reproducing the flavor in a processed food product. Therefore, we make no excuse for introducing you to many of the varied dairy foodstuffs of the world.

All dairy products start out as milk; their flavor components tend to be similar; the secret of their varied and unique characters is in the balance of those components.

10.1.1 *History of Animal Milks as a Human Food Source*

The earliest domestication of animals is believed to have been about 6500 BC, and with this came the widespread consumption of animal milks by humans. It is highly likely that the practice had begun long before this time, with the milk of wild animals hunted for their flesh. The nature of milk itself, with the influence of weather, probably gave rapid rise to a range of dairy products and this formed the basis of the wide range of milk products we have come to know today. We should be grateful that early humans discovered the variety of possible dairy products before they devised refrigeration and prevented many of them from developing!

10.1.2 *The Development of Flavor in Dairy Products*

Although there are significant species variations, animal milks are generally oil-in-water emulsions containing varying quantities of triglyceride fats, with the characteristic milk protein casein (plus other proteins at much lower levels), the milk sugar lactose, and a broad range of vitamins and minerals. In short, they contain just the necessary mix of nutrients to ensure the healthy development of the juvenile of the species until it is able to digest other foods.

These macroscopic components are not totally responsible for the varied flavors we associate with dairy products. They provide the raw materials for the development of aromatic compounds of an immense variety. Degradation of the protein, lactose and fat components directly yields many aroma compounds, some more desirable than others! But the great variety in dairy product flavors would not occur without the action of a range of microorganisms that selectively degrade the raw milk to give compounds responsible for the flavor characters we know so well. Further chemical interaction between these compounds increases the range of chemical species that contribute to dairy flavors.

Differences between different species' milks manifest themselves mainly as differences in lactose, protein and fat levels, and in particular in differences in the chemical make-up of the fat triglycerides. This in turn gives rise to differences in the types and balance of small aromatic molecules liberated in the degradation processes. Thus there is significant variation in flavor between dairy products derived from different species' milks, despite the broadly similar reactions taking place.

10.1.3 *Instrumental Analysis*

As in many areas of flavor science, understanding of dairy product flavors has advanced tremendously in the last thirty years. This has been largely driven by the increasing availability of instrumental analytical techniques that allow the detailed examination of the low-level components of complex mixtures. Gas chromatography and mass spectrometry have revolutionized the study of all flavors, and of dairy flavor in particular. Together, the two techniques enable the separation and identification of components in extracts from dairy products. However, they do not indicate which components are the most important contributors to the overall flavor of the dairy product. For this the trained human nose is still the most reliable, if not the only means of analysis. Sometimes the nose can usefully be used as an aroma-specific detector for the gas chromatograph, indicating which components should be further studied by mass spectrometry and other techniques. The result can be a list of the key compounds that together characterize the flavor of the studied product. But even this is of little use to the flavorist unless the compounds are available, or could be made available within commercial cost restraints. So, although analytical information is of great help, the skill and experience of the flavorist remain paramount in the development of commercially viable flavorings.

10.1.4 *The Development and Uses of Dairy Flavorings*

Dairy flavorings are used throughout almost all sectors of the manufacturing food industry. Particular dairy flavorings find uses from snack foods through to alcoholic beverages; from sugar confectionery to ready meals; from dairy products even to specifically non-dairy foods. The range is extremely wide, and this demands a certain approach from the dairy flavorist: It is vital to obtain as much information as possible about the application the customer has in mind. At least, each of the following considerations should be addressed.

10.1.4.1 *End-product*

- type of end-product
- processing
- ingredients—is the recipe available?
- pH of the end-product
- is unflavored base product available?
- packaging of end-product
- storage of end-product (freeze-thaw?)
- shelf life required
- countries of final sale

1. flavor "culture"
2. legislation
3. labeling

10.1.4.2 *Flavoring*

- flavor character required?
- is exclusivity required?
- status required (natural, NI, artificial)
- form required (powder, paste, liquid)
- solubility required
- strength required (handling problems?)
- any solvent or carrier restrictions
- packing required

10.1.4.3 *General*

- timescale—urgency?
- workload—priorities?

10.1.4.4 *Commercial*

- target flavoring price
- projected volume
- manufacturing capacity
- costs
  1. flavoring ingredients
  2. manufacture
  3. packing
  4. distribution
- potential earnings
  1. this customer
  2. other customers?
- credit terms
- competition

A flavorist must be a jack-of-all-trades: expert in whatever discipline is required! Intermittently, and concurrently, the skills of chemist, artist, food technologist, designer, marketer, consumer and even psychoanalyst may be needed in order to identify and satisfy the customer's needs. The customer has an idea of what he or she wants; the marketing department sets a brief; the food technologists research the practicalities. Ideally, the flavorist works alongside this whole process to achieve the desired flavor profile.

Flavorists can obviously benefit from a detailed knowledge of the compounds responsible for flavor in real dairy products, but this is only a part of the story.

The inherent flavor of the product will greatly affect the emphasis of the required flavoring. Many of the components of the "real thing" will be technically unavailable, or commercially unuseable. The flavor of the "real thing" may not even be what the customer really needs! Before commencing development work on any project, flavorists should satisfy themselves that they fully understand the customer's needs. The best way to do this is to question the customer directly. Such a direct approach is rarely rejected, because most food product developers realize this is the most effective way for them to achieve their goals.

10.2 MILK AND CREAM

10.2.1 *Whole Cow's Milk*

Cow's milk is by far the most commonly drunk milk in the U.K., so for the purposes of this section, we will confine our discussions to cow's milk, and future references to milk should be understood to imply that. Fresh milk is a stable emulsion of fat globules dispersed in a water phase containing both dissolved and suspended non-fat solids. The water content is around 87.5%. It has a slightly sweet taste due to the milk sugar lactose, and this is modified by the fatty acids, and their condensation and oxidation products derived from the milk fat. The most important fatty acid from the flavorist's viewpoint is butyric at 4% of the total fatty acids, with hexanoic and octanoic acids also significantly contributing to flavor.

Many factors affect the flavor of milk, some even before the milk leaves the cow (Figure 10–1). The cow's diet is of major significance (Forss, 1979), both in the development of desirable flavor, and in the responsibility for off-flavors. In particular, chlorophenols introduced into the diet through herbicides, pesticides and disinfectants give rise to well-known taints in milk. Pharmacologically active compounds in the diet may cause physiological changes in the cow, which are reflected in the composition and hence flavor of the milk (Dumont & Adda, 1979).

Even under relatively ideal refrigerated storage conditions, the flavor of milk changes further once it has been obtained from the cow. Oxidation can produce undesirable flavor compounds such as alcohols, alk-2-enals and alk-2,4-dienals. This even occurs in chilled milk, in which a common oxidation product is undeca-2,4-dienal (Boon et al., 1976). Alcohols, lactones, acids and hydrocarbons are also produced by oxidation of unsaturated fatty acids, but not all of these have a negative effect upon flavor.

Exposure to sunlight can cause the breakdown of methionine to methional, which can then react further to produce methanethiol and various methyl sulfides. Heating produces lactones from gamma- and delta-hydroxyacids, which are themselves derived from the triglyceride milk fat. In cow's milk delta-hydroxyacids predominate, explaining the importance of delta-lactones in milk flavor. The review of Forss (1979) gives much more detail of these reactions, and is recommended to those needing to study this area in more detail.

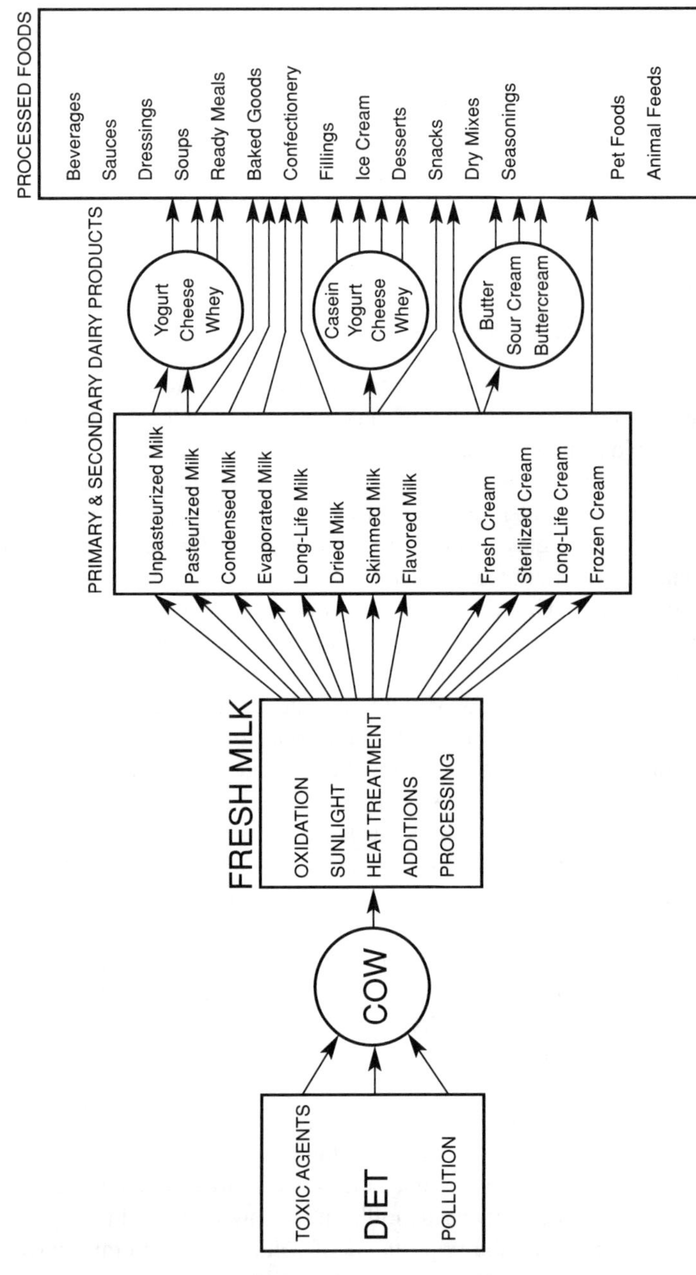

Figure 10–1 Milk products: their source and applications.

Badings reports that over 400 volatile aroma chemicals have been identified in milk, contributing in varying degrees to the overall flavor. He suggests that acceptable flavor is the result of the combination of these, many of which are at individual levels below their flavor thresholds. Off-flavors are thought to occur when the balance of these aroma chemicals is upset, resulting in some of them exceeding threshold levels.

Nearly all milk sold for drinking in the U.K. is pasteurized, as is most milk destined for further processing. Although fairly mild, the pasteurization conditions are sufficient to cause some changes in milk flavor. Most methods of further processing milk involve some degree of heating and hence oxidation. This effect is sometimes compounded by the addition of other materials, for example, sugars, which participate in the development of different flavor characters.

10.2.2 *Whole Milk Powder*

Whole milk is generally spray dried to form a powdered product. The process has several stages:

- Standardization
- Pasteurization
- Evaporation (to concentrate)
- Homogenization
- Drying

Each of these stages affects the flavor of the resulting powder, which has significantly impaired flavor compared to the fresh milk feedstock. The minimum fat content is 26% and this prevents the easy dissolution of whole milk powder. Coating the powder with about 0.2% lecithin can make an "instantized" version.

10.2.3 *Skimmed Milk*

Most of the fat is removed to produce skimmed milk. It has a much shallower and obviously less fatty taste than whole milk. In recent years increasing consumer awareness of the detrimental health effects of saturated fat in the diet have led to an increasing popularity for skimmed milk. It is widely used in both liquid and especially powder form in food processing. Key uses are in confectionery, ice cream, dessert mixes, comminuted meats, sauces, animal feeds, and further dairy products such as cottage cheese and yogurt.

10.2.4 *Sterilized Milk*

Sterilized milk is heated in the bottle to effect sterilization. Its flavor is definitely that of cooked milk. Key flavor formation occurs through the degradation of sulfur-containing whey proteins to give hydrogen sulfide and other thiols, and Maillard browning reactions with sugars present in the milk. The distinctive flavor shows great regional variations in popularity across the U.K.

10.2.5 *UHT Milk*

UHT (ultra-high temperature) or HTST (high temperature short time) treatment is a process used to effectively sterilize milk so that it can be aseptically packed to achieve long ambient shelf life. The milk, whole or skimmed, is heated to 122°C (minimum) for 1 second (minimum). Although a short duration, this high temperature treatment affects flavor significantly, especially in whole milk, where oxidation of the fat occurs. The flavor which results is much preferable to the distinctly caramellic notes of traditionally sterilized and bottled milk, but it is nevertheless very different from that of fresh or pasteurized milk, and is often referred to as "cooked" flavor. Consumer acceptance of "long-life" milk has generally been on grounds of convenience rather than taste, but many attempts have been made to reduce this cooked flavor (Lewis). As the unwanted characters seem to result from interactions of sulfides in the milk, the flavorist must be very cautious when using such notes in milk flavorings. The relative absence of fat in skimmed milk results in a UHT product which is more similar to its fresh counterpart, and this has formed the basis of a whole range of flavored milk drinks, in which fruit or chocolate flavorings are added prior to the UHT treatment.

10.2.6 *Evaporated and Sweetened Condensed Milk*

Evaporated and sweetened condensed milks undergo relatively severe heat processing. Not only are they evaporated to remove most of the water content of the milk, but they are also canned and sterilized at up to 120°C for 10 minutes! This gives rise to very pronounced cooked flavor characteristics. In sweetened condensed milk, sugar is added before the concentration process, and the resulting flavor is distinctly caramellic.

10.2.7 *Cream*

Cream is the fatty portion separated in the production of skimmed milk. It contains almost all the fat from the milk, and this is largely responsible for its characteristic flavor. Various standardized fat levels are available in the U.K. under the regulated names: single cream (18%), double cream (48%), whipping cream (35%). The Food Labelling Regulations (1996) regulate the composition and labelling of cream products in the U.K. Consumption of cream in the U.K. has been increasing in recent years, but still lags well behind that in other EC member states (Anon., 1989).

10.2.8 *Soured Creams*

Across Europe many different varieties of soured cream products exist. Different souring agents or organisms give rise to many subtle variations in flavor, linked only by the common factor of acidity, although even the degree of this can vary widely.

10.2.9 *Sterilized and UHT Cream*

Sterilized cream is produced in both cans and jars. Like sterilized milk, the heat treatment causes significant flavor changes and the character of sterilized cream can almost be regarded as a subcategory in its own right. The high fat content ensures that the effects of oxidation are quite marked. In UHT cream, the heating regime is less severe, but, because of the high fat content, the result is less acceptable than for UHT milk. The development of higher levels of lactones during the heating process contributes to this off-flavor, despite the fact that at lower levels lactones have an important part to play in the flavor of cream (Varnam & Sutherland).

10.2.10 *Clotted Cream*

In clotted cream, the fat content is increased to 55% by gentle evaporation. This pasteurizes the product and forms the characteristic "crust." As might be expected from the higher fat content and the further heat treatment, the flavor is stronger than that of fresh cream, but it is more rounded and subtle than sterilized cream.

10.2.11 *Casein*

Casein (milk protein) can be separated from milk either by the use of acid or rennet to coagulate the protein. It has little flavor and a large proportion is used for non-food purposes. In recent years rennet casein has found important use in the manufacture of cheese analogues, in which the milk fat is replaced by vegetable fat for both economic and health reasons. The other main food use of casein is as the sodium salt, which is used in meat products, coffee whiteners and whipped desserts for its emulsifying, stabilizing and water-binding properties (Burgess, 1988).

10.2.12 *Whey*

Whey is the byproduct of the coagulation and separation of casein from milk. There are four main types:

- Sweet cheese whey, largely from Cheddar production
- Acid cheese whey, largely from cottage cheese
- Acid casein whey, from acid casein production
- Rennet casein whey, from rennet casein production

Whey powders are widely used in food processing, mainly for their functional properties. However, cheese wheys in particular have more flavor and can be used to provide a cost-effective carrier for dairy flavorings. The subtle flavor acts as a useful base upon which to build a realistic dairy flavoring, but care must be taken to check its vegetarian status if this is important to the customer.

10.2.13 *The Applications of Milk and Cream Flavorings*

Milk and cream flavorings find two main uses in food processing:

- Products requiring distinct and identifiable dairy characters
- Products requiring the "body/creaminess/richness" which is associated with dairy products

In fulfilling these two needs, milk and cream flavors find use across the full spectrum of the food industry from animal feedstuffs to sauces, dips and dressings; from baked goods to coffee whiteners (van Eijk, 1986); from sugar confectionery to low-fat spreads. They are used for cost savings, to add milk character to non-dairy products; even to provide a stronger overall flavor character than is possible using the "real thing"!

The most commonly requested flavor types are milk, condensed milk and cream, both in their fresh and sterilized forms. Perhaps the most common requirement is to add "body" to products formulated with skimmed milk for cost and/or health reasons. This often results in a reduction in palatability compared to the full-fat product. Well-designed dairy flavorings can make a significant difference to such products.

10.2.14 *The Development of Milk and Cream Flavorings*

In approaching the problem of development of this type of flavoring, it is often best to consider the position(s) of the target character(s) within the spectrum of flavor types ranging from skimmed milk, through whole milk, cream and on to the heated characters of condensed milk. Skimmed milk is sweet, with relatively low flavor impact; whole milk adds some fatty character. Cream is fuller and fattier with lactonic and sulfurous notes; condensed milk has all these characters plus caramellic notes resulting from the heat treatment.

As in many other areas of flavoring development, there is a tradition of milk and cream flavoring types that bear little relationship to the real foods. They owe their existence to the easy availability of certain flavoring ingredients in the early days of the commercial flavoring industry. In this case these materials are typified by vanillin, maltol, ethyl butyrate and *p*-methyl acetophenone. Their use in dairy flavorings is more by association than a reflection of their occurrence, or, particularly, level in the foods they purport to emulate. Nevertheless the flavorist must recognize that there is still a demand for flavorings of this traditional type, not least because the consuming public has come to expect such flavor characters over a fairly long period of time.

Today, however, analytical instrumentation has given the flavorist a much better understanding of the compounds responsible for the flavor characters in real milk and cream. At the same time, many more of these compounds have become commercially available, thereby enabling the development of much more realistic flavorings. Ingredients of key importance in the creation of such realistic flavors include:

- delta-lactones, for creaminess
- carbonyls, for buttery notes
- short chain fatty acid, for cheesiness
- sulfur compound, for heated notes
- pyrazines (low levels), for nutty and green notes
- maltol (low levels), for "fullness"
- vanillin (low levels), for "fullness"

Many ingredients in these categories are commercially available in nature identical form. In recent years, in direct response to market demand, many have also become available in natural form, thereby opening up the possibilities for natural milk and cream flavorings.

However, a few characters still cannot be obtained as single natural flavoring substances. For these notes the flavorist must turn to flavor preparations from natural sources. The obvious ones are derived from dairy sources, and, for example, lipolyzed butterfat can be used to impart a useful base character to dairy flavors generally. Many other notes are available from natural vegetable and spice extracts.

In the now limited number of cases where artificial flavorings are acceptable, compounds such as butyl butyryllactate can be very useful for tenacious creamy body.

10.3 YOGURT AND FERMENTED PRODUCTS

Once ancient civilizations began to domesticate animals and obtain milk from them, the discovery of various means of keeping, storing and processing milk probably followed largely by accident and human curiosity. In ancient times the existence of bacteria was unknown, and many well-known products were the result of unguided empiricism.

Fermentation of milk involves the action of microorganisms, principally the lactic acid bacteria. These sour the milk by converting the milk sugar lactose to lactic acid. Many of these lactic acid bacteria occur naturally in milk and are, in fact, responsible for the natural souring process. Because large concentrations of lactic acid are generated, the development of other microorganisms that may cause spoiling is inhibited, and thus the fermented milk is preserved to some degree.

Action by a range of differing bacillus species results in curd formation in a range of products of varying solubility. In the U.K. the best known example of such fermented milk products is yogurt. However, in other parts of the world, many other types are common, and these form an important source of nutrition to much of the world's population. In this section, yogurt flavor will be covered in some detail, along with brief discussions of a representative number of the other types.

10.3.1 *Yogurt*

Yogurt is a white semi-solid with a clean acidic taste and a varying degree of creaminess which depends largely upon the fat content of the milk used in its production. In the U.K. cow's milk yogurt is the most common, but yogurts made with ewe's and goat's milk are growing in popularity, particularly in "health food" outlets. The source of the milk has a significant effect upon the taste of the yogurt produced, and both ewe's and goat's milk varieties have characteristic notes which are not present in the more familiar cow's milk product.

Yogurt is a very versatile foodstuff, which is eaten both as a food in its own right (often flavored with fruit, sweet brown, vegetable or spice/herb flavorings) or as an ingredient in other foods, both sweet and savory. It can be used in marinades, dips, sauces, dressings, baked goods, chilled and frozen desserts, to name just a few.

The two main bacteria involved in the manufacture of yogurt are *Streptococcus thermophilus* and *Lactobacillus bulgaricus*. Lactic acid is formed, and it is this compound that is largely responsible for the acidic "bite" and the bacteriological stability. The basic manufacturing procedure is very simple, and in fact it can be modelled quite well in the domestic kitchen. Figure 10–2 gives a simplified flow-process diagram of the domestic preparation. It is perhaps worth mentioning that cleanliness is vital, and it is essential to boil the milk before cooling. A more complete explanation of the processes involved, including commercial practices, is given by Tamine and Deeth (1980). The flavor of yogurt is produced by the action of the bacteria on the milk. Numerous biochemical changes occur during yogurt fermentation, and these are supported by chemical changes, which may involve products of thermal degradation, depending upon the degree of heating used in the manufacturing process. As already indicated, a major flavor contributor is lactic acid, which is derived by microbial fermentation of lactose. A further

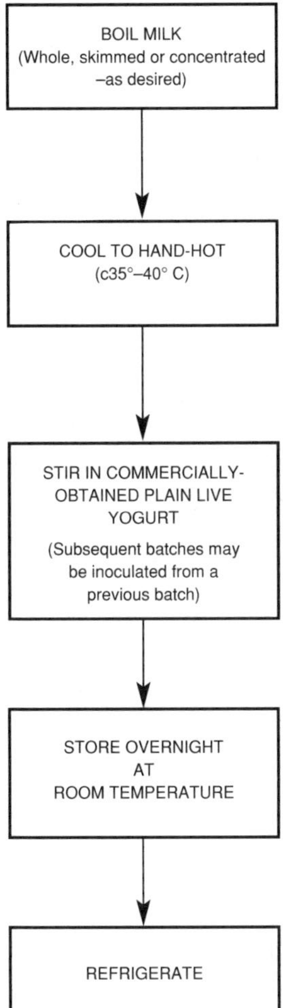

Figure 10–2 Simplified flow-process diagram of the domestic preparation of yogurt.

significant flavor contributor is acetaldehyde. The detailed biochemistry of yogurt flavor development is beyond the scope of this chapter, but more important for the creative flavorist is the range of compounds that are responsible for the characteristic flavor of yogurt. However, the review of Tamine and Deeth (1980) discusses yogurt biochemistry in detail, and would be a good starting point for those needing to study this area more closely.

Figure 10–3 gives a simplified diagrammatic representation of the flavor components and their sources in cow's milk yogurt, seen very much from an

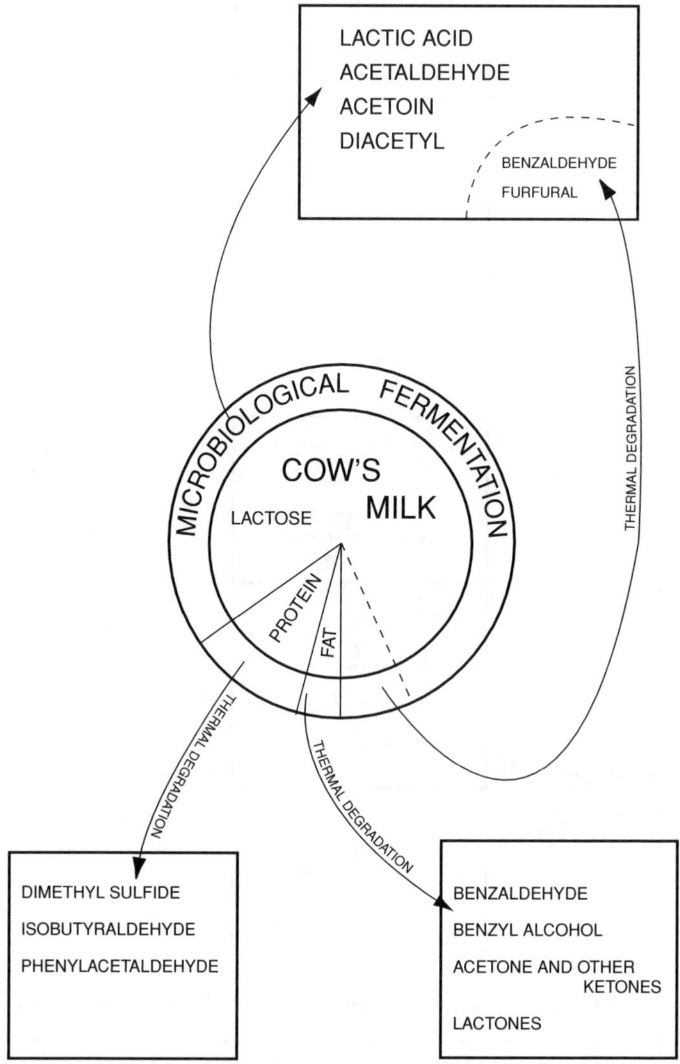

Figure 10–3 Flavor development in milk and cream.

organoleptic point of view. It is useful to utilize such a diagram when developing a new flavoring, as it helps to clarify the problem and identify the key components of the target flavor. Reference to the qualitative and quantitative lists of volatile compounds in food published by CIVO/TNO (Maarse & Visscher, 1983) reveals over 50 compounds that have been identified in yogurt and quantitative data on 13 of those compounds. However, not all these compounds have an equivalent role in characterizing yogurt flavor, so the flavorist must concentrate on those, which are:

- organoleptically most important
- readily available in the required form (natural or nature identical)
- cost-effective in use

In yogurt, lactic acid and acetaldehyde are very important, whereas other compounds (both those indicated in Figure 3–2 and others) vary in importance according to the geographical source/style of the yogurt.

So why use yogurt flavoring? Fresh yogurt is not easily used in products other than wet recipes, and even then it imposes limitations upon keeping quality, generally requiring chilled storage and short shelf life. Powdered yogurts are available, but, although they are generally acidic, they often fail to impart the full aromatic qualities of fresh yogurt. Further, yogurt powders are fairly expensive in use. Finally, and possibly most important, the average consumer's recollection, and therefore expectation, of yogurt flavor may differ markedly from the character of the fresh product. It is this expectation which the food product developer must aim to satisfy. Consequently there is a fairly high demand for high quality yogurt flavorings.

Yogurt flavorings can be used in snack food seasonings, dips, dessert mixes, cream fillings, dressings, soups, sauces and marinades. Consideration must be given to the overall acidity of the product to be flavored, and the exact characteristics the customer requires. Experience shows that customers often require a degree of creaminess which is not true to real yogurt, and they find a natural acidity level too high for their needs. It is important, therefore, to work with the customer to define the desired end product profile as completely and accurately as possible.

Undoubtedly the fresh character of acetaldehyde is appropriate in cow's milk yogurt flavor, but in flavoring formulations it must be used with caution as it is unpleasant and dangerous to handle, and in Europe it triggers a hazardous label on the flavoring when used at 1% or more.

A wide range of nature identical flavoring ingredients is available for the flavorist to work with:

- lactic acid, for acidity
- lactones, for creaminess
- short chain fatty acids, for cheesiness
- ketones, for buttery notes
- sulfur compounds, for off-milk notes
- heterocyclic compounds, for boiled milk notes

Many of the above materials are now also available in natural form, and those which are not can sometimes be substituted with natural flavoring preparations which are rich in the compounds needed.

If the customer's brief extends to the use of artificial materials (which is tending to be less and less common now), the flavorist's palette is extended by a few key compounds which can add a particular tenacity to the dairy character. The best known of these artificial materials is the compound ester butyl butyryllactate, the chemical components of which are themselves important dairy nature-identicals.

10.3.2 *Other Fermented Milk Products*

In other parts of the world, many different types of fermented milk product result from the use of different microorganisms. Although the demand for flavorings of these types is small compared to yogurt, they cannot be ignored; and many of them bear little similarity in flavor to yogurt.

10.3.2.1 *Cultured milks*

Cultured buttermilk is popular in the U.S.A. and in the Irish Republic, but is not commonly consumed in the U.K. Buttermilk is the liquid that separates when cream is churned to make butter; it is the traditional source for making cultured buttermilk. However, commercial "cultured buttermilk" is made by inoculating pasteurized skimmed milk with *Streptococcus cremoris* and/or *Streptococcus lactis*, then incubating at around 20°C to reach a pH of 4.5–4.6. The main flavor contributors are lactic acid and diacetyl, which produce a buttery character with a clean acidic background.

Cultured milks are becoming increasingly popular in the Scandinavian countries. There are several different types, some of which depend on acetaldehyde as well as diacetyl for the major flavor character (Marshall, 1982).

10.3.2.2 *Acidophilus milk*

Acidophilus milk is produced by souring milk with *Lactobacillus acidophilus*, an organism which is often found in the human gut. The product is easy to digest, and has been claimed to have therapeutic effects, supposedly neutralizing various disease-producing agents. For a full run-down, visit your local health food shop!

One of the major flavor contributors is again lactic acid, but the level is not as high as in yogurt, and the flavor is much more like that of sour milk.

10.3.2.3 *Kefir*

Kefir is a mildly alcoholic soured milk drink, which is generally accepted to have originated in the Soviet Union. It is now very popular throughout the Eastern bloc and the Middle East.

Kefir is made by the addition to cow's milk of "Kefir grains," small brownish "seeds" which are pre-swollen to double their size in warm water before being added to the milk. The resulting chemical reaction sours and ferments the milk.

The seeds are dried particles of clotted milk containing *Lactobacillus brevis*, *Streptococcus lactis* and various lactose fermenting yeasts.

Kefir is slightly effervescent and forms a head rather like beer. It has a creamy consistency and an acidic taste; the lactic acid content is between 0.6% and 0.9% and this is responsible along with a low level of acetic acid for a pH of about 4.3–4.4. The alcohol content varies widely between 0.2% and 1.0%. Other important flavor contributors are again acetaldehyde and diacetyl at a ratio of roughly 1:2. In the Soviet Union fresh Kefir is prescribed to treat gastrointestinal infections, because it is thought to destroy germs (Mann, 1989).

10.3.2.4 *Koumiss*

Koumiss is very similar to Kefir, but is usually made with mare's milk. It tends to have a slightly higher alcohol content of up to 3% because of the higher lactose content of mare's milk. Fermentation is effected by inoculation with a combination of thermophilic lactobacilli and yeasts.

10.3.2.5 *Labneh*

Labneh is concentrated Middle Eastern yogurt, which has the appearance of cream cheese. It is rolled into small balls and stored in olive oil. Regional variations include the addition of various herbs and/or paprika. It is generally eaten with bread as part of a main meal. It has good keeping qualities. The lactic acid content is 1.6–2.5%.

10.4 BUTTER

Butter is one of the most important manufactured dairy products in the U.K. and Western Europe. It is derived from the agglomeration of butterfat globules, with the release of watery buttermilk when cream is worked in the churning process. The resultant product is a water-in-oil emulsion containing not less than 80% butterfat.

There are two main types of butter: sweet cream and cultured cream. Either type may be salted at up to 2%.

10.4.1 *Sweet Cream Butter*

The cream for sweet cream butter manufacture is standardized to around 40% fat, pasteurized, cooled to 5°C and held to several hours to crystallize the fat. This crystallization is an important step in order to achieve efficiency in the subsequent churning (mixing) stage. The cooling period is known as the "aging period." After aging, the cream is warmed to 10°C for churning. In the churning process the fat globules are broken and the fat aggregates into grains. The fat binds together and water is released in the form of buttermilk. The fatty mass is then worked to achieve a good smooth texture and salted if required. In the U.K. the remaining water level must not exceed 16% (Spreadable Fats Regulations, 1966 to 1995).

10.4.2 *Cultured Cream Butter (Lactic Butter)*

Lactic butter is made from cream which has been "ripened" by lactic acid bacteria, yielding lactic acid and diacetyl, which are largely responsible for its characteristic taste. Most lactic butter is unsalted, but it may be salted at up to 0.5%. Lactic butter has long been very popular in France, Denmark and other continental countries; and recently it has increased its small share of the butter market in the U.K.

After pasteurization, the cream is inoculated with a starter culture, which may include a combination of *Streptococcus lactis, S. cremoris, S. lactis* subsp. *diacetylactis* and *Leukonostoc cremoris*. It is ripened for a few hours, and then the cream is cooled and held at 7° C for several hours before being churned. The acidity of the cream gives it different manufacturing qualities to sweet cream, which manifest themselves in faster churning and higher yields. However, lactic butter is more susceptible to oxidative rancidity than sweet cream butter, and the buttermilk produced is acid and less easily utilized. Modern manufacturing processes in which a culture concentrate is injected at the working stage of butter manufacture, after the separation of the buttermilk, have largely overcome these disadvantages.

10.4.3 *Buttermilk*

Buttermilk arising from the above processes has a similar composition to skimmed milk, but possesses good emulsifying properties due to its high phospholipid content. It is commonly used in dairy analogues, baked goods and ice cream.

10.4.4 *Ghee*

Ghee, commercial clarified butter, is made by heating butter to remove the water. The heating process causes Maillard-type reactions between fat and non-fat components of the butter, and this generates characteristically different flavor profiles from butter itself.

10.4.5 *The Flavor of Butter*

The flavor of any food is a combination of aroma, taste and mouthfeel. With an emulsion like butter, some flavor components reside mainly in the fat phase; others in the aqueous phase. The pH of the aqueous phase also affects this partition between the phases. This is obviously an important factor in cultured cream butter, where lactic acid is present in quite high concentration in the aqueous phase. Perception of individual flavor components also varies in different media. Generally, flavor compounds tend to taste stronger in aqueous media than in fatty media.

The flavor of butter is the result of a complex combination of the mouthfeel properties of the butterfat, with the varying release properties of each of the flavor components according to their partition between the phases. Much work has been done on flavor release in emulsions by McNulty (1975). Threshold values of individual flavor components, and synergistic effects between them are very important for flavor release in emulsions (Kinsella, 1975).

Many compounds have been identified in butters of various types. Useful qualitative and quantitative listings are given by CIVO/TNO (Maarse & Visscher, 1983). It is interesting to note that of the 233 different compounds listed by TNO, 61 are carbonyls, 21 are fatty acids and 24 are lactones; and these three groups represent by far the most important flavor contributors to butter character. Lactones in particular seem to be of great importance in butter flavor, and much work has gone into their identification (Forss et al., 1966). In general, a review of the literature would suggest that the flavor of cultured cream butter is well understood, whereas that of sweet cream butter has been studied to less depth. However, this is debatable; and the level of understanding of the development of flavor in the natural foodstuff is not necessarily directly related to the quality of the flavorings that the flavorist can develop. Cost effectiveness, raw material availability (with the right status) and an appreciation of the processing to which the flavoring is to be subjected are all also of vital importance in the development of a new flavoring.

Fatty acids are key materials in butter flavor, both in their own right and as precursors to other important compounds. The most important ones from a flavorist's point of view are the volatile acids—acetic, propionic, butyric, hexanoic and octanoic—and of course lactic acid in cultured cream butter.

The other important class of compounds that contribute to butter flavor is carbonyls; aldehydes, ketones and diketones. The best known of these is probably diacetyl, which along with acetaldehyde, is produced during the cream ripening stage of cultured butter manufacture. Both compounds also play a smaller part in the flavor of sweet cream butter. A number of other short chain aldehydes, particularly unsaturated ones, are also important in butter flavor.

Many alcohols, largely homologous with the ranges of acids and aldehydes present, have also been identified in butter. Their presence in equilibrium with the acids gives rise to low levels of esters which are often responsible for fruity off-flavors. Low levels of very low odor threshold compounds such as indole and dimethyl sulfide (Ramshaw, 1974) also have a key role in some types of butter.

Besides the basic distinction between sweet cream and cultured cream butter, there are many other geographical differences. In cultured cream butter, the exact strain(s) of bacteria in the starter culture greatly affect the development of flavor, so that Danish lactic butter is quite different from German sour cream butter.

There are similar differences between sweet cream butters, for example, Dutch butter is generally much creamier than English butter, which normally has a richer, more aged character.

When butter is heated, and therefore in virtually all processed food products in which it is an ingredient, the flavor becomes stronger. Aroma compound development is encouraged in both the fat and aqueous phases. The levels of volatile fatty acids increase in the fat phase, with a consequent increase in lactones (formed by cyclization of hydroxyacids). In the water phase, Maillard-type reactions occur, leading to typical products such as furfural and other conjugated furans.

So, the flavor of butter is complicated and cannot be reproduced with just a couple of key compounds. For example, diacetyl is important in all butters, but has been rather over-used; customers quite often ask for a "non-diacetyl type" butter flavoring!

10.4.6 *The Uses of Butter Flavorings*

Butter is used as an ingredient in many foods to provide both flavor and mouthfeel. Butter flavorings can be used in any application where butter might be used, but cannot either for technical or economic reasons. Some of the key applications are sauces, dressings, desserts, ice cream, baked goods, margarines and sugar confectionery. The diversity of applications requires a diversity of butter flavor types. Not only does the required butter character change with geography and processing (see above), but the required physical form, legal status, solubility and heat and freeze-thaw stability also vary from project to project. The range of common application dosages varies enormously, from as little as 0.001% (10 ppm) up to 1% or more.

Flavorings for margarine tend to be very low dosage (0.001–0.02%). They may need to be oil- or water-soluble (especially for low-fat spreads), and tend to be liquids, pastes or emulsions.

Baked goods must rely on less volatile flavoring materials. Common dosage for nature identical or artificial flavorings is around 0.2%, but it may be up to 1% or more for naturals. The physical form will depend upon the processing involved and the customer's convenience of handling.

High-acid dressings are often improved by the addition of a creamy butter flavor to offset the acidity without affecting the keeping quality of the product.

10.4.7 *Margarine and Low-Fat Spreads*

Margarine is an analogue of butter. It consists of an emulsion of animal and/or vegetable oils, with an aqueous phase that may contain dairy and non-dairy solids, and sometimes salt. It has no desirable inherent flavor, and thus must be flavored with suitable butter flavoring. In recent years there has been much development activity in this area, particularly in low-fat spreads, dairy blends and high-polyunsaturates products. This has put a new set of demands upon the flavor industry to supply new flavorings to suit these innovative products. For example, flavorings for low-fat spreads are often expected to mask the unpleasant gelatin character of the aqueous base. Also, customers often wish to apply flavorings to both the aqueous and fat phases, although the authors believe that whichever phase the flavor is applied to, the individual flavoring compounds will partition between the phases according to their relative solubilities, in an equilibrium distribution.

10.4.8 *The Development of Butter Flavorings*

The preceding discussions determine that it is essential to know the market into which the product to be flavored will be launched. In addition to geographical variations, in some countries there are restrictions upon the use of butter flavorings, particularly in respect of labeling. Also, as with all well-established flavor types, there are generally accepted, but unrealistic butter flavorings, which have gained credibility through many decades of use.

Choice of solvent system or powder carrier restricts the use of certain ingredients. Lactic acid will not dissolve in vegetable oils because of its water content. High levels of decanoic acid will not dissolve in propylene glycol. There are many other such examples of technical constraint.

A wide range of flavoring substances is available for the development of butter flavorings. They range from the now less commonly used artificials, through a wide range of nature identical synthetics, to an increasing number of natural substances and natural preparations.

When artificials are acceptable, the two most commonly used are butyl butyryllactate (for tenacious creamy body) and ethyl vanillin (for sweetness).

Among the wide range of nature identical and natural substances, the key types are:

- fatty acids, for bite, mouthfeel and roundness
- lactic acid, for acidity
- carbonyls, for creaminess, freshness and caramellic notes
- lactones, for creaminess and nutty characters
- sulfur compounds, for heated butter character
- alcohols and esters, for fruity notes

Lipolyzed butterfat, obtained either by enzymatic or acid hydrolysis of butterfat, is a useful building block for the development of natural butter flavorings. A huge range of other natural flavoring preparations find use in the development of natural butter flavorings.

As is the case for most dairy flavorings, the enormous number of ingredients that could find use in butter flavorings precludes the construction of a comprehensive list. Almost any aromatic raw material could be useful at some stage in the development of a flavoring to meet a particular market/processing need defined by the customer. The flavorist certainly should never be bound by the limits of those compounds known to exist in real butter.

10.5 CHEESE

The term cheese can be applied to dairy foods varying from a mild white acidic paste to an intensely aromatic brown waxy mass. Generally, cheeses are nutritionally rich (in calcium, protein, fat and vitamins) and organoleptically rich (in intensity and variety). Cheese is eaten cold or hot, and in savory and sweet dishes. Yet to some it is nothing more than rotten milk!

Technically, cheese is the product of coagulation of casein and removal of water from milk. The milk source may be whole, semi- or fully skimmed milk, cream or buttermilk, or any combination of these. The removal of water is what makes cheese significantly different from other coagulated milk products such as yogurt.

The history of cheese can be traced back a very long way. Archeological signs of domestication of dairy animals date back to 6500 BC. By 3500 BC artwork shows scenes of milking animals and the keeping of milk, and by 3000 BC definite proof of cheese-making is evident in Egypt, Greece and Italy (Christian, 1977). A rich history of the progression of cheese is well documented to the present day, when almost every country in the world makes some kind of cheese (Marquis & Haskell, 1985).

It is in the area of cheese that the variety of flavor characters that are derived from milk shows its true immensity. This variation itself determines that we can only scrape the surface of the subject in this chapter. Fortunately, reasonable classifications by basic flavor type are possible, and this will enable us to give an overview which should help the flavorist faced with the job of reproducing the character of a cheese of which he or she has never heard, let alone tasted!

10.5.1 *The Manufacture of Cheese*

The methods of cheese manufacture depend greatly upon the type of cheese being produced. However, certain basic steps are fairly common. The three fundamentals are coagulation of milk protein, separation of solids, and working upon the solids to modify flavor and texture. These basic steps are represented in Figure 10–4 by solid arrows. The steps indicated by broken arrows are undertaken for only some types of cheese, and their sequence in the production process

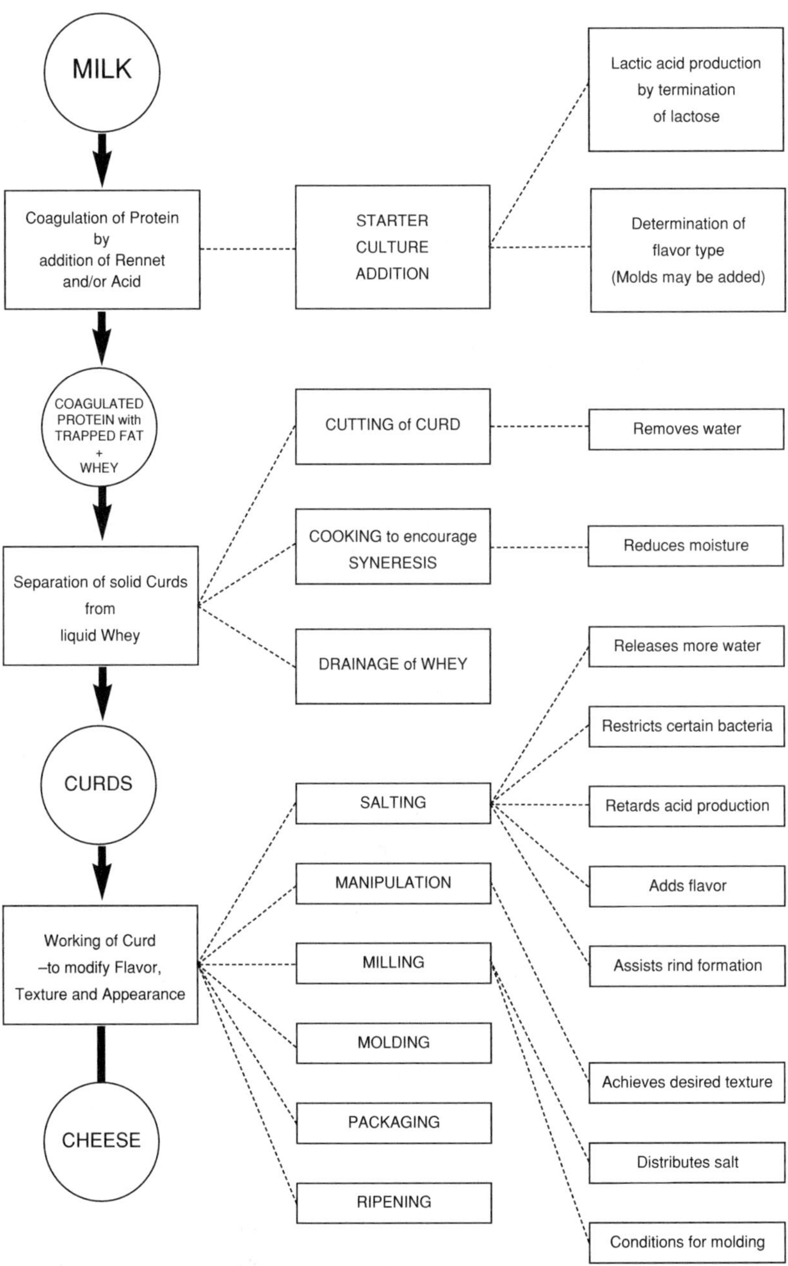

Figure 10–4 The processes involved in cheese-making.

varies depending on the type of cheese. Inoculation with bacteria, yeasts and molds is done at many different stages in the process, and this greatly affects flavor and texture development. In industrial scale manufacture, inoculation is scientifically controlled, whereas in the traditional farmhouse it may occur directly from the environment. In fact many cheeses owe their existence to such "accidental contamination."

10.5.2 *Classification of Cheese Types*

At this point, we intend to take the contentious step of offering a broad classification of cheeses by texture, ripening and flavor type. Such generalization will be certain to offend some readers, so we will again emphasize our concentration upon the flavor character of the individual cheeses (Table 10–1).

10.5.3 *The Development of Flavor in Cheese*

Complex biochemical and chemical processes, combining the breakdown of the curd by enzyme catalyzed reactions with further chemical interactions, give rise to the typical flavors, textures and appearances associated with particular types of cheese.

Lipolysis, proteolysis and fermentation each play significant roles in the development of cheese flavor. It should not be assumed that only volatile products are important in cheese flavor; the flavor of most cheeses is a complex blend of taste and aroma components, which results from the reaction, intermediate production and further reaction of many components. The perception of flavor is a result not only of the relative proportions of these components, but also of synergistic effects between them. In Figure 10–5 we have attempted to give a simplified overview of some of the chemical pathways in cheese flavor development. Those with a deeper interest are directed to the literature (Law, 1981, 1982, 1984).

Fatty acids in cheese are derived via a direct route from the milk, and it is in this area where the differences between species are particularly evident. Comparison of the saturated fatty acid contents of cow's, goat's and ewe's milks gives an interesting insight into possible explanation of the differences between their respective types of cheese (Ney, 1981). Figure 10–5 shows the importance of fatty acids as precursors of many of the compounds responsible for characteristic flavor in cheese.

10.5.4 *Review of a Range of Key Cheese Types*

As mentioned already, in this chapter it will be impossible to cover all, or even a large proportion, of the cheeses available in the world. However, a number of the most important types in the Western world will now be discussed. They are presented in alphabetical order by name for easy reference.

Table 10–1 Classification of cheeses by ripening, texture and flavor

Ripening	*Texture*	*Examples*	*Source*[a]	*Flavor type*[b]
Unripened	Soft	Cottage	Skimmed	Diacetyl/lactic
		Ricotta	Whey	Diacetyl/lactic
		Mascarpone	Cream	Creamy (heated)
		Neufchatel	Cream	Creamy/lactic
Unripened	Plastic	Mozzarella	Buffalo	Fresh/lactic
Ripened	Soft	Reblochon	Whole	Buttery
		Chabicou	Goat's	Goaty
		Feta	Mixed	Salty/ketones
Ripened	Semi-hard	Caerphilly	Whole	Diacetyl/lactic
		Gouda	Whole	VFA/waxy
		Saint-Paulin	Whole	Amino acids/bitter
Ripened	Hard	Cheddar	Whole	VFA/S-cpds:complex
		Lancashire	Whole	Diacetyl/lactic
		Cheshire	Whole	Diacetyl/lactic
		Parmesan	Whole	Strong VFA:complex
		Romano	Ewe's	Strong VFA:complex
		Emmental	Whole	VFA/nutty:complex
Surface mold	Soft	Camembert	Whole	VFA/NH_3:complex
		Brie	Whole	VFA NH_3:complex
Surface smear	Soft	Limburger	Whole	VFA/S-cpds:complex
		Munster	Whole	VFA/S-cpds:complex
Internal mold (blue)	Semi-hard	Stilton	Whole	VFA/ketones
		Roquefort	Ewe's	VFA/ketones
	Soft	Gorgonzola	Whole	VFA/ketones

[a]Source is cow's milk unless otherwise stated.

[b]Flavor type—broad indications only: VFA, volatile fatty acids; S-cpds, sulfur compounds; NH_3, amino compounds.

10.5.4.1 *Bel Paese*

Bel Paese is a soft, uncooked Italian cow's milk cheese with an 80-year history (Christian, 1984). Its aroma is slightly sulfurous, acidic and reminiscent of sweaty feet! The taste is acidic and creamy, with a pleasant slightly bitter aftertaste. The better quality the cheese, the less bitterness is evident.

10.5.4.2 *Blue-veined cheeses*

"Blue-veined" describes the appearance of these cheeses, the body of which is veined with growths of blue-green *Penicillium* molds. In the manufacturing process, the curds are inoculated with spores, then when air is admitted to the

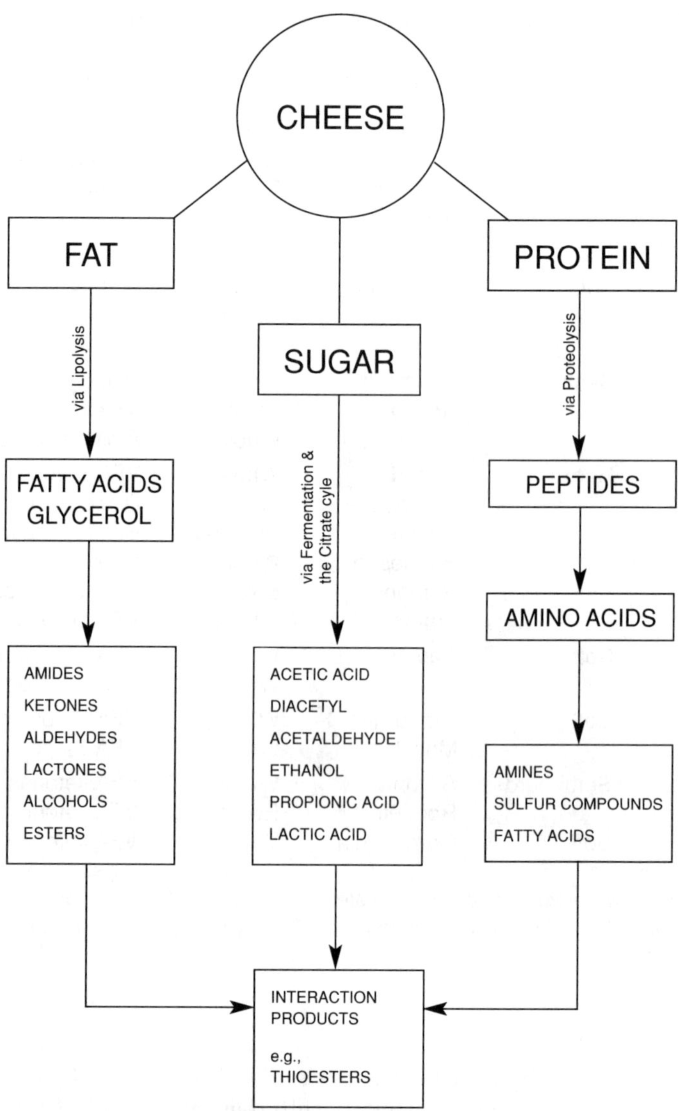

Figure 10–5 Flavor development in cheese.

fresh cheeses by piercing, the spores germinate and spread throughout the cheese. The mold assists in flavor development, being strongly proteolytic and lipolytic, and adding the effect of the starter culture.

The characteristic flavor of blue-veined cheeses is a combination of short-chain fatty acids and methyl ketones. The action of the mold releases fatty acids, which are further degraded to methyl ketones, which are largely responsible for the fruity character of such cheeses. Extreme proteolysis is also common in such mold-ripened cheese, and this gives characteristic bitterness (Fernandez-Salguero et al., 1989).

10.5.4.3 *Brie and Camembert*

These are French soft cheeses made using mesophilic starter cultures. The curd is uncooked, leaving its moisture content high. The characteristic flavor is due to the white mold that grows on the surface. Spores of this mold (*Penicillium candidum, P. caseicolum* and/or *P. camemberti*) are seeded onto the surface of the cheeses after dry salting, and flavor development occurs by proteolysis and lipolysis from the surface inwards. In addition, naturally occurring environmental yeasts contaminate the surface and add to flavor development.

Brie has a creamy, slightly acidic, nutty flavor with characteristic musty, mushroom-like ammoniacal overtones.

Camembert is matured longer than Brie, and develops a secondary surface flora of *Brevibacterium linens*, which gives rise to other strongly aromatic compounds including methanethiol.

An important volatile contributor to the mushroom-like notes of both these cheeses is 1-octen-3-ol (Karahadian et al., 1985).

10.5.4.4 *Cabrales (Picon)*

Cabrales is a hard Spanish blue-veined cheese made with cow's milk. The cheese is matured for up to six months in limestone caves in the Asturias, where it develops a sharp and distinctive blue-veined flavor, which is almost peppery.

10.5.4.5 *Caerphilly*

Caerphilly is a crumbly, semi-hard white cheese originating in Wales. Made from whole cow's milk, it has an acidic flavor, with slight reminiscences of cowsheds.

10.5.4.6 *Chaumes*

A French unpasteurized cow's milk cheese, which is soft-textured, golden creamy in color, with a brown rind. The strong aroma is of old socks, but the taste is milder, light and creamy.

10.5.4.7 *Cheddar*

Cheddar is probably manufactured in more countries throughout the world than any other type of cheese. Consequently, there is much literature on the flavor of Cheddar cheese, but there are few really good Cheddar-type cheese flavorings available. Cheddar has a very complex flavor profile that is not easily summarized in terms of individual aromatic compounds. In application, its flavor has quite different characters. Cooked Cheddar develops a significantly changed flavor during the heating process. The common availability of Cheddar at various stages of maturation also adds to the variation in flavor, and the consequent difficulty in characterization.

The literature indicates the importance to Cheddar of fatty acids, methyl ketones, amines and sulfur compounds, especially methanethiol, but there is great debate about the mechanisms that give rise to typical Cheddar flavor. This debate is well summarized by Lawrence et al. (1993) but it is brought to no conclusion that is useful to the bench flavorist. It even questions whether methanethiol is itself important to Cheddar flavor, or whether its presence is merely an indicator that the required processes have taken place. From a flavorist's perspective the debate is rather sterile; either methanethiol is required in Cheddar flavorings or it is not. Typical Cheddar flavor has a creamy, acidic, sulfurous character. Some of the important volatiles *may* be methanethiol, hydrogen sulfide, diacetyl, butanone, pentan-2-one, dimethyl disulfide, acetone and ethanol (Manning, 1978). Some esters may be present, but high levels are detrimental to the flavor. None of these compounds is, in itself, characteristic of the complex flavor of Cheddar, and this is reflected by the difficulty in developing realistic Cheddar flavorings.

10.5.4.8 *Cheshire*

White and red Cheshire, the latter simply colored with annatto, are hard cow's milk cheeses, which have a mild acidic, almost buttery, slightly salty flavor. Blue-veined Cheshire is stronger in flavor with an additional bitter acidic tang.

10.5.4.9 *Cottage cheese*

Cottage cheese is a soft white, lumpy, watery cheese, which is sold and eaten in the fresh, unripened state. It is made from skimmed milk that is acidified with a mixed starter culture of lactic acid bacteria. Rennet may be added to provide the clotting agent chymosin, but it is not fundamentally responsible for flavor in unripened cheese. The flavor is derived from fermentation products.

The acidity of the cheese is provided by lactic acid; diacetyl is produced by *Streptococcus diacetylactis* from milk citrate. Acetaldehyde is also a product of fermentation, but in a well-flavored cottage cheese the level of acetaldehyde should be well below that of diacetyl, the reverse situation to that pertaining in cow's milk yogurt.

10.5.4.10 *Cream Cheese*

Cream cheese is made from a blend of milk and cream. It has a smooth and buttery texture, with a creamy taste. It is often sold with added flavor and textural ingredients, such as pineapple and chives.

10.5.4.11 *Danablu*

Danablu is the Danish trade name for their cow's milk equivalent of Roquefort. This whole cow's milk blue-veined semi-hard cheese has a strong, savory, typically ketonic/acid blue cheese character.

10.5.4.12 *Double Gloucester*

This pale orange colored smooth textured hard cheese has a milder flavor than much Cheddar, but is broadly similar to Cheddar in both production and flavor.

10.5.4.13 *Doux de Montagne*

This French cheese originating from the Pyrenees region is semi-hard, mild and creamy. It has a slightly lactic, creamy aroma with a hint of sourness. The taste is creamy, buttery, lactonic, slightly caramellic with a hint of bitterness.

10.5.4.14 *Feta*

Feta is a cheese of Greek origin, which is sold in blocks stored in brine. The taste of this white cheese is sharp and salty. Originally made from ewe's milk, Feta is now made in many countries from many animals' milks. The flavor varies in profile according to the milk used (Horwood et al., 1981), largely influenced by the varying balance of fatty acids in the milk fat. Sheep milk fat has a higher concentration of C8 and C10, but a lower concentration of C14 and C16 fatty acids when compared to cow's milk. Thus Feta made from cow's milk lacks some of the characteristic pungency associated with these shorter chain acids.

10.5.4.15 *Fromage Frais*

This term is applied to a group of very soft moist cheeses made from pasteurized cow's milk. The taste is mild, fresh and acidic, and these cheeses are often eaten with fruit, and even sold fruit-flavored. Quarg, which is common across northern Europe, is a member of this group. These cheeses can be made with whole or skimmed milk, resulting in different fat contents with associated differences in flavor and mouthfeel.

10.5.4.16 *Gjetost*

Gjetost is made from goat's milk whey in Scandinavia. There are hard and soft varieties, but neither is commonly available in the U.K. Strictly, it is not a true cheese, the whey being cooked and evaporated prior to cheese production. Gjetost has a caramelized note and resembles sweetened condensed milk or fudge.

10.5.4.17 *Gorgonzola*

Gorgonzola is an Italian soft blue-veined cheese, made with uncooked whole cow's milk. The bluish-green internal veining is the result of growth of *Penicil-*

lium glaucum, a different strain from that used in Roquefort and most other blue cheeses. Gorgonzola has a very salty, tangy, almost "moldy" blue flavor, with a slightly dirty aroma. It is an excellent cheese when at its best.

10.5.4.18 *Gouda and Edam*

These are Dutch semi-hard cow's milk cheeses. Both are available in relatively mildly flavored young form, or more mature. In both cases, the rindy, "waxy" notes increase with age. Gouda has a fuller flavor due to its higher fat content, while Edam has a salty note derived from its brine treatment. The characteristic Dutch cheese flavor is the result of controlled lipolysis, which yields C14–C18 fatty acids, and hence the waxy characters.

10.5.4.19 *Havarti*

This is a Danish semi-hard cow's milk cheese which has small eyes, with a fairly mild flavor reminiscent of mild Dutch cheeses.

10.5.4.20 *Italian Romano, Provolone and Parmesan*

The efficiency of degradation of triglycerides in cheese manufacture greatly affects flavor development. In these three Italian cheeses, distinctive flavor is generated by the deliberate addition of animal lipases, which generate low molecular weight free fatty acids. Some other important volatiles that also contribute to the characteristic Italian hard cheese flavor are ethyl hexanoate, 2-heptanone and pentan-2-ol (Meinhart & Schreier, 1986).

Provolone is a plastic spun curd cheese. Unless well aged it is a mild, but tangy cheese with a slight phonemic hint.

Pecorino Romano is a cooked curd ewe's milk hard cheese. It is full flavored and sharp, with typical ewe's milk aftertaste. Romano is the name used in the U.S.A. for the domestic equivalent hard cow's milk cheese, made by the Pecorino Romano technique.

10.5.4.21 *Jarlsberg*

A softer, sweeter variation of Emmental, which is made in Norway.

10.5.4.22 *Lancashire*

A pale creamy soft cheese which is moist and crumbly with a delicate salty, creamy flavor. A good toasting cheese.

10.5.4.23 *Leicester*

A hard cheese with a grainy texture, made by a process similar to that of Cheddar. It has a mellow, sweet and nutty Cheddar-type flavor.

10.5.4.24 *Limburger*

Limburger is a semi-soft cheese originating from Belgium. Its flavor is influenced greatly by the growth of microorganisms on its surface during ripening. The principal microorganisms are yeasts, micrococci and *Brevibacterium linens*. Major flavor contributors are phenol (which arises from microbial decomposition of tyrosine), dimethyl disulfide and indole (Parliament et al., 1982). A very smelly cheese!

10.5.4.25 *Manchego*

A Spanish ewe's milk hard cheese, whose flavor is distinctive and sharp, and varies in strength with age.

10.5.4.26 *Mascarpone*

An Italian soft cream cheese. It looks like whipped double cream. The taste is like slightly sweetened double cream, with a hint of cooked milk character.

10.5.4.27 *Mozzarella*

Mozzarella is the traditional plastic curd cheese used on Italian pizza. Traditionally made from buffalo milk, it has a very mild, slightly acidic buttery flavor.

10.5.4.28 *Munster*

Munster is a French cow's milk soft cheese. Its flavor is influenced by the growth of microflora on its orange/yellow rind. It has a delicate salty flavor when young, which develops into a full tangy flavor after maturation.

10.5.4.29 *Neufchatel*

Neufchatel is a soft cheese, which is eaten young when its flavor is like tangy cream cheese.

10.5.4.30 *Pont l'Eveque*

This is a French cow's milk soft cheese, which has the typical strong smell associated with washed rind cheeses. It has a rich creamy taste, characterized by methyl thioesters, phenol, creosol and acetophenone (Dumont et al., 1976).

10.5.4.31 *Port-Salut*

Virtually indistinguishable from Saint Paulin.

10.5.4.32 *Quarg*

Although historically linked with Germany, Quarg is already widespread across most of Europe, and is gaining popularity in the U.K. It is a soft acid-curd cheese with a bland but acidic flavor.

10.5.4.33 *Queso Blanco*

This is a white cheese usually eaten fresh in Latin America. It can also be vacuum packed for a one-year shelf life. Queso Blanco can be made in a variety of ways: starter and rennet, or by addition of acids with heating.

There are several varieties of Queso Blanco, determined by variations in salting or addition of different microorganisms. High temperature storage results in butyric notes from fermentation of lactose. It can also be flavored at various stages in manufacture by addition of flavoring ingredients such as fatty acids and enzyme-modified cheese (Siapantas, 1981).

10.5.4.34 *Reblochon*

Reblochon is a French cow's milk soft cheese with a particularly creamy, buttery flavor.

10.5.4.35 *Ricotta*

This is not strictly a cheese at all! Whey is re-cooked to precipitate the proteins that form the basis of the cheese. Ricotta has some similarity to cottage cheese, but is slightly fuller in both texture and flavor.

10.5.4.36 *Ridder*

Ridder is a semi-hard cow's milk cheese made in Scandinavia. It smells of sweaty feet in the Gouda/Edam sense, with a creamy diacetyl, New Zealand butter character. In taste it is rather like a young Gouda, with a hint of caramel or cane syrup.

10.5.4.37 *Roquefort*

The famous ewe's milk blue-veined cheese matured in the limestone caves of Cambalou at Roquefort. The curd is inoculated with spores of *Penicillium roqueforti* that grow throughout the cheese to form bluish-green veins. Roquefort has a distinctive ketonic blue cheese taste, with buttery undertones, and a characteristic ewe's milk tang. The flavor is very powerful.

10.5.4.38 *Saint Paulin*

A French cheese which is inoculated with a lactic acid bacteria culture, which increases the acidity and leads to its distinctive character. It smells and tastes of hydrolyzed casein.

10.5.4.39 *Samso*

A Danish cheese with a mild aromatic, sweet flavor.

10.5.4.40 *Smoked cheese*

When smoke flavoring is introduced into the cheese during manufacture the resulting product is often referred to as smoked cheese. There are many such examples of smoked cheese from around the world.

10.5.4.41 *Stilton*

The most famous English blue-veined cheese, Stilton has a sharp acidic, slightly bitter character, with a pungent, almost cowshed-like aroma. *Penicillium roqueforti* is used in its manufacture.

10.5.4.42 *Swiss Emmental and Gruyere*

Emmental and Gruyere are the best-known Swiss cheeses in the U.K. They are hard ripened cheeses, similar to, but differing significantly from Cheddar. Bacteria involved during ripening include propionic acid bacteria, which generate propionic acid. This acid plays an important role in the aroma of these cheeses, along with acetic and lactic acids. A further product of propionic acid bacteria fermentation is carbon dioxide, which is responsible for the large holes in these cheeses. A good Emmental cheese has a characteristic nutty flavor. Several components are said to be important: acetic acid, propionic acid, butyric acid and diacetyl (Mitchell, 1981). Steffen et al. (1993) also identify among this *volatile* group: primary and secondary alcohols, methyl ketones, aldehydes, esters, lactones, and sulfur- and nitrogen-containing compounds. In addition a *non-volatile* group, including peptides, peptones, free amino acids, amines, salts and fat contribute to the characteristic taste and mouthfeel of these cheeses.

Gruyere generally has a "dirtier" flavor than Emmental, especially near the surface. The mountain varieties of Gruyere have flavor derived from surface flora growth such as lactate-utilizing yeasts. These lift the pH at the surface, so allowing other microorganisms, for example *Brevibacterium linens*, an orange pigmented coryneform, to grow. The mechanisms involved in the development of flavor are very complex. The putrid character is partly attributed to methanethiol and thioesters of acetic acid and propionic acid. The flavor does vary considerably through the cheese from rind to center, as demonstrated by the relative abundance of aromatic compounds analyzed in different regions of the cheese (Liardon et al., 1982; Bosset & Blanc, 1984). This variation makes it particularly important for the flavorist to establish exactly what type of Gruyere character a customer is requesting; the description "Gruyere" is insufficient.

10.5.4.43 *Taleggio*

This is a surface-ripened, uncooked curd soft Italian cheese. It has a delicate mildly acidic, buttery flavor.

10.5.4.44 *Tilsit*

This is a semi-hard cow's milk cheese, originating from Germany. Many aromatic compounds have been identified in Tilsit. Key flavor contributors include the branched chain fatty acids: isobutyric, isovaleric and isohexanoic (Ney, 1985).

10.5.4.45 *Wensleydale*

A white crumbly cheese with a nutty, creamy taste—best eaten fairly young.

10.5.5 *Related Products*

A number of other related products should be mentioned at this stage.

In processed cheese, typically, hard and/or semi-hard cheeses such as Cheddar and Gouda are blended with skimmed milk powder, butter and other ingredients, then heated by direct steam injection with emulsifying salts to produce a homogeneous cheese-like mass. The proportions can be varied to achieve the desired consistency for spreads, slices, etc. If the temperature/time treatment is sufficient, pasteurization of the product can be achieved, giving valuable extension to shelf life, even under ambient conditions. Remelting of processed cheeses for culinary uses is also achieved easily and without fat separation, because of the emulsifying salts. Processed cheeses cannot generally be flavored under current legislation in most parts of the world, but often enzyme-modified cheeses are used because they can provide extra flavor without the need to label.

The expense of milk fat in some parts of the world has led to the development of cheese analogues, in which a cheese-like product is made using casein (usually derived by rennet coagulation) and cheaper vegetable fats. Normal cheese flavor development does not occur in such products, because the key fatty acids and other dairy precursors are not present. Therefore cheese flavorings are often added to achieve a desirable product. Such materials find use in food processing, especially for pizzas, where there are cost and handling problems in using real cheese (e.g., Mozzarella).

A specialist area is that of totally non-dairy cheese substitutes, where an analogue is made without even the use of rennet casein. Such "cheeses" clearly must rely totally upon flavoring to achieve an acceptable product—the absolute challenge for the dairy flavorist?

The development of spray drying has enabled the production of dehydrated cheese powders and cheese powder compounds (blends of cheese, vegetable fat and other ingredients including flavorings). Such products are of great value to the food industry because they eliminate the refrigeration and handling problems associated with "wet" cheese.

10.5.6 *Applications of Cheese Flavorings*

The most obvious applications for cheese flavorings are sauces, dips, baked goods, fillings and snack foods; but the range of products in which cheese flavorings find use continues to surprise even those working in the industry. Cheese flavorings are not permitted in processed cheeses in many countries, but in others this is an important application area, whereas cheese analogues could not exist without flavorings. There is also some application in cheese powder compounds.

In the U.K., one of the most often-requested cheese types is (unsurprisingly) Cheddar, but this is followed closely by Italian and blue-veined types. With the increasing integration of the U.K. into the European Union, the range of other cheese types requested is growing rapidly.

Two areas of recent special interest are worth elaborating: bake stability and vegetarian products.

10.5.6.1 *Bake stability*

When bake-stable flavorings are mentioned, the applications which spring to mind are biscuits, cakes, breads, pastries, and, perhaps to a lesser extent, breakfast cereals. There are some very delicious examples where the flavor profile relies entirely on the food ingredients used, such as oats, dried fruit, brown sugar, nuts, butter, yeast, cheese and eggs. These are traditional ingredients in the Western world, used both for their dietary properties and textural function, as much as for the desirable flavor characters they can help to produce. Other ingredients added specifically for their flavor and "medicinal powers" include herbs, spices, citrus zest, cocoa, coffee and vanilla. As "synthetic flavorings" evolved they often incorporated or mimicked these traditional flavor types, and today form the basis of the range of "bake-stable" flavoring applications: chocolate biscuits, garlic bread sticks, coconut cakes and cheese straws being examples. All these types of ingredient or flavoring addition fall into two distinct groups; some are truly bake-stable and are not changed by the cooking process, whereas the others change during the process, interacting with the food substrate to develop their desirable flavors—they are not bake-stable.

In some applications the use of real cheese is just not possible if the price, texture and dietary claims specified by the marketing brief are to be achieved. A good bake-stable cheese flavoring, especially when no real cheese is to be used, needs to mimic the flavor changes which cheese undergoes when it is baked. The flavoring has to have both the heavier long chain fatty acid attributes and the bitter peptide baked taste, but also needs the elusive mouth-watering volatile sulfurous components. The flavoring system therefore needs to remain dynamic during the processing, only reaching equilibrium as the product is completed.

Dynamic cheese flavoring systems work with the process and cook *with* the product, to reach their best only when baked—they really are "bake-unstable"!

10.5.6.2 *Vegetarian products*

The last few years have seen a massive increase in demand for vegetarian versions of many types of flavoring, but cheese is probably the area where this has been most marked. Many products are now made especially for the vegetarian market, or more interesting for the flavoring manufacturer; manufacturers like to be able to flash their mainstream products with "Suitable for Vegetarians" in order to draw in an extra sector of the market. Although this provides an extra segmentation of the market for cheese flavorings, it also sets a formidable challenge to the flavorist. Cheese flavorings do not normally perform very well in the total absence of real cheese. These days this is not much problem if the food manufacturer is using one of the increasingly wide range of vegetarian cheeses (made with microbial rennet) that are now available. However, sometimes even this is not desirable and the flavoring is called upon to provide all the flavor character. This is perhaps the ultimate challenge for the cheese flavorist, but it is remarkable what has been achieved under these difficult conditions.

10.5.7 *The Development of Cheese Flavorings*

There is much in the literature about the flavor of cheese, but with the odd exception (Rondenet & Ziemba, 1970), little of value about the development of cheese flavorings. This is almost certainly because of the commercial sensitivity of flavoring formulations. This chapter will be no exception to this general principle, but we shall try to highlight the key points for consideration by the flavorist embarking on the creation of a cheese flavoring.

As with any flavor area it is useful for the flavorist to determine a set of essential characteristics and then decide on the relative abundance of each. Figure 10–6 sets out one possible way of profiling a cheese flavor. This assumes that the individual flavorist chooses a compound or set of compounds for each flavor element (e.g., "sicky acidic" = butyric acid).

So, in this example, Danablu cheese has a predominantly soapy and acid character, with emphasis on the sharp and sicky acidic notes. Soapiness (and some fruity notes) can be achieved with methyl ketones; acidic character with fatty acids; and fruitiness with esters. The saltiness should not be forgotten, and, in order to achieve a realistic effect in the end product, salt will either have to be incorporated in the flavoring or added to the end product.

Important flavoring materials for use in cheese flavorings are cited in the literature: fatty acids, ketones (especially methyl ketones), aldehydes, lactones, alcohols, esters, amines, sulfur compounds. These are the volatile components. Also vitally important in cheese flavor are the lipids, lipoproteins, peptides and amino acids. By having some knowledge of cheese manufacture and its relationship to the cheese type produced, it is even possible to "guess" a flavoring formulation without sampling the cheese itself. This is clearly an important factor for the fla-

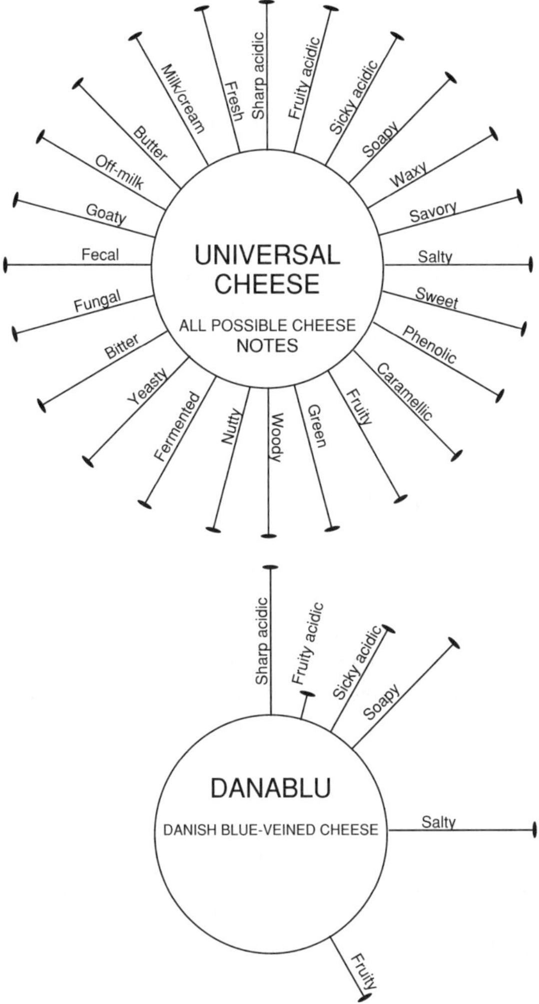

Figure 10–6 Flavor profile diagrams for a universal cheese and for Danablu.

vorist who is asked to service the cheese flavoring needs of distant parts of the world, whose local cheeses he or she is not familiar with.

The choice of ingredients open to the flavorist is very wide, because many of the compounds identified in cheeses are available in nature identical form, and, increasingly, in natural form. The premium for naturals is also falling. For example, natural butyric acid is currently about 15 times the price of the nature identical material, whereas when this chapter was first written around 1990 the multiplier was 25!

Despite this wide range of available single chemicals, the flavorist is unlikely to be able to formulate realistic and cost-effective cheese flavorings with these alone. A wide range of natural flavoring preparations can also be used in cheese flavorings. These range from foods themselves, such as cheese powders, through many extracts, to microbiological preparations such as EMC (enzyme-modified cheese: conventionally prepared cheese to which lipases have been added to accelerate ripening), and other enzymatic preparations of dairy and other materials. EMCs are particularly interesting in that they provide a source of natural cheese flavor components that is relatively cheap. However, most EMC has a characteristic flavor that is not always appropriate on its own for the intended use. So the early promise of EMCs replacing all other cheese flavorings because of their cheapness and ability to be used without the declaration "flavoring" has not been realized in most cases. In practice, EMCs tend to be used as important building blocks for the formulation of natural cheese flavorings, combined with many other natural ingredients, but they are rarely the *sole* ingredients.

10.6 MANUFACTURING CONSIDERATIONS

When developing any new flavoring the flavorist has two main customers to consider. The product development technologist in the customer company is the obvious and first priority. However, the flavor house's own production department is also a vital customer of the flavorist's work. It is no good the flavorist developing a flavoring which totally fulfils the needs of the brief, in terms of character, dosage, price and effectiveness, if the production department is unable to scale the product up from the bench level at which it was developed. The flavorist must have a good appreciation of the needs, methods, equipment and capacity of the production department in order to provide them with a formulation which can turn into a saleable product.

The developed flavor should be organoleptically repeatable when scaled-up, and as simple as possible to compound. The formulation should be as concise as possible, using only reliable and approved suppliers, and without redundant ingredients. The flavoring must be easy to produce repeatably within practical and sensible specifications, as well as meeting the total requirements of the customer company.

Particular problems to be considered in the production of dairy flavors include:

- unpleasant smell
- difficult and hazardous materials to handle (e.g., fatty acids, lactones and acetaldehyde)
- flammability (e.g., diacetyl)

- possible biological activity in dairy ingredients
- many ingredients and solvents are oily; spills can lead to very slippery floors

Generally, these problems and hazards (to both the product and the operatives) can be overcome by strict adherence to good manufacturing practice (IFST, 1989) and normal health and safety rules.

10.7 CONCLUSIONS

The area of dairy flavorings is no longer as specialized as it was. Many companies produce good dairy flavorings, though few offer as broad a range in this area as in the more traditional areas. It remains a profitable niche in the flavor market, sharing the opportunities and constraints of the rest of the industry, but providing an interesting challenge to the flavorist. Understanding of dairy flavor is now almost as well developed as that of fruit flavor, but the relative complexity of the flavors and the multiphase systems in which they develop, and are often used, add to the intellectual challenge. On a business level the market place is still less saturated with a wide variety of competitive flavorings than in many other flavor areas.

REFERENCES

Anon. (1989). *Dairy Industries International* **54**(8), 19.

H.T. Badings, in *Dairy Chemistry and Physics,* ed. P. Walstra and R. Jenness, 342.

P.M. Boon, A.R. Keen & N.J. Walker. (1976). *N. Z. J. Dairy Sci. Techol.* 11, 189.

J.O. Bosset & B. Blanc. (1984). *Lebensm. Wiss. Tecnol.* **17,** 359.

K.J. Burgess. (1988). Ch. 3 in *Food Industries Manual* (22nd ed.), ed. M..D. Ranken, Blackie, Glasgow.

G. Christian. (1977). *Cheese and Cheese-making,* Macdonald, London.

G. Christian. (1984). *World Guide to Cheese,* Ebury Press, London.

J.P. Dumont & J. Adda. (1979). Ch. 21 in *Proc. 2nd Weurmann Symposium* (eds. D.G. Land and H. Nursten).

J.P. Dumont, C. Degas & J. Adda. (1976). *Le Lait* 56, 177.

The Food Labelling Regulations (1996).

The Food Regulations (1970). S.I. 1499.

J. Fernandez-Salguero, A. Marcos, M. Alcala & M.A. Esteban. (1989). *J. Dairy Res.* 56, 141.

IFST. (1989). *GMP: A Guide to its Responsible Management* (2nd ed.), IFST (UK), London.

D.A. Forss. (1979). *J. Dairy Res.* 46, 691.

D.A. Forss, G. Urbach & W. Stark. (1966), *Proc. 17th Int. Dairy Congress* C:2, 211.

J.F. Horwood, G.T. Lloyd & W. Stark. (1981). *Aust. J. Dairy Technol.* 36(1), 34.

C. Karahadian, D.B. Josephson & R.C. Lindsay. (1985). *J. Agric. Food Chem.* 33, 339.

J.E. Kinsella. (1975). *Food Technol.* 29(5), 82.

B.A. Law. (1981). *Dairy Sci. Abstracts* 43(3), 143.

B.A. Law. (1982). *Perfum. Flavor.* 7, October/November 9.

B.A. Law. (1984). Ch. 7 in *Proc. Dairy Symposium,* Elsevier.

R.C. Lawrence et al. (1993). In *Cheese: Chemistry, Physics and Microbiology,* Vol. 2, ed. P.F. Fox, Chapman & Hall, London.

M.J. Lewis, in *Modern Dairy Technology, Advances in Milk Processing,* ed. P.K. Robinson, Elsevier.

R. Liardon, J.O. Bosset & B. Blanc. (1982). *Lebensm. Wiss. Tecnol.,* 15, 143.

H. Maarse & C.A. Visscher, eds. (1983). *Volatile Compounds in Food.* Nutrition and Food Research TNO, Zeist.

E.J. Mann. (1989). *Dairy Ind. Int.* 54(9), 9.

D.J. Manning. (1978). *Dairy Ind. Int.* 43(4), 37.

V. Marquis and P. Haskell. (1985). *The Cheese Book,* Simon & Schuster, New York.

V.M. Marshall. (1982). *Perfum. Flavor.* 7, 27.

P.B. McNulty. (1975). *Proc. SCI Symposium,* Sept. 236.

E. Meinhart & P. Schreier. (1986). *Milchwissenschaft* 41(11), 689.

G.E. Mitchell. (1981). *Aust. J. Dairy Technol.* 36(1), 21.

K.H. Ney. (1981). In *The Quality of Foods and Beverages,* ed. G. Charalambous, Academic Press, New York.

K.H. Ney. (1985). *Fette Seifen Anstrichm.* 87, 289.

T.H. Parliament, M.G Kolor, & D.J. Rizzo. (1982). *J. Agric. Food Chem.* 30, 1006.

E.H. Ramshaw. (1974). *Aust. J. Dairy Technol.* 29(3), 110d.

E.L. Rondenet & J.V. Ziemba. (1970). *Food Eng.* October.

L. Siapantas. (1981). In *The Quality of Foods and Beverages,* ed. G. Charalambous, Academic Press, New York.

The Spreadable Fats (Marketing Standards) Regulations 1995 S.I. No. 3116 as amended by S.I. 1998 No. 452 H.M.S.O. London.

C. Steffen, et al. (1993). In *Cheese: Chemistry, Physics and Microbiology,* Vol. 2, ed. P.F. Fox, Chapman & Hall, London.

A.Y. Tamine & H.C. Deeth. (1980). *J. Food Protec.* 43(12), 939.

T. van Eijk. (1986). *Dragoco Rep.* 3, 63.

A.H. Varnam & J.P. Sutherland. *Milk and Milk Products, Technology, Chemistry and Microbiology,* Chapman & Hall, London.

Chapter 11

Flavor Modifiers

Günter Matheis

11.1 INTRODUCTION

The flavor of a food is composed of taste, odor (aroma) and trigeminal perceptions in the oral and nasal cavities (see Chapter 5). The sensation that our brain registers as flavor is initiated by the simultaneous stimulation of our gustatory (taste), olfactory (odor) and trigeminal senses from an array of components present in our food. In addition to food components that trigger taste, odor and trigeminal impressions, there are some components that are capable of supplementing, enhancing, decreasing or modifying the flavor of foods, although they have little or no flavor of their own at typical usage levels. These substances are commonly known as "flavor enhancers" or "flavor potentiators" (Maga, 1983, 1994; Heath & Reineccius, 1986; Sugita, 1990; Nagodawithana, 1992, 1994a, 1994b, 1995a).

Kemp and Beauchamp (1994) criticized the failure to distinguish flavor enhancer from flavor potentiator. In their view, a flavor potentiator is a substance that increases the perceived intensity of the flavor of another substance, whereas the term flavor enhancer should be used for a substance that increases the pleasantness of the flavor of another substance. Thus, the term flavor enhancer should be restricted to hedonic improvement. Kemp and Beauchamp (1994) also point out that some substances, commonly termed flavor enhancers or flavor potentiators, may suppress the flavor of other substances (e.g., monosodium glutamate suppresses sweetness, bitterness and the cooling effect of menthol but not sourness or butteryness). They propose that substances giving such effects be termed flavor modulators.

The International Organisation of the Flavour Industry (IOFI, 1990) defines flavor enhancer as "substance with little or no odor at the level used, the primary purpose of which is to increase the flavour effect of certain food components well beyond any flavour contributed directly by the substance itself." The EU

(1995) definition of flavor enhancers is "substances that enhance the taste and/or odour of a food."

Flavor enhancers, potentiators or modulators may affect the taste, odor and/or trigeminal impressions of foods. Usually, only taste and/or odor are affected, although maltol and ethylmaltol have been reported to be effective in improving the mouthfeel in low fat food systems (Murray et al., 1995). Because of the lack of a firm definition of flavor enhancer, potentiator or modulator, the term flavor modifier will be used for substances that enhance, suppress or otherwise modify the flavor of foods. Flavor modifiers have been classified into the five categories shown in Table 11–1. Based on this classification, monosodium glutamate and purine 5′-ribonucleotides, for example, are flavor enhancers and flavor suppressors at the same time, because they exhibit taste enhancing and odor and taste (bitter) suppressing effects.

In the following chapter, the most important flavor modifiers, together with some of lesser importance, will be discussed.

11.2 MONOSODIUM GLUTAMATE, PURINE 5′-RIBONUCLEOTIDES AND RELATED SUBSTANCES

11.2.1 *Historical Background*

From time immemorial, cooks around the world have known how to prepare good soup and other food using vegetables and meat or bones. Since more than 2,000 years ago, the Japanese culture has traditionally used the seaweed kombu (*Laminaria japonica*), dried fermented bonito (katsuobushi, a mackerel-type fish), the dried mushroom shiitake (*Lentinus edodes*) and other natural material to improve the quality of their food preparations (Sugita, 1990; Maga, 1994; Nagodawithana, 1995a). Kombu connotes delight and katsuobushi means victory. Experience has taught cooks what scientists discovered only in the early 1900s.

Although glutamic acid was first isolated from wheat gluten and named after it by the German scientist Ritthausen in 1866, it was in 1908 that the Japanese scientist Ikeda first attributed the flavor-improving effect of dried kombu or sea tangle (a type of seaweed) to glutamic acid (Ikeda, 1909). The importance of Ikeda's discovery was soon evident, because the commercial production of the monosodium salt of glutamic acid (monosodium glutamate, MSG) for the intentional addition to foods began shortly thereafter. Apparently, Ikeda was also the first to propose the name umami for the flavor-improving effect of MSG. Umami means deliciousness, palatability, savoriness or succulence in Japanese (Nagodawithana, 1994b, 1995a; Maga, 1994; Imafidou & Spanier, 1994). In China, the word xianwei, which represents the taste common to fish and meat, corresponds to umami (Sugita, 1990).

Table 11–1 Categories of Flavor Modifiers

Category	*Examples*	*Remarks*
1. Flavor enhancers exhibiting little or no flavor at typical usage level	Monosodium glutamate, purine 5′ribonucleotides	Enhance sweet and salty taste impressions and beef stock odor impression
2. Flavor enhancers exhibiting flavor at typical usage	Vanillin[a], ethyl vanillin[a]	Enhance odor impressions (e.g., fruity, chocolate)
	Maltol[b], ethyl maltol[b], 4-hydroxy-2, 5-dimethyl-3(2H)-furanone[b],	Enhance odor impressions (e.g., fruity, creamy)
	4-hydroxy-5-methyl-3(2H)-furanone[b]	Improve the mouthfeel of low fat foods (maltol and ethyl maltol)
	3-Methyl-2-cyclopentene-2-ol-1-one[b]	Enhances odor impressions (e.g., nutty, chocolate)
3. Flavor suppressors exhibiting little or no flavor at typical usage level	Monosodium glutamate, purine 5′-ribonucleotides	Mask or suppress odor impressions (e.g., sulfurous, hydrolysate notes) and sour and bitter taste
4. Flavor suppressors exhibiting flavor at typical usage level	Sucrose[a]	Suppresses unpleasant odor impressions in fruit juices
5. Other flavor modifiers	Miraculin[c]	Sour tasting substances are perceived as sweet tasting for approx. two hours

[a]Although vanillin, ethyl vanillin and sucrose are normally not considered as flavor modifiers, they have flavor modifying properties.

[b]Also used as flavoring substances.

[c]Glycoprotein from miracle fruit, the fruit of the West African shrub *Richadella dulcifera* (*Synsepalum dulcificum*).

A few years later, in 1913, Kodama (1913) discovered that the histidine salt of 5′-inosinic acid, a component present in dried bonito, also exhibits the umami effect. It was later concluded that it was not the histidine part of this substance but the inosinic acid, which is responsible for the flavor-improving effect. This confirmed Justus von Liebig's mid-19th century work on beef broth which also contains inosinic acid. In the early 1960s, the flavor-improving component of shiitake was identified as guanosine 5′-monophosphate (Kuninaka, 1960; Nakajima et al., 1961; Shimazono, 1964).

After these discoveries, other umami substances have been identified, including peptides, amino acids and amines. Today, the most commonly used umami substances are MSG, inosine 5′-monophosphate (IMP) and guanosine 5′-monophosphate (GMP). They are commercially available world-wide.

11.2.2 *Monosodium Glutamate and Glutamic Acid*

Monosodium glutamate is the monosodium salt of L-(+)-glutamic acid (Figure 11–1). Only the completely dissociated form of L-(+)-glutamic acid exhibits the umami effect. The percentages of dissociation at various pH values are shown in Table 11–2 and the pH dependent ionic forms of glutamic acid in Figure 11–2. It is apparent from the data in Table 11–2 that only at pH 6 to 8 does glutamic acid show its optimal umami effect.

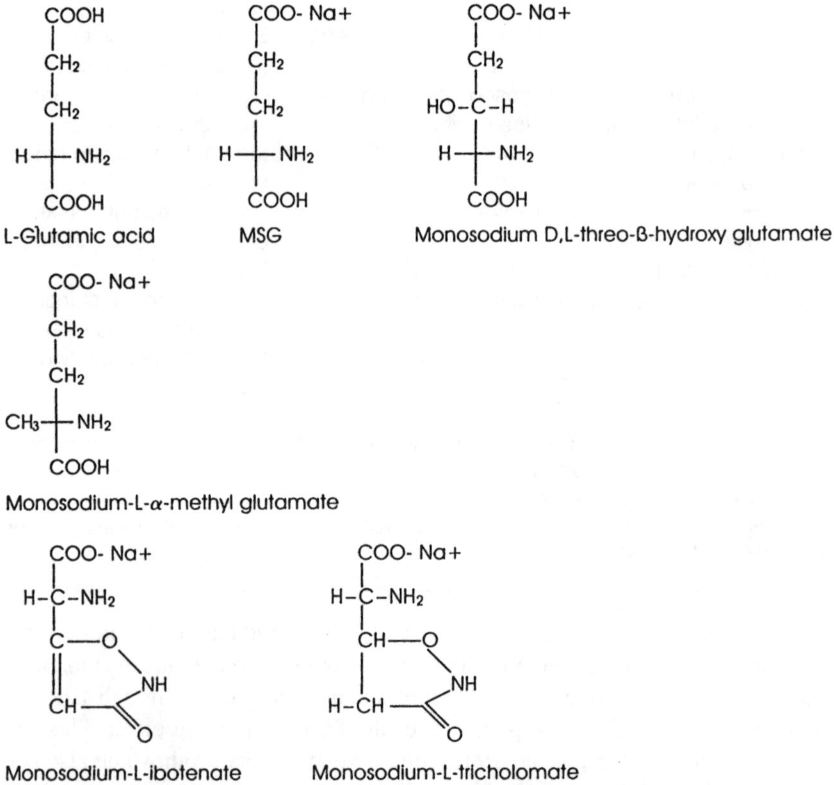

L-Glutamic acid

MSG

Monosodium D,L-threo-β-hydroxy glutamate

Monosodium-L-α-methyl glutamate

Monosodium-L-ibotenate

Monosodium-L-tricholomate

Figure 11–1 Chemical structures of glutamic acid, MSG and selected related substances.

Table 11–2 Percentages of Dissociation of Glutamic Acid at Various pH

pH	*% of dissociation*
3.0	5.3
3.5	15.1
4.0	36.0
4.5	64.0
5.0	84.9
5.5	94.7
6.0	98.2
7.0	99.8
8.0	96.9

Source: Data from J.A. Maga, Umami Flavour of Meat, in *Flavor of Meat and Meat Products,* F. Shahidi, ed., pp. 98–115, © 1994, Blackie Academic and Professional.

In practice, MSG (which corresponds to the completely dissociated form of the acid) is used almost exclusively. MSG is permitted world-wide, although in some countries and some types of foods, maximum concentration limits apply. Other salts of glutamic acids that are occasionally used are potassium, calcium and ammonium glutamates. These salts are not permitted world-wide.

Originally, glutamic acid and MSG were isolated from natural sources. Glutamic acid is ubiquitous in nature and occurs as building block of all proteins. It is also the most abundant amino acid in almost all proteins. To isolate glutamic acid from natural sources would not cover the quantities currently required by the food industry. Current estimated global demand for MSG is about 500,000 metric tons annually (Nagodawithana, 1995a). Presently, the vast majority of MSG is produced through fermentation processes. Among the microorganisms used, bacteria of the genera *Corynebacterium* and *Brevibacterium* are widely employed.

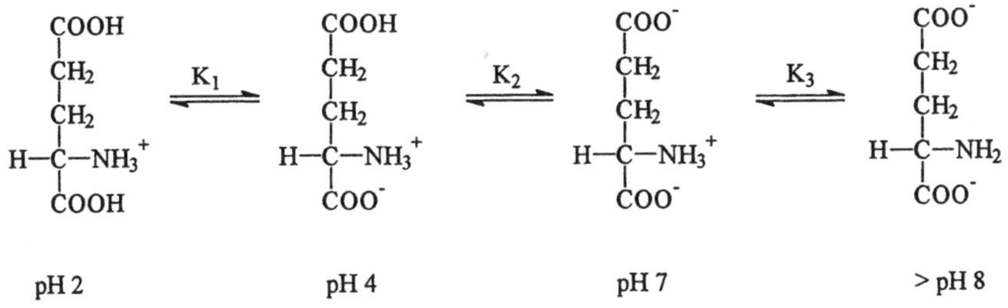

COOH		COOH		COO^-		COO^-
CH_2	K_1 ⇌	CH_2	K_2 ⇌	CH_2	K_3 ⇌	CH_2
CH_2		CH_2		CH_2		CH_2
H—C—NH_3^+		H—C—NH_3^+		H—C—NH_3^+		H—C—NH_2
COOH		COO^-		COO^-		COO^-
pH 2		pH 4		pH 7		> pH 8

Figure 11–2 Ionic forms of glutamic acid.

For example, special strains of *Corynebacterium glutamicum* (formerly reported as *Micrococcus glutamicum*) with a reduced α-ketoglutarate dehydrogenase activity are widely used. The reduced conversion of α-ketoglutarate to succinate causes the accumulation of α-ketoglutarate which is converted to glutamate by reductive deamination (Figure 11–3). Starch, cane and beet molasses or sugar are employed as the carbon source. Ammonium chloride, ammonium sulfate and urea are suitable nitrogen sources. The ammonium ion is detrimental to both cell growth and product formation. Therefore, its concentration must be kept at a low level. Gaseous ammonia has a great advantage over the ammonium salts in maintaining the pH at 7.0 to 8.0, the optimum pH for glutamic acid formation. From glutamic acid, MSG of more than 99% of purity is obtained by neutralization and purification (Figure 11–4). It consists of odorless white crystals.

The taste of MSG has been described as sweet-salty with some tactile properties capable of providing a feeling of "mouth satisfaction" (Nagodawithana, 1995a). Others have described the taste impression as savory, beefy, brothy, meaty and mouthwatering. The detection threshold of MSG in aqueous solution has been reported to be 100 to 300 ppm (Maga, 1983, 1994; Sugita, 1990; Oberdieck, 1980; Niederauer, 1995).

MSG is not hygroscopic and does not change in quality during storage. It is not decomposed during normal food processing or in cooking at pH 5 to 8 (Table 11–3). It is less stable in acidic conditions (below pH 4) and at high temperatures, where cyclization occurs to form 5-pyrrolidone-2-carboxylate (Figure 11–5). At very high temperatures and particularly under alkaline conditions, glutamate tends to racemize to D, L-glutamate. MSG, like other amino acids, also has the capability to undergo Maillard-type reactions in the presence of reducing sugars.

Typical applications for MSG are soups, sauces, ready-to-eat meals, meat and fish products, snacks and vegetable products (with the exception of pickled products because of their low pH of 2 to 3). Typical usage levels are shown in Table 11–4. They range from 0.1 to 0.6% in finished food. There appears to be some variability from one person to another with respect to the preferred optimum level of use.

MSG has no flavor-improving effect on some foods such as confectionery and dairy products, soft drinks, fruit juice drinks, desserts and others. In fact, the addition of MSG to such products may even have an adverse effect on their flavor.

It should be noted that many foods naturally contain free glutamic acid. Examples are given in Tables 11–5 and 11–6. Protein-bound glutamic acid, which occurs in virtually all proteins, has no umami effect.

In the late 1960s, a condition became apparent through the literature, which was called the "Chinese restaurant syndrome" (Maga, 1983; Taliaferro, 1995). Administration of MSG to "normal" individuals was reported to result in various symptoms such as burning sensation in the back of the neck, facial pressure, chest pain, sweating, nausea, weakness, thirst and headache.

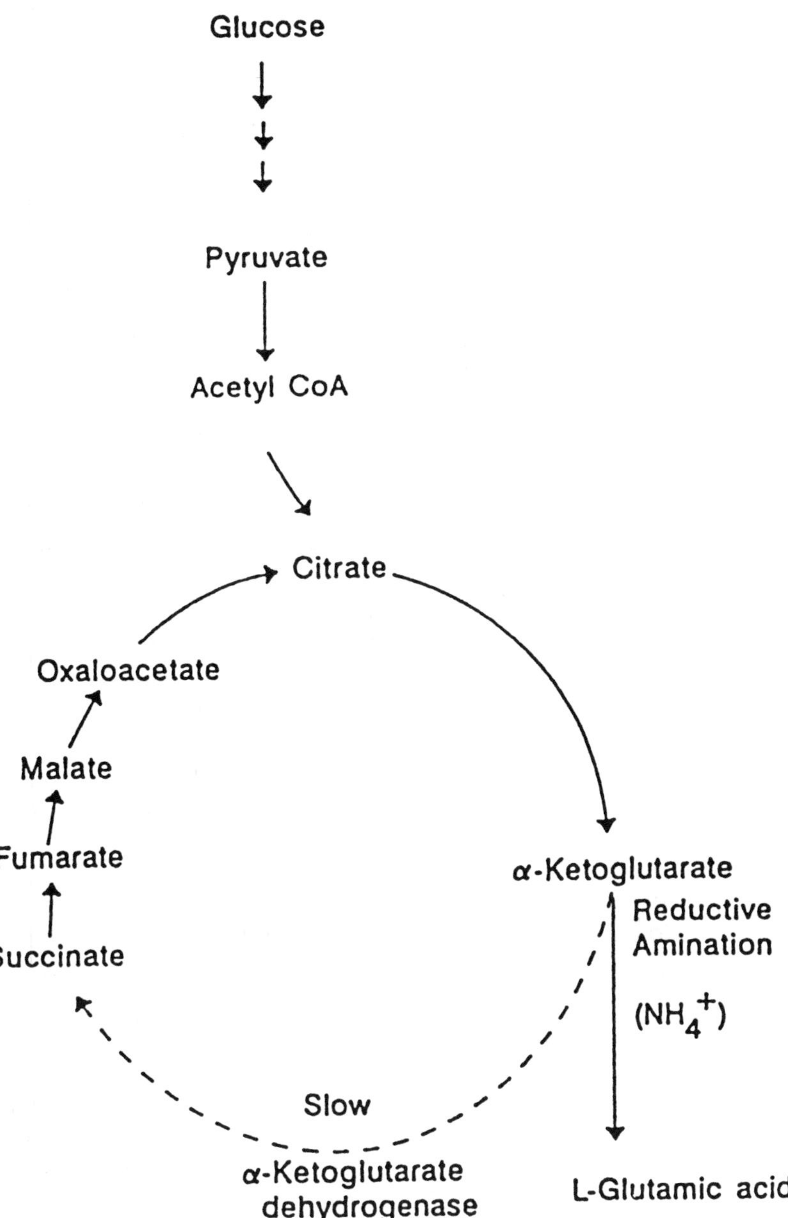

Figure 11–3 Glutamic acid accumulation by *Corynebacterium glutamicum*. *Source:* Reprinted with permission from T.W. Nagodawithana, *Savory Flavors,* p. 321, © 1995, Esteekay Associates.

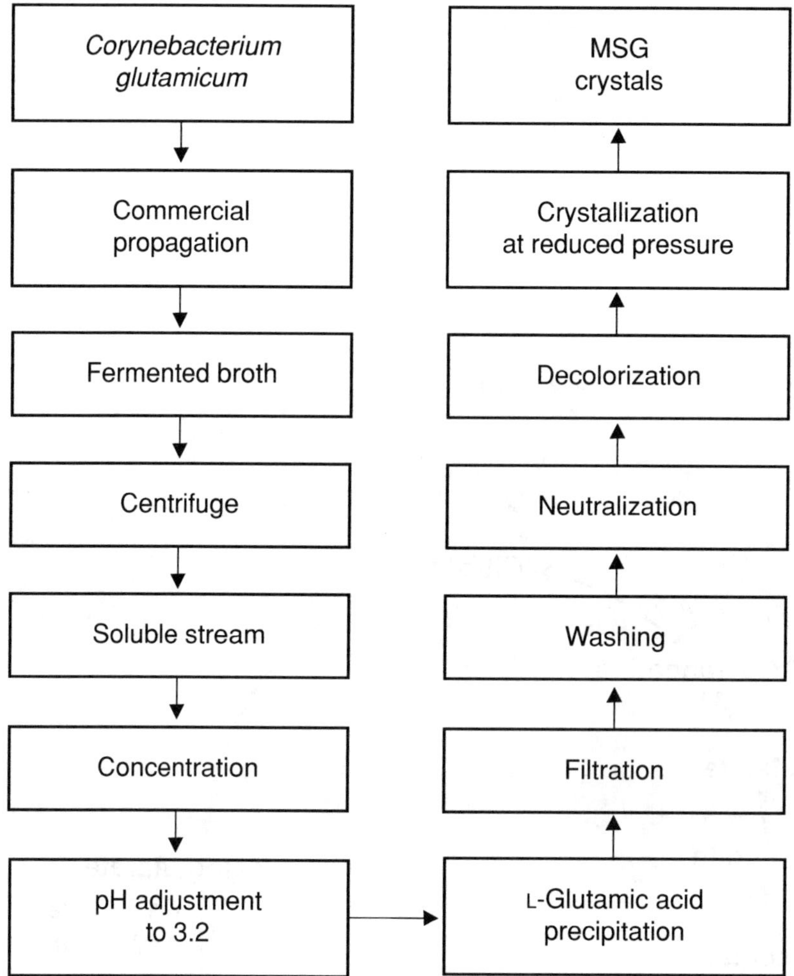

Figure 11–4 Flow chart for the commercial production of MSG by fermentation.

The Joint Expert Committee on Food Additives (JECFA) of the Food and Agricultural Organisation (FAO) of the United Nations and the World Health Organization (WHO) considered the issue of MSG hypersensitivity ("Chinese restaurant syndrome"). They concluded in 1987 that "studies have failed to demonstrate that MSG is the causal agent in provoking the full range of symptoms of Chinese Restaurant Syndrome. Properly conducted double-blind studies among individuals who claimed to suffer from the syndrome did not confirm MSG as the causal agent" (Sugita, 1990). In 1993, the conclusions of the JECFA were confirmed by

Table 11–3 Some Physical Data for MSG, IMP and GMP

	MSG	*IMP*	*GMP*	*IMP/GMP*[a]
pH—optimum for application	5–8[b]	5–7	5–7	5–7
Solubility at 20°C				
• in water	67%	24%	20%	
• in 1% NaCl solution		23%	16%	
Stability				
• after 1 h at 100°C and pH 5.6		95%	89%	
• after 1 h at 100°C and pH 7.0		97%	97%	
• after 2 h at 100°C and pH 5.6		93%	84%	
• after 2 h at 100°C and pH 7.0		93%	94%	
• after 30 min. in corned beef at 100°C		98%	98%	
• after 30 min. in corned beef at 120°C		94%	94%	
• after 3 min. of frying a fish dish at 170–180°C		99%	99%	
• after 50 days at 30°C and pH 5	60%	95%	95%	
• after 40 min. at 115°C and pH 3				71%
• after 40 min. at 115°C and pH 6				77%

[a]Mixture of IMP/GMP (1:1), marketed under the trade name Ribotide®.

[b]5–8 is the pH range of most foods, except for example pickled food (pH 2–3).

Source: Data from J.A. Maga, Flavor Potentiators, *CRC Critical Reviews in Food Science Nutrition,* Vol. 18, pp. 231–312, © 1983; T. Nagodawithana, Yeast-Derived Flavors and Flavor Enhancers and Their Probable Mode of Action, *Food Technology,* Vol. 46, No. 11, pp. 138–144, © 1992; and S. Fuke and T. Shimizu, Sensory and Preference Aspects of Umami, *Trends in Food Science Technology,* Vol. 4, pp. 246–251, © 1993.

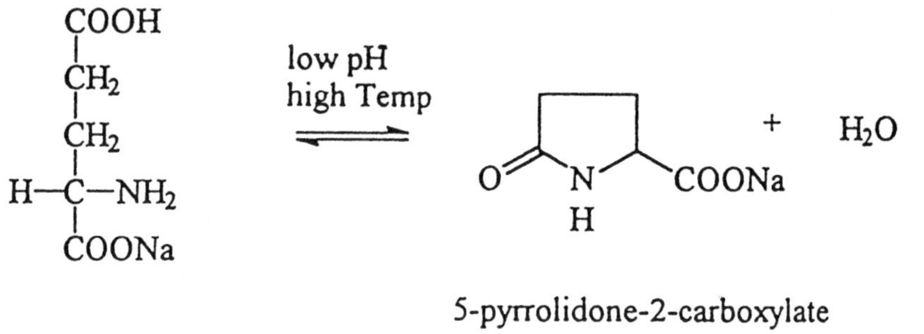

$$\begin{array}{c} COOH \\ | \\ CH_2 \\ | \\ CH_2 \\ | \\ H-C-NH_2 \\ | \\ COONa \end{array} \xrightleftharpoons[\text{}]{\text{low pH, high Temp}} \text{5-pyrrolidone-2-carboxylate (COONa)} + H_2O$$

5-pyrrolidone-2-carboxylate

Figure 11–5 Formation of 5-pyrrolidone-2-carboxylate from MSG. *Source:* Reprinted with permission from T.W. Nagodawithana, *Savory Flavors,* p. 313, © 1995, Esteekay Associates.

---

the University of Western Sydney, Australia (Tarasoff & Kelly, 1993). It appears that histamine, allergenic proteins, preservatives, food colorings or high levels of salt could be responsible for provoking the Chinese restaurant syndrome (Tarasoff & Kelly, 1993; Dayton, 1994). In 1995, the Federation of American Societies for Experimental Biology issued a report that concludes that MSG is safe for the general population at levels normally consumed (Raiten et al., 1995; Institute of Food Technologists, 1995). The panel that prepared the report used the term "MSG symptom complex" to describe possible reactions to MSG. The old term Chinese restaurant syndrome was deemed misleading and pejorative. The United States Food and Drug Administration (FDA) has placed MSG on the list of food additives which are considered "generally recognized as safe" (GRAS). In fact, MSG is cited as an example in the definition of GRAS substances, along with other common food ingredients such as salt, pepper and sugar.

In 1991, the European Union's Scientific Committee for Food (SCF) placed MSG in the safest category for food additives when it determined that it was not necessary to allocate an acceptable daily intake (ADI) for MSG (Glutamate Information Service, 1995).

11.2.3 *Purine 5′-Ribonucleotides*

5′-Ribonucleotides may contain purine or pyrimidine bases. Only the purine 5′-ribonucleotides have flavor-modifying properties. They are building blocks of ribonucleic acid (RNA) and consist of a purine base (e.g., hypoxanthine, guanine, adenine), ribose and phosphoric acid linked to the 5′-position of ribose (Figure 11–6). The most commonly used purine 5′-ribonucleotides are inosine 5′-monophosphate (also known as inosinic acid, 5′-inosine acid, inosine 5′-phosphate or disodium 5′-inosinate) and guanosine 5′-monophosphate (also known as

Table 11–4 Usage Levels of Important Umami Substances in Selected Foods

	Usage level (%)			
Food	*MSG*	*IMP*	*GMP*	*IMP/GMP*[a]
Asparagus, canned	0.08–0.16			0.003–0.004
Cheese, processed	0.40–0.50			0.005–0.010
Crab, canned	0.07–0.10			0.001–0.002
Dressings	0.30–0.40			0.010–0.150
Fish, canned	0.10–0.30			0.003–0.006
Ham, canned	0.10–0.20			0.003–0.010
Hamburgers, frozen	0.10–0.15	0.002–0.004	0.001–0.002	0.001–0.002
Ketchup	0.15–0.30			0.010–0.020
Mayonnaise	0.40–0.60			0.012–0.018
Meat and fish products, canned	0.07–0.30	0.010–0.015	0.004–0.007	0.001–0.010
Poultry, canned	0.10–0.20			0.006–0.010
Sauces	1.00–1.20			0.010–0.030
Sausages	0.30–0.50			0.002–0.014
Sausages, canned	0.10–0.20			0.006–0.010
Snacks	0.10–0.50	0.005–0.010	0.002–0.004	0.003–0.007
Soups and sauces, canned	0.12–0.18	0.004–0.005`	0.002	0.002–0.003
Soups and sauces, dehydrated	5.00–8.00	0.200–0.260	0.090–0.110	0.100–0.200
Soup powders for instant noodles	10.00–17.00			0.300–0.600
Soy sauce	0.30–0.60			0.030–0.050
Vegetable juice	0.10–0.50			0.005–0.010

[a]Mixture of IMP/GMP (1:1), marketed under the trade name Ribotide®.

Table 11–5 Free Glutamic Acid Content of Selected Animal Foods

Food	*Free glutamic acid (%)*[a]
Meat and meat products	0.001–0.064
Beef	0.033–0.042
Chicken	0.044–0.056
Duck	0.064
Lamb	0.003
Mutton	0.008
Pork	0.023–0.029
Sausage	0.001–0.004
Milk and milk products	0.002–2.755
Milk, cow	0.002–0.003
Milk, human	0.022–0.024
Cheese, Camembert	0.390–0.760
Cheese, Danish Blue	0.850
Cheese, Emmental	1.155
Cheese, Gouda	0.584–1.244
Cheese, Gruyere	1.050–1.333
Cheese, Parmesan	0.600–2.755
Cheese, Roquefort	1.605
Cheese, Stilton	1.041
Fish and seafood	0.004–0.508
Abalone	0.138
Carp	0.009–0.022
Clam	0.031–0.316
Cod	0.011
Crab	0.032–0.072
Eel	0.013
Halibut	0.012–0.013
Herring	0.009
Lobster	0.009
Mackerel	0.024–0.075
Octopus	0.037
Oyster	0.335
Pilchard	0.356
Prawn	0.065
Salmon, canned	0.025
Scallop	0.191
Sea bream	0.012–0.024
Sea urchin	0.381–0.508
Squid	0.004–0.056

continues

Table 11–5 continued

Food	*Free glutamic acid (%)*[a]
Tuna	0.005–0.011
Tuna, canned	0.025
Eggs	0.023–0.029

[a]Protein-bound glutamic acid has no umami effect.

Source: Data from J.A. Maga, Flavor Potentiators, *CRC Critical Reviews in Food Science Nutrition,* Vol. 18, pp. 231–312, © 1983 and Y.-H. Sugita, Flavor Enhancers, in *Food Additives,* A.L. Branen, P.M. Davidson, and S. Salminen, eds., pp. 259–296, © 1990, Marcel Dekker.

guanylic acid, 5′-guanylic acid, guanosine 5′-phosphate or disodium 5′-guanylate). In order to show the flavor-improving effect, the 5′-ribonucleotide must have a hydroxyl group or an amino group in the 6-position of the purine base (Figure 11–6), preferably a hydroxyl group such as in IMP and GMP. An amino group in the 6-position of the purine base, such as in adenosine 5′-monophosphate (AMP; Figure 11–6) decreases the flavor-improving effect. Another prerequisite is the phosphate in the 5′-position of the ribose moiety (Figure 11–6). 2′- and 3′-phosphates have no flavor-modifying effects.

As in the early days when MSG was produced from glutamic acid rich proteins, such as gluten, purine 5′-ribonucleotides were also initially isolated from natural sources (e.g., meat, fish, dehydrated mushrooms). Fresh muscles of marine fish, for example, served as a satisfactory source for the production of IMP. In the 1950s, production techniques involving microbial fermentation came into widespread use for both IMP and GMP, because of their higher level of productivity. Today, several processes are being employed and the most widely used methods are summarized in Table 11–7. Almost the entire supply of IMP and GMP available is produced by applying the direct fermentation techniques (Maga, 1983), using, for example, *Brevibacterium ammoniagenes* (Nagodawithana, 1995a). The direct fermentation techniques are outlined in Figure 11–7; the hydrolysis of RNA in Figure 11–8. AMP offers less flavor-enhancing properties than IMP and GMP, but it can serve as precursor of IMP (Figure 11–8). Cytosine monophosphate (CMP) and uracil monophosphate (UMP) exhibit no flavor-improving effect but have found use in the pharmaceutical industry.

IMP and GMP are non-hygroscopic salts. They are stable under heat (up to 120°C) and acidity (optimum pH for the flavor-improving effect is 5–7) (Table 11–3). Taste thresholds in aqueous solution have been reported to be 25–250 ppm and 12–200 ppm for IMP and GMP, respectively (Maga, 1983, 1994; Sugita, 1990; Nagodawithana, 1995a; Oberdieck, 1980). The tastes of IMP and GMP

Table 11–6 Free Glutamic Acid Content of Selected Plant-Based Foods

Food	*Free glutamic acid (%)*[a]
Fruits	Traces–1.219
Apple	0.005
Grape	0.044
Lemon	0.009
Nectarine	1.219
Orange	0.015
Pear	0.020
Strawberry	0.055
Vegetables	0.001–0.724
Asparagus	0.051–0.076
Bean	0.005–0.025
Beet, canned	0.038
Broccoli	0.176–0.213
Carrot	0.004
Corn	0.051–0.165
Cucumber	0.001
Egg plant	0.001
Garlic	0.002
Mushroom, shiitake, dried	0.177–0.635
Mushroom, *Agaricus bisporus*	0.175–0.685
Mushroom, *Agaricus bosporus*, dried	0.571
Mushroom, *Boletus edulis*, dried	0.025
Mushroom, *Tricholoma nudum*	0.393
Mushroom, *Tricholoma protentosum*	0.535
Onion	0.001
Pea	0.051–0.254
Potato	0.045–0.254
Pumpkin	0.004
Radish	0.002
Spinach	0.005–0.049
Tomato	0.005–0.724

[a]Protein-bound glutamic acid has no umami effect.

Source: Data from J.A. Maga, Flavor Potentiators, *CRC Critical Reviews in Food Science Nutrition,* Vol. 18, pp. 231–312, © 1983 and Y.-H. Sugita, Flavor Enhancers, in *Food Additives,* A.L. Branen, P.M. Davidson, and S. Salminen, eds., pp. 259–296, © 1990, Marcel Dekker.

have been described as beefy and oak-mushroom, respectively (Nagodawithana, 1995a).

The use of IMP and GMP in liquid foods or foods with high water content may present some problems. Many animal and plant-based foods contain phosphomo-

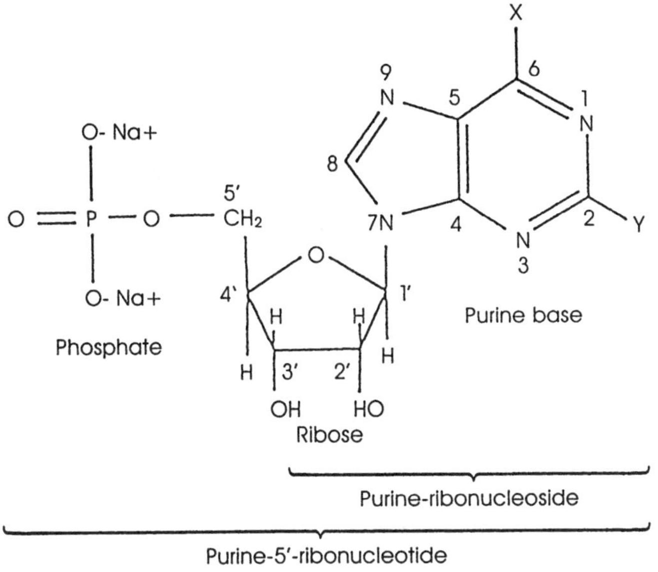

X = OH, Y = H: Inosine monophosphate (IMP) = disodium-5′-inosinate
X = OH, Y = NH_2: Guanosine monophosphate (GMP) = disodium-5′-guanylate
X = NH_2, Y = H: Adenosine monophosphate (AMP) = disodium-5′-adenylate

Figure 11–6 Examples of disodium salts of purine 5′-ribonucleotides. *Source:* G. Matheis, Flavor Modifiers, *Dragoco Report,* Vol. 42, pp. 5–27, © 1997, Dragoco.

noesterases. These enzymes can easily split the phosphomonoester linkage of the ribonucleotides, and the flavor-improving effect is lost. These enzymes should be inactivated by heating the food to 85°C before adding IMP or GMP, or the foods should be frozen to slow down enzyme activity.

Typical applications for IMP and GMP are soups, sauces, dressings, ready-to-eat meals, meat and fish products, snacks and vegetable products. Usage levels range from 0.001 to 0.150% in finished food (Table 11–4). A 1:1 mixture of IMP and GMP is commercially available under the tradename Ribotide® and extensively used (Table 11–4). The taste threshold of Ribotide® has been reported to be 63 ppm (Maga, 1983, 1994; Sugita, 1990). IMP and GMP have no flavor-improving effects on a range of foods, including confectionery, dairy products, soft drinks and desserts.

Many foods naturally contain purine 5′-ribonucleotides. Examples are given in Tables 11–8 and 11–9. In most meats, IMP results from the decomposition of AMP which in turn results from adenosine triphosphate (ATP) (Maga, 1983; Oberdieck, 1980). In fish and chicken, for example, the enzymatic conversion of ATP to IMP occurs within a few hours after death (Maga, 1983). Thus, the IMP

Table 11–7 Production Methods for Purine 5′-Ribonucleotides

Type of process	*Brief description*
Direct fermentation	Conversion of sugars into IMP and GMP
Direct fermentation	Conversion of sugars into purine ribonucleotides with subsequent phosphorylation into the purine 5′-ribonucleotides
Hydrolysis of RNA	Degradation of yeast RNA to purine 5′-ribonucleotides (GMP, AMP, UMP[a], CMP[b])
	Subsequent conversion of AMP into IMP
Combination	Any combination of the above three procedures

[a]Uracil 5′-monophosphate
[b]Cytosine 5′-monophosphate
Source: Data from T.W. Nagodawithana, *Savory Flavors,* pp. 297–333, © 1995, Esteekay Associates.

content of fresh raw meat is usually rather high. After several days of refrigerated storage or in processed raw meat, however, it is relatively low, because raw meat contains phosphomonoesterase, and the IMP in raw meat is easily lost in processes as thawing, washing and salting.

11.2.4 *Umami Effect*

Experimental data on receptor mechanisms of umami substances in animals have shown that receptor sites for umami are independent of those for the four

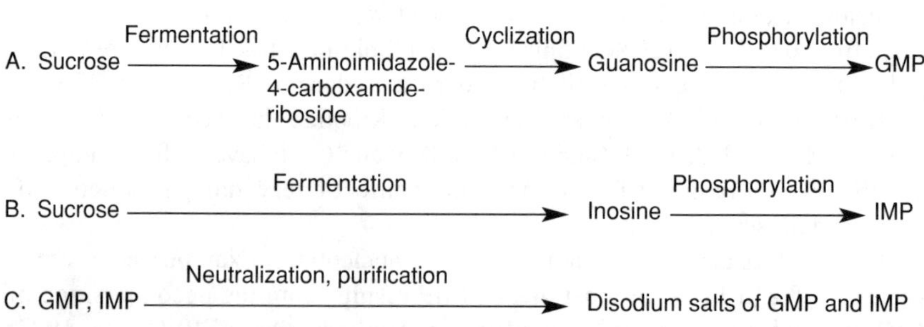

Figure 11–7 Production of IMP and GMP by direct fermentation. *Source:* Reprinted with permission from G. Matheis, Flavour Modifiers, in *Flavourings,* E. Ziegler and H. Ziegler, eds., pp. 310–327, © 1998, Wiley-VCH.

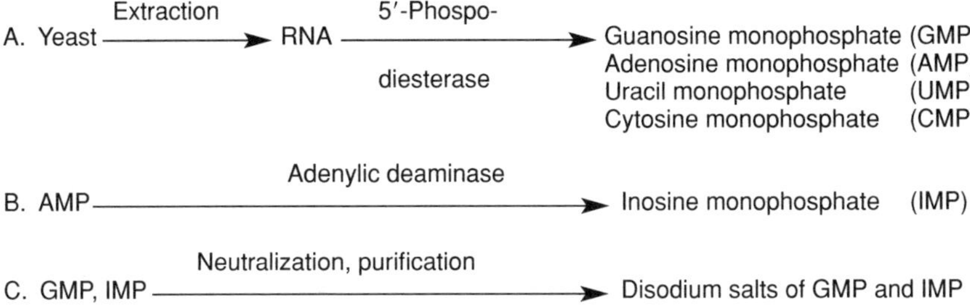

Figure 11–8 Production of IMP and GMP by hydrolysis of RNA. *Source:* Reprinted with permission from G. Matheis, Flavour Modifiers, in *Flavourings,* E. Ziegler and H. Ziegler, eds., pp. 310–327, © 1998, Wiley-VCH.

---

basic tastes (sweet, sour, salty and bitter). Applying psychological and physiological studies on human subjects, Yamaguchi (1987) developed a three-dimensional model, where the four basic tastes are on the edges, the faces, on the inside or in the near vicinity of a tetrahedron, and umami is outside the tetrahedron (Figure 11–9). Thus, various authors postulate the idea that umami is the fifth primary taste impression. This is contradictory to early literature reports that in some manner MSG modifies the four basic taste sensations (Maga, 1983) and to the observed facts that umami substances enhance or suppress taste impressions and mask odor impressions (Table 11–1). Moreover, MSG, best known for its umami effect, possesses structural features that could elicit all the basic taste qualities sweet, sour, salty and bitter (Figure 11–10). This approach has led Birch (1987) to propose that MSG not only exhibits the umami effect but also has a salty taste. Whether umami represents an independent and uniquely different entity of oral sensation or a part of the four basic tastes remains an unresolved question. The fundamental understanding of the way that umami substances improve the palatability of a whole range of foods remains elusive.

Nevertheless, notable advances have been made in recent years on the probable mechanism by which MSG, IMP and GMP enhance taste impressions (Nagodawithana, 1994a). This is illustrated in Figure 11–11. Figure 11–11a shows schematically the binding of a flavoring substance A with a taste stimulus A to the surface of a taste receptor located on the tongue. The stimulus A initiates the taste signal A. Figure 11–11b shows weak binding of MSG to the MSG binding site of the receptor. Because of the weak binding of MSG, stimulus-receptor interaction is still weak. As a result, the taste signal is still weak, although stronger than in the absence of MSG. Figure 11–11c illustrates the proposed stimulus-receptor interaction when purine 5′-ribonucleotides are bound in addition to MSG. GMP or IMP are thought to allosterically alter the MSG binding site. Both MSG and IMP or GMP are strongly bound and force the flavoring substance to be more strongly bound, as well. This results in a stronger taste signal (or taste enhancement). A

Table 11–8 IMP, GMP and AMP Content of Selected Animal Foods

	Purine 5'-ribonucleotide (%)		
Food	*IMP*	*GMP*	*AMP*
Meat and meat products	0.075–0.443	0.001–0.004	0.006–0.013
Beef	0.106–0.443	0.002	0.006–0.008
Chicken	0.075–0.122	0.001–0.002	0.007–0.013
Pork	0.186	0.004	0.009
Fish and fish products	0.000–1.310	0.000–0.023	0.000–0.184
Abalone	0	0	0.081
Bonito	0.285		0.007
Bonito, dried	0.630–1.310	Trace	Trace
Clam	0	0	0.012–0.098
Cod	0.043		0.023
Crab	0.000–0.076	0.000–0.023	0.010–0.059
Eel	0.165	Trace	0.020
Herring, dried	0.072		
Lobster	0	0	0.082
Mackerel	0.014–1.300	Trace	0.006–0.007
Octopus	0	0	0.026
Oyster	0	0	0.021
Salmon	0.002–0.235	Trace	0.006–0.007
Sardine	0.192		0.006
Scallop	0	0	0.116
Sea bass	0.188	Trace	0.009
Sea bream	0.214–0.421	Trace	0.012
Shrimp, dried	0.099		
Squid	0	0	0.184
Squid, dried	0.023		
Swordfish	0.019	0	0.003
Trout	0.117–0.187		0.004–0.014
Tuna	0.286	Trace	0.005
Whale	0.214–0.385	0.005	0.002
Milk			
Cow	0.115–0.326	0.002–0.005	0.002–0.013

Source: Data from Maga, 1983; Sugita, Y.-H., 1990; Maga, 1994; Oberdieck, 1980; Niederauer, 1995; and Fuke & Shimizu, 1983.

similar reaction for a flavoring substance with a taste stimulus B that initiates a taste signal B is depicted in Figure 11–11d.

The common structural element of umami substances is the presence of two negative charges in the molecule, spaced 3 to 9 carbon or other atoms apart, and preferably 4 to 6 atoms. MSG, IMP and GMP fit well into this concept (Figure

Table 11–9 IMP, GMP and AMP Content of Selected Vegetables

	Purine 5′-ribonucleotide (%)		
Vegetable	*IMP*	*GMP*	*AMP*
Asparagus		Traces	0.004–0.027
Corn	0	0	0.006
Cucumber			0.001–0.002
Bean			0.001
Mushroom, shiitake	0	0.045–0.103	0.030–0.175
Mushroom, shiitake, dried		0.033–0.216	0.063–0.321
Mushroom, *Agaricus bisporus*		0.019–0.041	0.015
Mushroom, *Agaricus bisporus*, dried		0.011	
Mushroom, *Boletus edulis*, dried		0.001	
Mushroom, *Tricholoma nudum*		0.004	
Mushroom, *Tricholoma protentosum*		0.001	
Onion		Traces	0.001
Pea	0	0	0.002
Potato		0.001	
Tomato		0.001	0.010–0.012

Source: Data from Maga, 1983; Sugita, Y.-H., 1990; Maga, 1994; Oberdieck, 1980; Niederauer, 1995; and Fuke & Shimizu, 1983.

11–12). AMP, which is less effective as a umami substance, has one negative charge at one end of the molecule and an electronegative atom (i.e., nitrogen) at the other (Figure 11–12).

11.2.5 *Other Umami Substances*

Other umami substances include amino acids (such as cysteine, homocysteine, cysteine S-sulfonic acid, aspartic acid, α-methyl glutamic acid, tricholomic acid, ibotenic acid), sulfur containing substances (e.g., 3-methyl thiopropylamine) and peptides. They are of less commercial interest than MSG, IMP and GMP. Chemical structures of some of these substances are shown in Figure 11–1. Relative umami effects of some are shown in Tables 11–10 and 11–11. Tricholomic acid and ibotenic acid have been found in the mushrooms *Tricholoma muscarium* and *Amanita strobiliformis*, respectively.

11.2.6 *Synergism*

A fascinating aspect of umami substances is their ability to act synergistically in foods. Synergism is defined as the cooperative action of two (or more) components of a mixture whose total effect is greater than the sum of their individual ef-

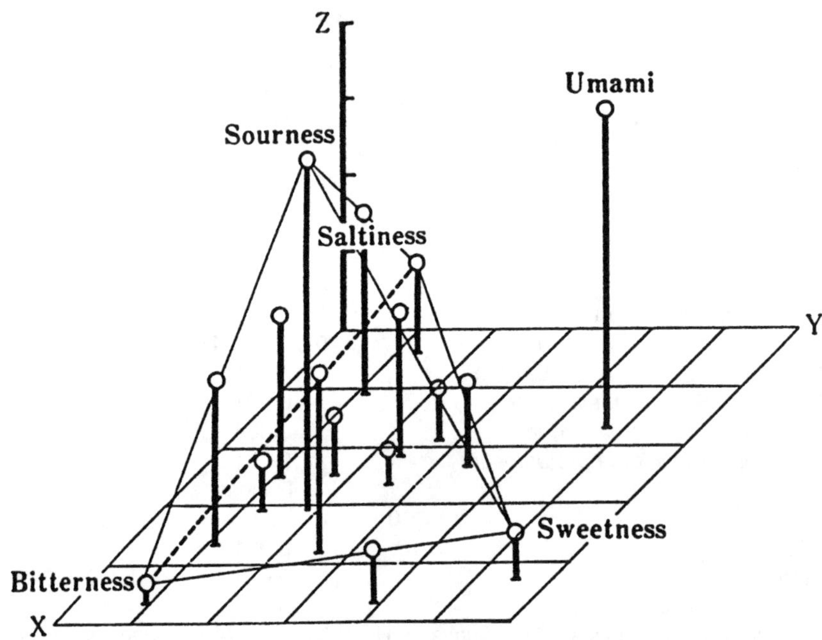

Figure 11–9 Three-dimensional model for the stimuli sweet, sour, salty, bitter and umami. *Source:* Reprinted from Y.-H. Sugita, Flavor Enhancers, in *Food Additives*, A.L. Branen, P.M. Davidson, and S. Salminen, eds., pp. 259–296, © 1990, Marcel Dekker.

---

fects. Umami substances show pronounced synergistic effects between themselves and with other substances (Maga, 1983, 1994; Nagodawithana, 1992, 1995a; Heath & Reineccius, 1986; Sugita, 1990; Oberdieck, 1980; Fuke & Shimizu, 1993). Table 11–12 illustrates such effects in MSG/IMP and MSG/GMP combinations. Note that a 1:1 mixture of MSG/GMP produces a thirtyfold increase in umami intensity over MSG alone and that the intensities go through a maximum with increasing amounts of MSG. It appears, therefore, that the most effective flavor potentiator system is the 1:1 mixture of MSG/GMP. Due to the relatively high cost of purine 5′-ribonucleotides, however, they are seldomly used at 1:1 ratio with MSG. Typically a 95:5 ratio of MSG/purine 5′-ribonucleotides is used in the food industry, the purine 5′-ribonucleotides being a 1:1 mixture of IMP/GMP (Ribotide®). This combination yields a synergistic effect of approximately sixfold over MSG. This means, in practice, that MSG can be replaced by 17 to 20% of a mixture of 95% MSG and 5% Ribotide® with a cost reduction of approximately 50%. By using IMP/MSG mixtures, even more cost reduction can be achieved (Table 11–13).

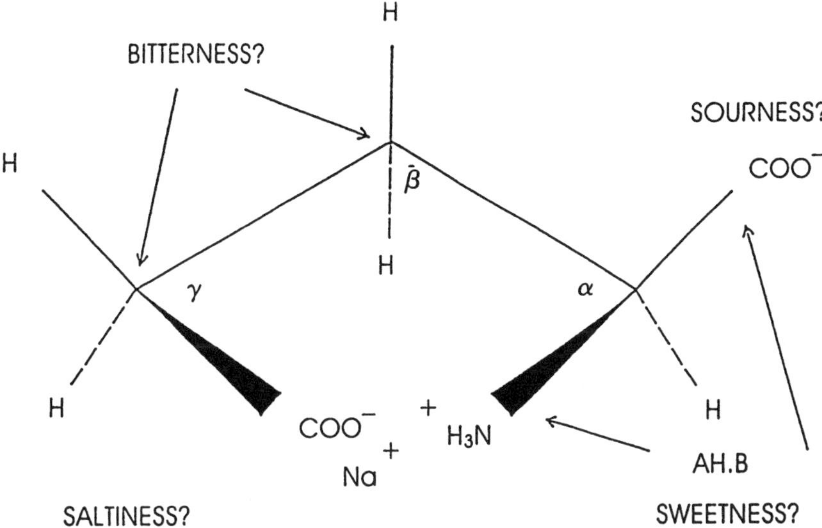

Figure 11–10 Structure of MSG showing the structural features that could elicit the four primary taste qualities. *Source:* Reprinted with permission from T. Nagodawithana, Yeast-Derived Flavors and Flavor Enhancers and Their Possible Mode of Action, *Food Technology,* Vol. 46, No. 11, pp. 138–144, © 1992, Institute of Food Technologists.

One of the approaches to quantify synergistic effects is the measurement of taste thresholds of individual and mixed systems. Studies have demonstrated a dramatic decrease in the taste threshold of one flavor modifier as a result of the synergistic influence by another flavor modifier. The threshold values of IMP and GMP, for example, are 25–250 ppm and 12–200 ppm, respectively, when tasted individually, whereas the taste threshold of Ribotide® has been reported to be 63 ppm. When used in combination with 0.8% MSG, the taste threshold of Ribotide® decreases to 0.3 ppm (Maga, 1983, 1994; Nagodawithana, 1995a), a dramatic reduction due to the strong synergism between these flavor modifiers.

Synergistic effects between MSG and sodium chloride have also been reported (Birch, 1987). For maximum palatability of a clear soup, more MSG must be added if only small amounts of salt are used, and vice versa. Optimum amounts of salt and MSG are 0.8% and 0.4%, respectively. Thaumatin, an intensely sweet tasting protein, has been reported to have a similar synergistic effect to that of purine 5′-ribonucleotides on MSG, but at significantly lower dosages (Van Eijk, 1987). Other umami substances with synergistic effects on each other include various amino acids (e.g., cysteine, homocysteine, cysteine S-sulfonate), cycloalliin and histamine (Van Eijk, 1987).

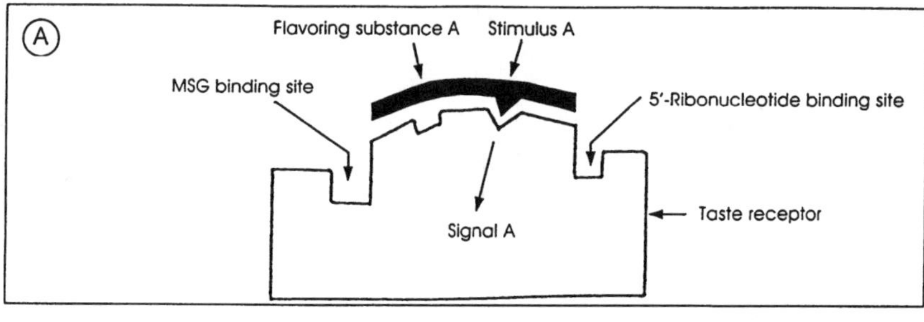

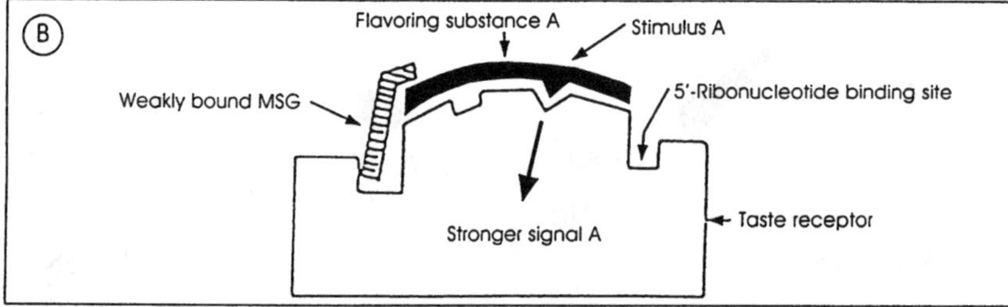

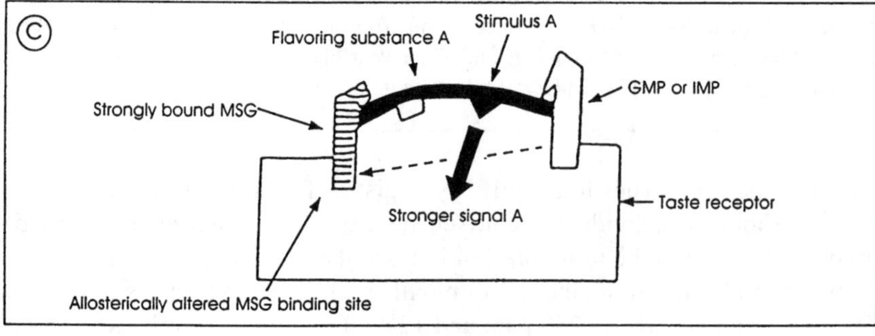

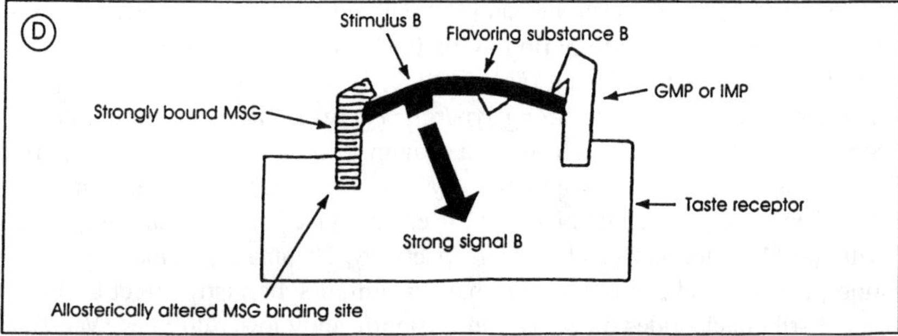

Figure 11–11 Proposed interactions of flavoring substances at the taste receptor surface in the absence and presence of MSG, IMP and GMP. *Source:* G. Matheis, Flavor Modifiers, *Dragoco Report,* Vol. 42, pp. 5–27, © 1997, Dragoco.

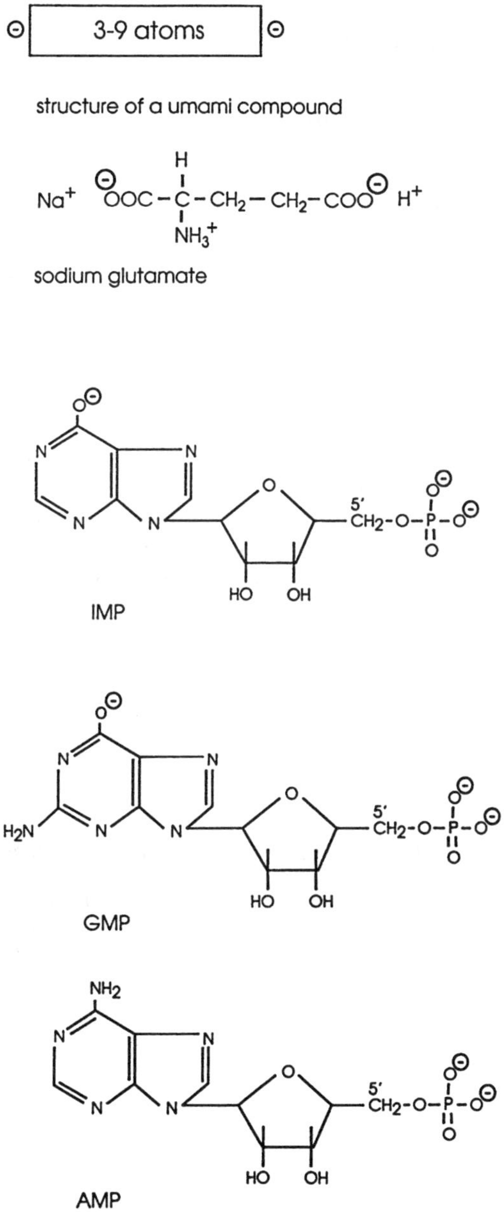

⊖ 3-9 atoms ⊖

structure of a umami compound

$Na^+ \ ^{\ominus}OOC-CH(NH_3^+)-CH_2-CH_2-COO^{\ominus} \ H^+$

sodium glutamate

IMP

GMP

AMP

Figure 11–12 Common structural element of umami substances. *Source:* G. Matheis, Taste, Odor, Aroma, and Flavor, *Dragoco Report,* Vol. 39, pp. 50–65, © 1994, Dragoco.

Table 11–10 Umami Effect of Selected Substances Relative to MSG

Umami substance	*Relative effect*
MSG	1.00
Monosodium-D,L-threo-β-hydroxyglutamate	0.86
Monosodium-D,L-homocystate	0.77
Monosodium-L-aspartate	0.01
Monosodium-L-α-aminoadipate	0.01
Tricholomic acid	5–30
Ibotenic acid	5–30

Source: G. Matheis, Flavor Modifiers, *Dragoco Report,* Vol. 42, pp. 5–27, © 1997, Dragoco.

11.2.7 *Yeast Extracts (or Autolysates) and Hydrolyzed Vegetable Proteins*

It should be mentioned that not only single chemical components but also complex materials are used as umami substances. These include yeast extracts and autolysates (due to high glutamic acid and purine 5′-ribonucleotides content) (Tables 11–14 and 11–15), as well as hydrolyzed vegetable proteins (due to high glutamic acid content; Table 11–14) (Nagodawithana, 1995a, 1995b). Hydrolyzed vegetable proteins (HVP) have been known for a long time in the food industry, but only since the 1930s have these products gained prominence as flavor modifiers. They may be obtained by hydrolysis with mineral acid (e.g., HCl)

Table 11–11 Umami Effect of Selected Substances Relative to IMP

Umami substance	*Relative effect*
IMP	1.00
GMP	2.30
AMP	0.18
2-Methyl 5′-inosinate	2.30
2-Ethyl 5′-inosinate	2.30
2-Methylthio 5′-inosinate	8.00
2-Ethylthio 5′-inosinate	7.50
2-Furfurylthio 5′-inosinate	17.00

Source: Data from J.A. Maga, Flavor Potentiators, *CRC Critical Reviews of Food Science Nutrition,* Vol. 18, pp. 231–312, © 1983; Y.-H. Sugita, Flavor Enhancers, in *Food Additives,* A.L. Branen, P.M. Davidson, and S. Salminen, eds., pp. 259–296, © 1990, Marcel Dekker; and T.W. Nagodawithana, *Savory Flavors,* pp. 297–333, © 1995, Esteekay Associates.

Table 11–12 Umami Effect Intensities of MSG/IMP and MSG/GMP Combinations

Ratio of		*Relative umami*	*Ratio of*		*Relative umami*
MSG	*IMP*	*intensity*	*MSG*	*GMP*	*intensity*
1	0	1.0	1	0	1.0
1	1	7.0	1	1	30.0
10	1	5.0	10	1	18.8
20	1	3.5	20	1	12.5
50	1	2.5	50	1	6.4
100	1	2.0	100	1	5.4

Source: G. Matheis, Flavor Modifiers, *Dragoco Report,* Vol. 42, pp. 5–27, © 1997, Dragoco.

or by enzymatic hydrolysis (Table 11–14). They are used, for example, in soups, gravies, savory snacks, sauces and ready-to-eat meals. They are stable under varying process conditions, e.g., canning, freezing and heating up to 180° C. HVP preparations now gaining popularity have salt concentrations sometimes less than half of the levels found in traditional HVP. This is achieved by including several crystallization steps after the concentration of the material or by partial replacement of HCl with H_2SO_4.

Yeast extracts are concentrates of soluble material derived from yeast following hydrolysis of the cell material, particularly the proteins, carbohydrates and nucleic acids. Hydrolysis is carried out by use of the yeast's own hydrolytic enzymes (autolysis) or by other methods (hydrolysis or plasmolysis). Yeast extracts are commercially available as powders and pastes. Many types of cheese flavored products such as crackers are commonly improved by the incorporation of yeast extracts.

Table 11–13 Cost Reduction by Using MSG/IMP Mixtures

MSG/IMP ratio	*Cost reduction (%)*
100:0	
98:2	59
96:4	63
94:6	62
92:8	61

Source: G. Matheis, Flavor Modifiers, *Dragoco Report,* Vol. 42, pp. 5–27, © 1997, Dragoco.

Table 11–14 Examples of Complex Materials Used as Flavor Modifiers

Material	*Remarks*
Yeast autolysates	Contain up to 6% 5′-ribonucleotides, especially GMP
Hydrolyzed vegetable proteins (hydrolyzed with mineral acid)	Contain up to 17% MSG; 30 to 50% of dry matter is sodium chloride
Hydrolyzed vegetable proteins (enzymatically hydrolyzed)	Contain up to 35% MSG; practically free of sodium chloride
Hydrolyzed vegetable proteins (hydrolyzed with organic acid, e.g., acetic acid)	No commercial importance at present

Source: G. Matheis, Flavor Modifiers, *Dragoco Report,* Vol. 42, pp. 5–27, © 1997, Dragoco.

Although neither the FDA in the United States nor the European Commission have identified reasons to restrict MSG consumption, there is evidence for a negative public perception of MSG, probably due to the above mentioned "Chinese restaurant syndrome." In Europe, MSG as well as the purine 5′-ribonucleotides have an additional negative image because of the labeling requirement with an E-number.

It has been shown that the production of HVP using HCl can lead to the formation of chloropropanols which are known to be carcinogenic (Nagodawithana, 1995b). These compounds appear as a result of the reaction between HCl and traces of lipids. Various countries have set maximum levels of chloropropanols in HVP and in food. Some countries are currently reviewing this issue.

Yeast extracts or autolysates have not shown adverse reactions in humans and hence have, thus far, received the "clean label" status of being natural. Products containing glutamic acid (such as HVP and yeast extracts) are considered to be GRAS. In addition to HVP and yeast extracts, various oligopeptides are occasionally used as MSG replacers. These will not be discussed here, except for the so-called beefy meaty peptide (BMP) which will be referred to in section 11.6.

Table 11–15 Ribonucleotide Content of Yeast Autolysates

	GMP (%)	*IMP (%)*	*AMP (%)*	*Other (%)*[a]	*Total (%)*
Yeast autolysate	1.5	0–1.5	0.1–1.6	2.4	5.5

[a]UMP (uridine monophosphate) and CMP (cytidine monophosphate) contain the pyrimidine bases uracil and cytosine and have no umami effect.

Source: G. Matheis, Flavor Modifiers, *Dragoco Report,* Vol. 42, pp. 5–27, © 1997, Dragoco.

11.3 MALTOL AND ETHYL MALTOL

11.3.1 *Historical Background*

In 1861, maltol (Figure 11–13) was first isolated from the bark of the larch tree by Stenhouse. He noticed the pleasant odor of maltol and its slightly bitter and astringent taste. In 1894, Brand isolated maltol from roasted malt.

The first synthesis of maltol started from pyromeconic acid, which itself was an expensive material, and gave maltol in low yields. Today, there are two routes for industrial production. One is fermentation combined with chemical synthesis (Oberdieck, 1980; Le Blanc & Akers, 1989). Kojic acid, a fermentation product of an *Aspergillus* fungus, is oxidized. After oxidation, formaldehyde is added. Adding acetaldehyde instead of formaldehyde yields ethyl maltol (Figure 11–13). The other route starts from furfural and results in maltol or ethyl maltol (Murray et al., 1995).

11.3.2 *Maltol*

Maltol occurs naturally in many foods (e.g., baked goods, cocoa, chocolate, coffee, caramel, malt, condensed milk, chicory, cereals, soy sauce and beer) (Murray et al., 1995; Oberdieck, 1980; Le Blanc & Akers, 1989; Belitz & Grosch, 1982). It is formed when carbohydrates are heated (Figure 11–14). Maltol is a white, crystalline powder with a caramel-like odor. Its taste threshold in water at 20° C has been reported to be 35 ppm (Belitz & Grosch, 1982).

Maltol has long been known to enhance the flavor of sweet foods. The addition of 5 to 75 ppm maltol may permit a 15% sugar reduction in some sweet foods (Heath & Reineccius, 1986). Table 11–16 shows some usage levels of maltol in selected sweet foods.

11.3.3 *Ethyl Maltol*

Ethyl maltol has not yet been found occurring naturally in foods. It is 4 to 6 times stronger than maltol and can replace maltol in sweet foods. Typical usage levels are shown in Table 11–16.

11.3.4 *Maltol and Ethyl Maltol*

It has been reported recently that maltol and ethyl maltol are also capable of improving the flavor of savory (or spicy) foods (Murray et al., 1995). In salad dressings, they round the spiciness and decrease the "bite" or acid sensation contributed by acetic acid. Both substances have also been reported to be highly effective in improving the perception of low-fat food systems (Murray et al., 1995).

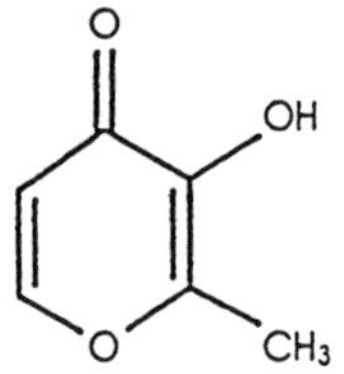

Maltol

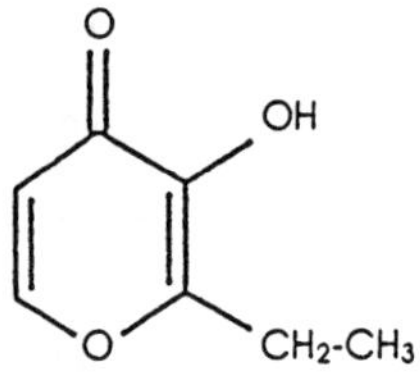

Ethyl maltol

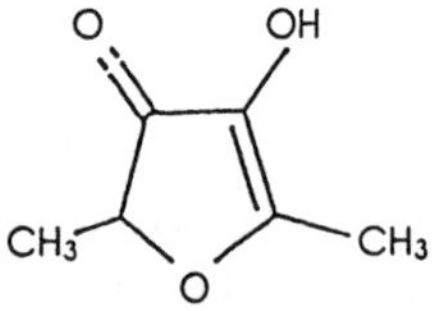

4-Hydroxy-2,5-dimethyl-3(2H)-furanone

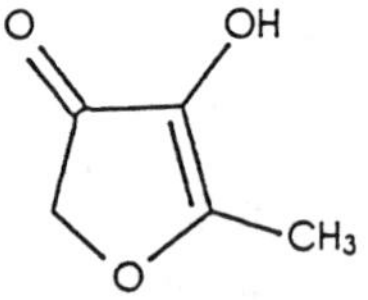

4-Hydroxy-5-methyl-3(2H)-furanone

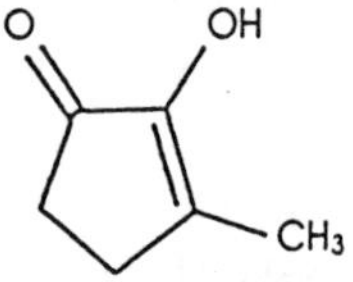

3-Methyl-2-cyclopentene-2-ol-1-one

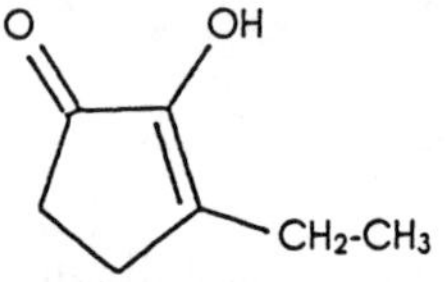

3-Ethyl-2-cyclopentene-2-ol-1-one

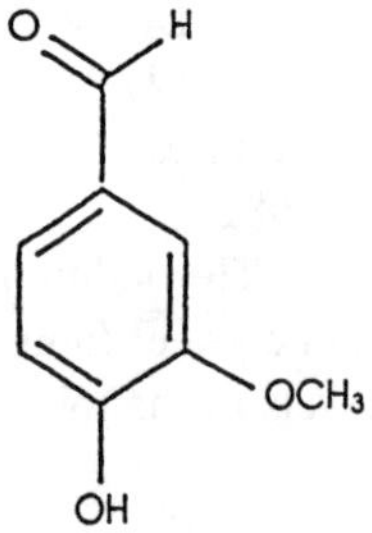

Vanillin

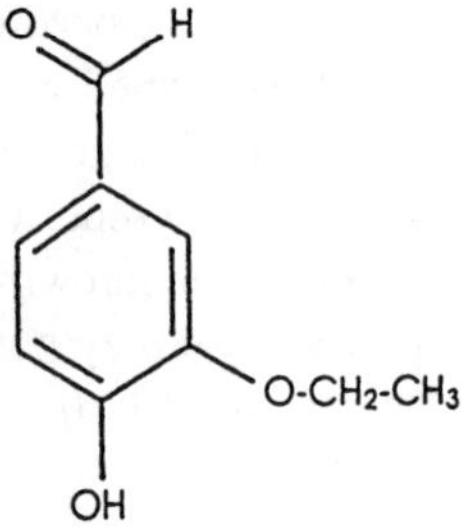

Ethyl Vanillin

Figure 11–13 Chemical structures of some important flavor modifiers. *Source:* G. Matheis, Flavor Modifiers, *Dragoco Report,* Vol. 42, pp. 5–27, © 1997, Dragoco.

Low-fat yogurt, ice cream and salad dressings taste richer, fuller and creamier with the addition of ppm levels of maltol or ethyl maltol. In other words, their mouthfeel is improved.

The JECFA concluded that up to 2 mg/kg/day (120 mg/day for a 60-kg person) is an acceptable level of consumption of both maltol and ethyl maltol for humans (Murray et al., 1995; Le Blanc & Akers, 1989). This value is many times greater than the current average consumption level for both substances.

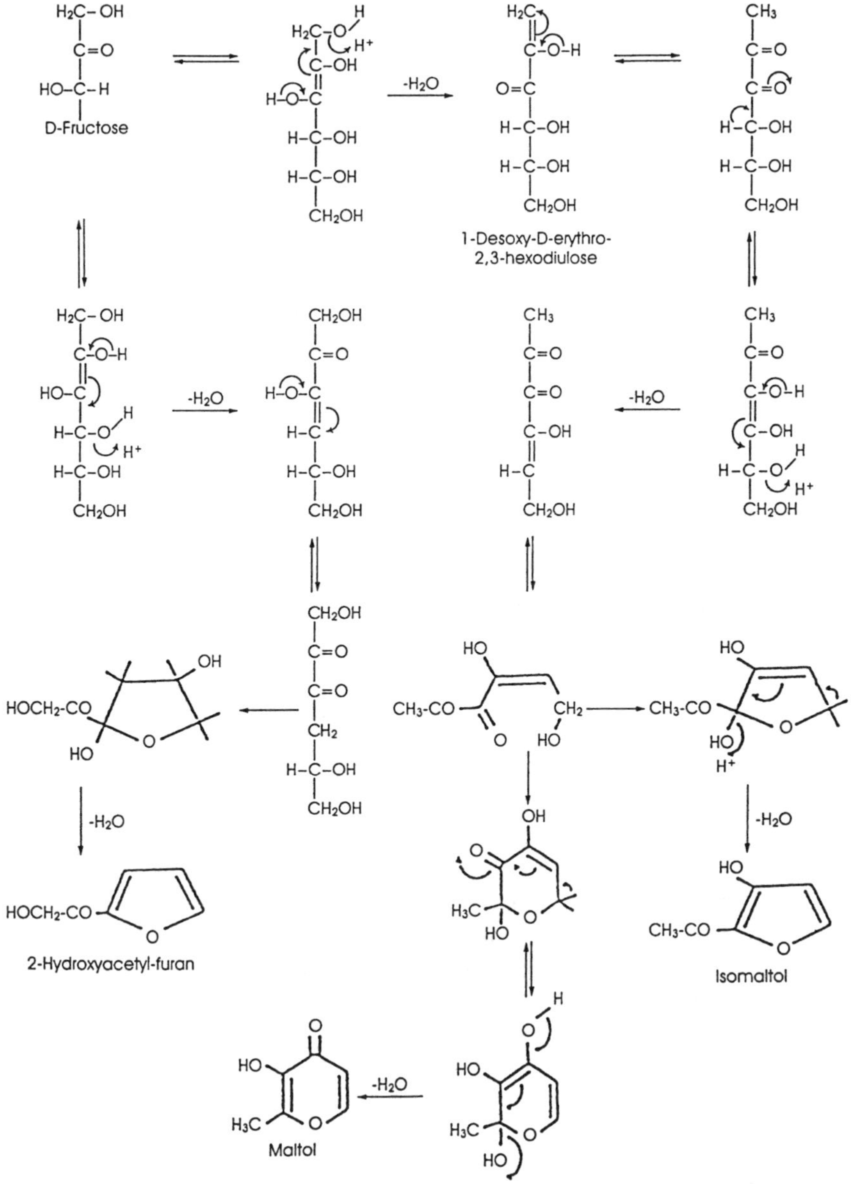

Figure 11–14 Formation of maltol from carbohydrates. *Source:* Reprinted with permission from H.O. Belitz and W. Grosch, *Lehrbuchder Lebensmittelchemie,* p. 219, © 1982, Springer-Verlag GmbH & Co. KG.

Table 11–16 Typical Usage Levels of Maltol, Ethyl Maltol and Cyclotene® in Selected Foods

	Flavor modifier (ppm)		
Food	*Maltol*	*Ethyl maltol*	*Cyclotene®*
Baked goods	75–250	25–150	15–100
Beverages	2–250	1–100	10–50
Candies	3–300	1–100	15–100
Chewing gum			5–30
Chocolate foods	30–200	5–40	
Dairy products	10–150	5–50	
Desserts	30–150	10–75	
Ice cream			5–50
Jams	40–200	15–60	

Source: Data from R. Oberdieck, *Geschmacksverstärker,* pp. 156–164, © 1980, Alkohol-Industrie and K.H. Ney, *Lebensmittelaromen,* © 1987, Behr's Verlag.

11.4 FURANONES AND CYCLOPENTENOLONES

11.4.1 *Furanones*

Both 4-hydroxy-2,5-dimethyl-3(2H)-furanone (trade name Furaneol®) and 4-hydroxy-5-methyl-3(2H)-furanone (Figure 11–13) contribute to the flavor at normal usage level. They are, however, also enhancers of fruity and creamy odor impressions. Both furanones have a caramel-like odor. Furaneol® possesses an additional burnt pineapple odor. The odor threshold of Furaneol® has been reported to be 0.00004 ppm in water at 20° C (Belitz & Grosch, 1982).

Furaneol® is formed when rhamnose is heated in the presence of a substance containing an amino group through Maillard reaction (Figure 11–15). 4-Hydroxy-5-methyl-3(2H)-furanone results from heating fructose (Figure 11–16). Both furanones occur in a variety of foods. Furaneol® has been found in pineapple, strawberry and popcorn. Both Furaneol® and 4-hydroxy-5-methyl-3(2H)-furanone have been identified in meat broth. Both furanones are applied as flavor modifiers in foods where maltol and ethyl maltol are used.

11.4.2 *Cyclopentenolones*

3-Methyl-2-cyclopentene-2-ol-1-one (trade name Cyclotene®) is an enhancer of nutty and chocolate odor impressions. Cyclotene® (Figure 11–13) has been isolated from birch wood tar and found in various foods (e.g., maple syrup). It is formed when sugars are heated at pH 8–10 (Figure 11–17) and has a caramel-like odor. Typical usage levels in selected foods are shown in Table 11–16. Cyclotene®

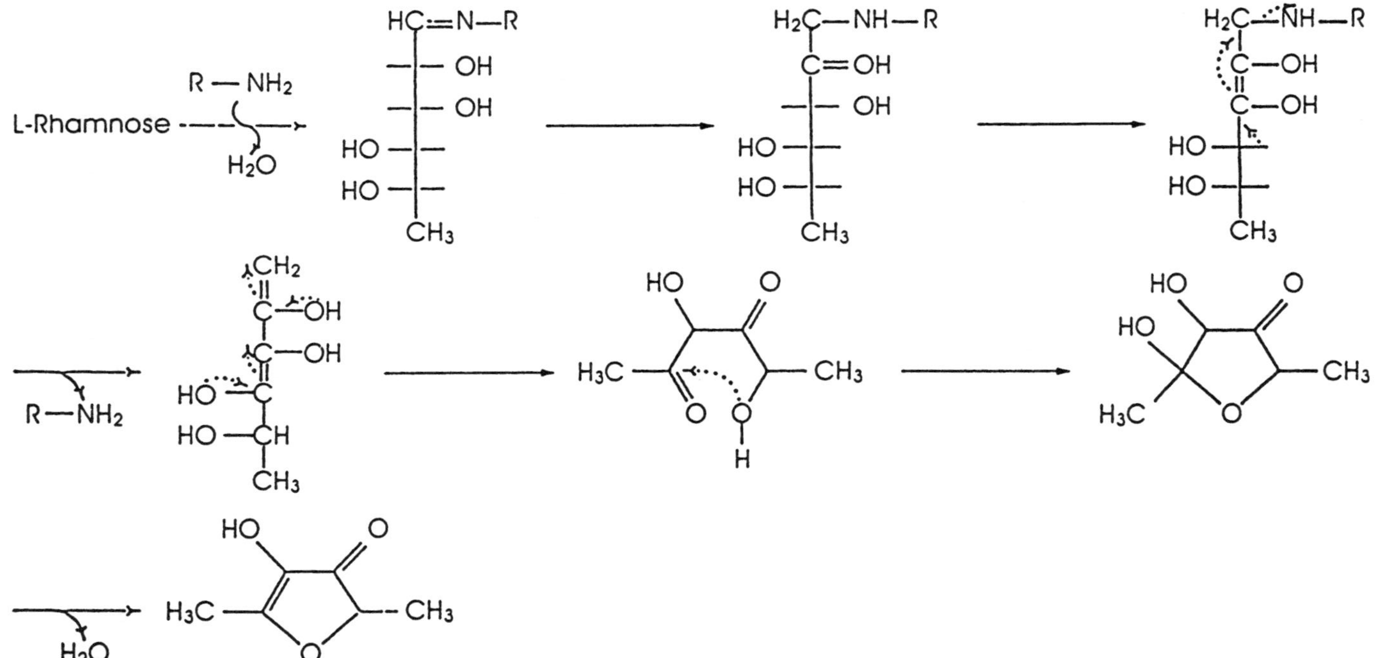

Figure 11–15 Formation of Furaneol® from rhamnose. *Source:* Reprinted with permission from H.O. Belitz and W. Grosch, *Lehrbuchder Lebensmittelchemie,* p. 273, © 1982, Springer-Verlag GmbH & Co. KG.

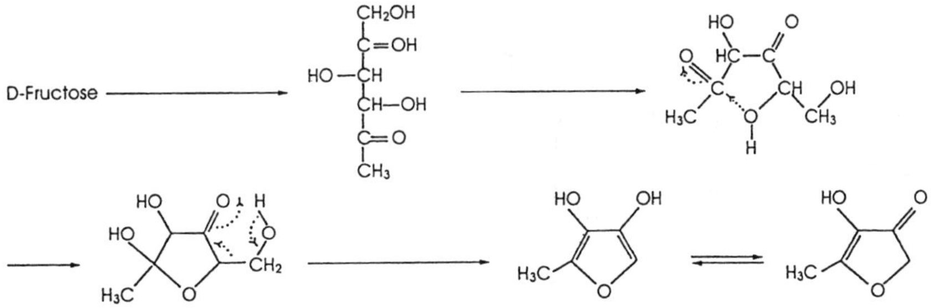

Figure 11–16 Formation of 4-hydroxy-5-methyl-3(2H)-furanone from fructose. *Source:* Reprinted with permission from H.O. Belitz and W. Grosch, *Lehrbuchder Lebensmittel-chemie,* p. 272, © 1982, Springer-Verlag GmbH & Co. KG.

---

has also been reported to mask the salty taste impression (10–20 ppm masks 1–2% of salt) (Ney, 1987).

3-Ethyl-2-cyclopentene-2-ol-1-one (Figure 11–13) may be used to replace Cyclotene®.

11.5 VANILLIN AND ETHYL VANILLIN

11.5.1 *Historical Background*

Vanillin (Figure 11–13) has been known as a flavoring substance since about 1816. By 1858, the pure chemical had been obtained from ethanolic extract of vanilla pods. In 1876, Reimer and Tiemann synthesized vanillin from guaiacol. For many years, the most important source of vanillin was eugenol, from which it was obtained by oxidation.

11.5.2 *Vanillin*

Today, the majority of commercial vanillin is obtained either by processing waste sulfite liquors or by fully synthetic processes starting from guaiacol (Belitz & Grosch, 1982). Lignin, present in the sulfite waste from the cellulose industry, is treated at elevated temperatures and pressures with alkalis in the presence of oxidants (Figure 11–18). The vanillin formed in this way is separated from by-products by extraction, distillation and crystallization processes. Condensation of guaiacol (obtained from catchol) with glyoxylic acid, followed by oxidation and decarboxylation of intermediates, also results in crude vanillin (Figure 11–19). Various commercial grades of vanillin are obtained by distillation and subsequent recrystallization.

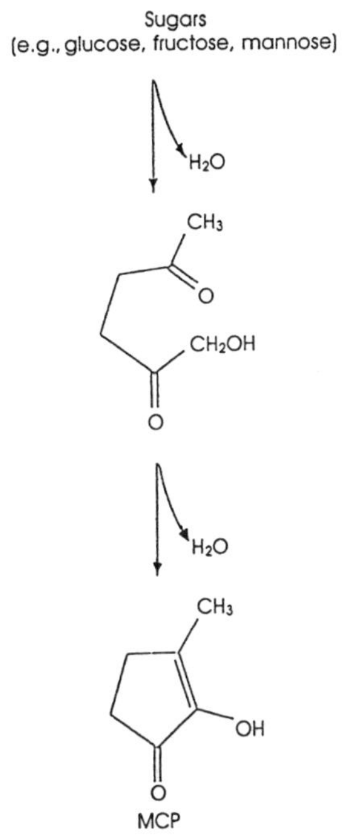

Sugars
(e.g., glucose, fructose, mannose)

H_2O

CH_3

O

CH_2OH

O

H_2O

CH_3

OH

O

MCP

Figure 11–17 Formation of Cyclotene® from sugars at pH 8–10. *Source:* G. Matheis, Flavor Modifiers, *Dragoco Report,* Vol. 42, pp. 5–27, © 1997, Dragoco.

Vanillin has a vanilla-like odor with a threshold of 0.02 ppm in water at 20° C (Belitz & Grosch, 1982). In addition to being a very important flavoring substance, it enhances fruity and chocolate odor impressions.

11.5.3 *Ethyl Vanillin*

Ethyl vanillin (Figure 11–13) has not yet been found occurring naturally in foods. It is 2 to 4 times stronger than vanillin. Ethyl vanillin is synthesized from eugenol, isoeugenol or safrol. As vanillin, it enhances fruity and chocolate odor impressions.

11.6 OTHER FLAVOR MODIFIERS

There are several flavor modifiers which are of minor industrial importance at present. These include natural and synthetic materials. Examples are given in Table 11–17. Studies aiming at using miraculin and curculin as low-calorie sweeteners are in progress (Kurihara, 1992).

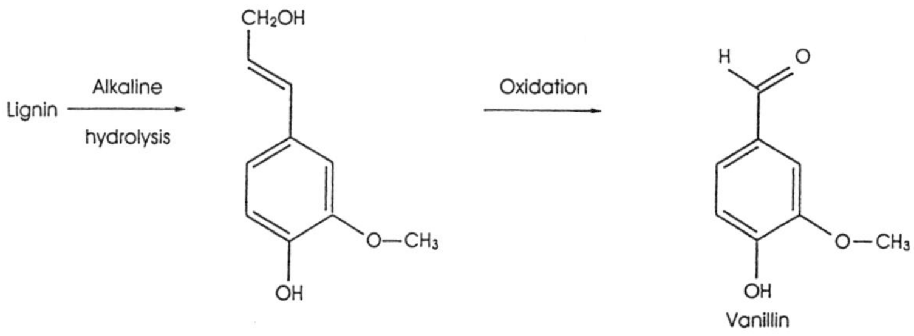

Figure 11–18 Synthesis of vanillin from lignin. *Source:* G. Matheis, Flavor Modifiers, *Dragoco Report,* Vol. 42, pp. 5–27, © 1997, Dragoco.

BMP may become a savory peptide and a flavor modifier of the future (Nagodawithana, 1995a). It occurs naturally in beef. One method of production of BMP already initiated by several concerns is biotechnology (Spanier et al., 1995). The taste threshold of BMP has been reported to be 1,600 ppm (Spanier et al., 1995). There are also reports in the literature claiming that BMP cannot be considered as a flavor modifier and that its occurrence in beef is highly unlikely (Van Wassenaar et al., 1995; Hau et al., 1997).

11.7 SODIUM CHLORIDE

Sodium chloride (table salt, salt) is generally classified as a taste substance. In addition, it functions as a flavor modifier at usage levels below and above its taste threshold, which is reported to range from 370 to 5,000 ppm. Many foods (both sweet and savory) without salt have a flat taste. Salt may enhance sweetness and mouthfeel and decrease bitter, sour and metallic sensations (Kemp & Beauchamp, 1994; Gilette, 1985). Even in sweet foods such as cakes, candies and

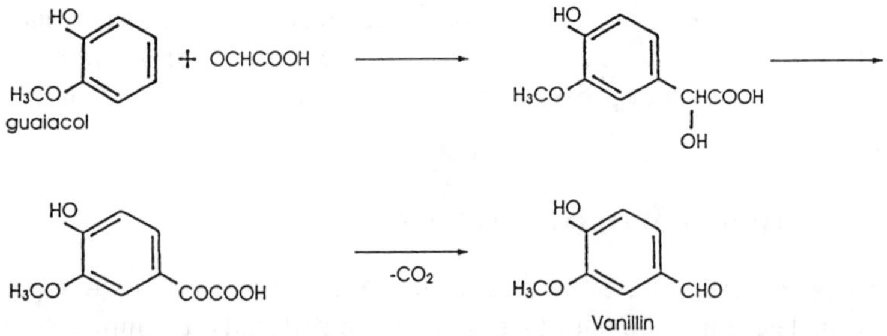

Figure 11–19 Synthesis of vanillin from guaiacol. *Source:* Reprinted with permission from K. Bauer and D. Garbe, *Common Fragrance and Flavor Materials,* © 1985, Wiley-VCH.

Table 11–17 Selection of Commercially Less Important Flavor Modifiers

Name	*Effect*	*Remarks*
Gymnemic acid	Inhibits sweet taste for several hours (powdered sugar tastes like sand and sugar solution tastes like plain tapwater). Salty, bitter and sour tastes are not affected.	Various triterpene glycosides isolated from the leaves of *Gymnema sylvestre*.
Miraculin (trade name: Mirlin®)	Modifies sour into sweet taste. After rinsing the mouth with a solution of miraculin, lemon juice tastes like sweetened lemon juice. Salty, bitter and sweet tastes are not affected.	Tasteless glycoprotein (molecular weight: 40,000–48,000; sugar moiety: 14%) isolated from the miracle fruit, i.e., the fruits of *Richadella dulcifica* (*Synsepalum dulsificum*).
Curculin	The sweet taste of curculin disappears in a few minutes after holding it in the mouth. Then, water elicits the sweet taste again and black tea tastes like sweetened tea. In addition, curculin modifies sour into sweet taste (like miraculin).	Sweet tasting protein (molecular weight: 28,000) isolated from the fruits of *Curculigo latefolia*.
Ziziphin	Inhibits sweet taste.	Triterpene glycoside isolated from the leaves of *Ziziphus jujuba*.
Hodulcin	Inhibits sweet taste.	Glycoside isolated from the leaves of *Hovenia dulcis*.
Sodium-2-(4-methoxyphenoxy) propionate (common name: lactisol; trade name: Cypha®)	Decreases sweet taste.	Synthetic compound. Has been identified in roasted coffee.

continues

Table 11–17 continued

Name	*Effect*	*Remarks*
Thaumatin (trade name: Talin®)	Masks bitter taste.	Sweet tasting protein (molecular weight: 22,000) isolated from the fruit of the West African perennial plant *Thaumatoccus danielli.* Although the primary taste property of thaumatin is sweetness, its main use is as a flavor modifier.
Lecithin	Masks bitterness and harsh off-notes in mint or menthol flavored chewing gum.	The primary use of lecithin is as an emulsifier.
Neohesperidin dihydrochalcone	Enhances fruity notes and reduces sharp or spicy notes	Primary use as intense sweetener.
Beefy meaty peptide (BMP), also known as "delicious peptide"	Enhances meat flavor.	Octapeptide of the sequence Lys-Gly-Asp-Glu-Glu-Ser-Leu-Ala.
Gurmarin	See gymnemic acid.	Polypeptide isolated from the leaves of *Gymnema sylvestre*. Has also been synthesized.

Source: Data from Nagodawithana, 1995a; Belitz & Grosch, 1982; Kurihara, 1992; Lindley, 1993; Kurihara & Nirasawa, 1994; Salminen & Hallikainen, 1990; Glass et al., 1986; Lindley et al., 1993; Spanier et al., 1995; Ota et al., 1996.

toffees salt has its place. However, its presence is most critical for savory foods. Salt is often referred to as "the poor man's flavor enhancer." It is difficult, if not impossible, to make a general statement about the most appropriate salt concentration needed in various food products. This is due to the wide variation in consumer preference and to the individual differences in threshold detection.

11.8 CONCLUSIONS

Flavor modifiers are extremely important components of many foods and are a valuable means of improving flavor quality, particularly of savory foods. Table 11–18 classifies flavor modifiers based on type of foods they are used in.

Table 11–18 Classification of Flavor Modifiers Based on the Types of Foods They Are Used in

Type of food	*Examples*
Savory foods	MSG, GMP, IMP
Sweet foods	Maltol, ethyl maltol, Furaneol®, 4-hydroxy-5-methyl-3(2H)-furanone, Cyclotene®, vanillin, ethyl vanillin, Cypha®
Sour foods	Miraculin
Low fat foods	Maltol, ethyl maltol
Sweet and savory foods	Sodium chloride

REFERENCES

K. Bauer and D. Garbe. 1985. *Common Fragrance and Flavor Materials*. VCH Verlagsgesellschaft, Weinheim.

H.O. Belitz and W. Grosch. 1982. *Lehrbuch der Lebensmittelchemie.* Springer, New York.

G.G. Birch. 1987. Structure, Chirality, and Solution Properties of Glutamates in Relation to Taste. In: *Umami: A Basic Taste* (Y. Kawamura and M.R. Kare, eds.). Dekker, New York, pp. 173–184.

L. Dayton. 1994. Why MSG Myth Is a Lot of Chop Suey. *New Scientist,* January 1994, p. 15.

European Union, 1995. Council Directive on Additives other than Colours and Sweeteners (95/2/EEC). *Off. J. Eur. Comm.* 38(L61), 1–40.

S. Fuke and T. Shimizu. 1993. Sensory and Preference Aspects of Umami. *Tr. Food Sci. Technol.* 4, 246–251.

M. Gilette. 1985. Flavour Effects of Sodium Chloride. *Food Technol.* 39(6), 47–52, 56.

Glutamate Information Service. 1995. Myths and MSG. *Food Engin. Int.* October 1995, p. 24.

M. Glass, V. Corsello, D.A. Orlandi and A. Guzowski. 1986. Process for Preparing a Chewing Gum Composition with Improved Flavor Perception. U.S. Patent 4 604 288.

J. Hau, D. Cazes and L.B. Fay. 1997. Comprehensive Study of the "Beefy Meaty Peptide." *J. Agric. Food Chem.* 45, 1351–1355.

H.B. Heath and G. Reineccius. 1986. *Flavor Chemistry and Technology.* Avi Publishing, Westport, CT.

Institute of Food Technologists 1995. Monosodium Glutamate. A Statement of the Institute of Food Technologists. *Food Technol.* 49(10), 28.

K. Ikeda. 1909. On a New Seasoning. *J. Tokyo Chem.* 30, 820–826.

G.I. Imafidou and A.M. Spanier. 1994. Unraveling the Secret of Meat Flavor. *Tr. Food Sci. Technol.* 5, 315–321.

International Organisation of the Flavour Industry. 1990. *Code of Practice for the Flavour Industry.* IOFI, Geneva.

S.E. Kemp and G.K. Beauchamp. 1994. Flavor Modification by Sodium Chloride and Monosodium Glutamate. *J. Food Sci.* 59, 682–686.

S. Kodama. 1913. On a Procedure for Separating Inosinic Acid. *J. Tokyo Chem.* 34, 751–755.

A. Kuninaka. 1960. Studies on Taste of Ribonucleic Acid Derivatives. *J. Agric. Chem. Soc. Jpn.* 34, 487–492.

Y. Kurihara and S. Nirasawa. 1994. Sweet, Antisweet and Sweetness-Inducing Substances. *Tr. Food Sci. Technol.* 5, 37–42.

Y. Kurihara. 1992. Characteristics of Antisweet Substances, Sweet Proteins, and Sweetness-Inducing Proteins. *CRC Rev. Food. Sci. Nutr.* 32, 231–252.

D.T. Le Blanc and H.A. Akers. 1989. Maltol and Ethyl Maltol: From the Larch Tree to Successful Food Additives. *Food Technol.* 43(4), 78–84.

M.G. Lindley. 1993. Sweetness Antagonists. In: *Flavor Science. Sensible Principles and Techniques* (T. Acree and R. Teranishi, eds.). American Chemical Society, Washington, DC, pp. 117–133.

M.G. Lindley, P.K. Beyts, I. Canales and F. Borrego. 1993. Flavor Modifying Characteristics of the Intense Sweetener Neohesperidin Dihydrochalcone. *J. Food Sci.* 58, 592–594, 666.

J.A. Maga. 1983. Flavor Potentiators. *CRC Crit. Rev. Food Sci. Nutr.* 18, 231–312.

J.A. Maga. 1994. Umami Flavor of Meat. In: *Flavor of Meat and Meat Products* (Shahidi, F., ed.). Blackie Academic and Professional, London, pp. 98–115.

P.R. Murray, M.G. Webb and G. Stagnitti. 1995. Advances in Maltol and Ethyl Maltol Applications. *Food Technol. Int. Eur.,* pp. 53–55.

T. Nagodawithana. 1992. Yeast-Derived Flavors and Flavor Enhancers and Their Probable Mode of Action. *Food Technol.* 46(11), 138–144.

T. Nagodawithana. 1994a. Flavor Enhancers: Their Probable Mode of Action. *Food Technol.* 48(4), 79–85.

T. Nagodawithana. 1994b. Flavour Enhancers and the Probable Way They Work. *Food Technol. Int. Eur.,* pp. 119–121.

T.W. Nagodawithana. 1995a. *Savory Flavors.* Esteekay Associates, Milwaukee, WI, pp. 297–333.

T.W. Nagodawithana. 1995b. *Savory Flavors.* Esteekay Associates, Milwaukee, WI, pp. 225–262.

N. Nakajima, K. Ishikawa, M. Kamada and E. Fujita. 1961. Food Chemical Studies on 5′-Ribonucleotides. 1. On the 5′-Ribonucleotides in Foods. Determination of the 5′-Ribonucleotides in Various Stocks by Ion Exchange Chromatography. *J. Agric. Chem. Soc. Jpn.* 35, 797–804.

T. Niederauer. 1995. Aroma- and Geschmacksverstärker. Grundlagen der Anwendung. *Fleischwirtschaft.* 75, 28–31.

K.H. Ney. 1987. *Lebensmittelaromen.* Behr's Verlag, Hamburg.

R. Oberdieck. 1980. *Geschmacksverstärker.* Alkohol-Industrie, pp. 156–164.

M. Ota, K. Tonosaki, K. Miwa, T. Fukuwatari and Y. Ariyoshi. 1996. Synthesis and Characterization of the Sweetness-Suppressing Polypeptide Gurmarin and ent-Gurmarin. *Biopolymers* 39, 199–205.

D.J. Raiten, J.M. Talbot and K.D. Fisher (eds.). 1995. *Analysis of Adverse Reactions to Monosodium Glutamate (MSG).* Prepared by the Life Sciences Research Office, Federation of American Societies for Experimental Biology, for the Center for Food Safety and Applied Nutrition, FDA/HHS.

S. Salminen and A. Hallikainen. 1990. Sweeteners. In: *Food Additives* (A.L. Branen, P.M. Davidson and S. Salminen, eds.). Dekker, New York, pp. 297–326.

H. Shimazono. 1964. Distribution of 5′-Ribonucleotides in Foods and Their Application to Foods. *Food Technol.* 18, 294–298.

A.M. Spanier, J.M. Bland, J.A. Miller, J. Glinka, W. Wasz and T. Duggins. 1995. BMP: A Flavor Enhancing Peptide Found Naturally in Beef. Its Chemical Synthesis, Descriptive Sensory Analysis, and Some Factors Affecting its Usefulness. In: *Food Flavors: Generation, Analysis and Process Influence* (G. Charalambous, ed.). Elsevier, Amsterdam, pp. 1365–1378.

Y.-H. Sugita. 1990. Flavor Enhancers. In: *Food Additives* (A.L. Branen, P.M. Davidson and S. Salminen, eds.). Dekker, New York, pp. 259–296.

P.J. Taliaferro. 1995. Monosodium Glutamate and the Chinese Restaurant Syndrome: A Review of Food Additive Safety. *J. Environ. Health* 57(10), 8–12.

L. Tarasoff and M.F. Kelly. 1993. Monosodium L-Glutamate: A Double-Blind Study and Review. *Food Chem. Toxicol.* 31, 1019–1035.

T. Van Eijk. 1987. Umami Substances—Flavor Enhancers or Modifiers? *Dragoco Rep.* 32, 3–17.

P.D. Van Wassenaar, A.H.A. van den Oord and W.M.M. Schaaper. 1995. Taste of "Delicious" Beefy Meaty Peptide Revised. *J. Agric. Food Chem.* 43, 2828–2832.

S. Yamaguchi. 1987. Fundamental Properties of Umami in Human Taste Sensation. In: *Umami: A Basic Taste* (Y. Kawamura and M.R. Kare, eds.). Dekker, New York, pp. 41–73.

Chapter 12

Flavorings for Pharmaceutical Products

Günter Matheis

12.1 INTRODUCTION

The acceptance of pharmaceutical preparations is significantly influenced, like that of foodstuffs, by odor and taste. This applies particularly to children, but it is true of adults as well. We all associate unpleasant childhood memories with the concept of "taking our medicine"; think, for example, of a course of cod-liver oil in the winter. Understandably, children find it hard simply to suppress the unpleasant sensations. This is why, from very early times, people have tried to make some of the preparations in children's medicine more attractive. Many, for instance, have taken worm pills mixed with chocolate in their childhood.

Adults like to pretend that these unpleasant impressions are of no consequence to them, and are very proud of their ability to overcome them. In particular in the case of elderly people, who frequently take more medicinal products than middle-aged adults, an attractive preparation can act as a positive stimulus on the psyche and thus assist the healing process.

The flavor industry offers a wide range of products with which almost every pharmaceutical preparation can be made sensorially attractive.

The flavor of a pharmaceutical preparation, like that of a foodstuff, is composed of taste, odor (or aroma) and trigeminal impressions in the nose and mouth cavities (Figure 12–1). In accordance with the guidelines of the International Organisation of the Flavour Industry (IOFI, 1990), the terms flavor and flavorings will not be used as synonyms. Flavour will be used for the combined effects of the sensory impressions of taste, odor and trigeminal perceptions. Flavoring will be used for a product that is produced by the flavor industry to impart flavor to pharmaceutical products or foodstuffs (Table 12–1). Trigeminal perceptions are also known as somatosensory impressions, common chemical sense or chemesthesis (Lawless & Lee, 1993).

From the definitions of the IOFI (1990) and the Council of the European Communities (1988) it is apparent that, in general, flavorings consist of flavor-impart-

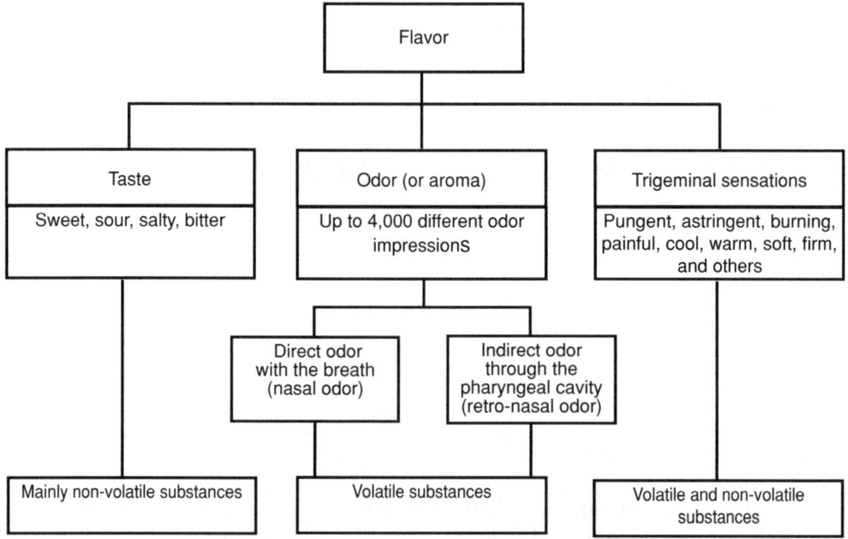

Figure 12–1 Flavor of a pharmaceutical preparation.

ing ingredients and ingredients that do not contribute to the flavor (Table 12–2). The latter are flavor adjuncts and include, for example, solvents, carriers, preservatives, and other additives. Generally, the flavor-imparting ingredients amount to 10 to 20% of the flavoring.

12.2 THE CHOICE OF FLAVORINGS

If pharmaceutical products are to be flavored, a whole series of factors must be taken into consideration. Examples are shown in Table 12–3. It can already be seen here that general rules are very difficult to adduce. Almost every preparation has to be considered individually.

Before using flavorings, an attempt should be made to tone down any inherent unpleasant taste or odor by adding flavor chemicals and/or other substances. Examples of this are given in Table 12–4. It is also possible to deliberately involve the inherent taste of the preparation in the flavoring. A harmonious chocolate impression can, for instance, be given to a bitter preparation in conjunction with a sweet cocoa and vanilla flavoring and sugar. In the case of bitter alcoholic preparations, the addition of herb liqueur flavoring can give the overall impression of bitters.

Depending on the nature of the preparation (see Table 12–3), liquid, dry or emulsion flavorings have to be used (Table 12–5); viscous flavorings hardly play

Table 12–1 Definitions of Taste, Odor, Aroma, Flavor and Flavoring

Designation		*Definition*
English	*German*	
Taste	Geschmack	Sensory impression which is perceived mainly by the taste receptors on the tongue[a]
Odor	Geruch	Sensory impression which is perceived by the odor receptors (directly through the nose and retro-nasally)
Aroma	Aroma[b]	Pleasant odor impression
Flavoring	Aroma[b]	Product to impart flavor to pharmaceutical products or foodstuffs
Flavor	Flavour	Combined effect of the sensory impressions of taste, odor and trigeminal perceptions

[a]In everyday language, taste usually has a wider meaning and includes retro-nasal odor and trigeminal perceptions.

[b]In the German language, aroma has two different meanings.

any role in pharmaceutical preparations. The liquid flavorings include essential oils, distillates and, above all, compositions. These are mixtures of a number of raw materials, usually in a solvent. In the case of aqueous and aqueous-alcoholic preparations and aqueous suspensions, a hydrophilic solvent is required, whereas in oily products a lipophilic solvent is necessary. Examples of hydrophilic sol-

Table 12–2 Definitions of Flavoring (*IOFI*, 1990; *CEC*, 1988)

International Organisation of the Flavour Industry	*Council of the European Communities*
Concentrated preparation, with or without flavor adjuncts,[a] used to impart flavor, with the exception of only salty, sweet or acid tastes.	Flavoring means flavoring substances, flavoring preparations, process flavorings, smoke flavorings or mixtures thereof.
It is not intended to be consumed as such.	Flavorings may contain foodstuff as well as other substances.[b]

[a]Food additives and food ingredients necessary for the production, storage and application of flavorings as far as they are nonfunctional in the finished food.

[b]Additives necessary for the storage and use of flavorings, products used for dissolving and diluting flavorings, and additives for the production of flavorings (processing aids) where such additives are not covered by other CEC provisions

Table 12–3 Examples of Factors That Play a Part in the Flavoring of Pharmaceutical Preparations

Factor	*Examples*	
Structure	Aqueous	Oily
	Aqueous alcoholic	Oil-in-water emulsion
	Aqueous suspension	Oily suspension
	Capsule	Pill
	Drops	Powder
	Lozenge	Pressed tablet
		Tablet
Inherent odor	Odorless	Unpleasant
Inherent taste	Bitter	Sweet
	Salty	Tasteless
	Sour	
Inherent trigeminal impression in mouth and pharynx	Alkaline	Burning
	Astringent	Metallic
Target group	Adults	Children
How to be taken	In cocoa	On sugar
	In fruit juice	In water
Potential interactions between flavoring substances and ingredient of the preparation		
pH value		
Costs of flavoring		

Source: G. Matheis, Flavoring Pharmaceutical Products, *Dragoco Report*, Vol. 38, pp. 22–33, © Dragoco.

vents include ethanol, isopropanol and propylene glycol. Lipophilic solvents include plant oils, plant oil fractions, triacetin and benzyl alcohol. Figure 12–2 shows some parameters which the flavorist must take into account when creating a flavoring.

Emulsion flavorings, which are themselves oil-in-water emulsions, are suitable for flavoring oil-in-water emulsions. Frequently, the oil phase is represented by essential oils or oleoresins; they are emulsified with the help of emulsifiers and in some cases stabilizers are also added.

Dry flavorings are used extensively in powder preparations, tablets, pills and pressed tablets. In principle they can be made in dried form in practically all taste and odor directions. As already mentioned (see Table 12–5), dry flavorings are made from liquid or paste-like precursors, or they may be mixtures of two or more dry products. Examples of dry flavorings produced in various ways are given in Table 12–6.

Table 12–4 Examples of Toning Down the Inherent Taste and Odor of Pharmaceutical Preparations

Inherent taste or odor	*Toned down by*
Bitter	Addition of common salt alone or in conjunction with sugar and/or sweetener
Sour	Sugar and/or sweetener
Unpleasant odor	Neutral tasting and neutral smelling thickening agent (some of the volatile odor chemicals become bound, so that they are no longer available to the odor receptors)

Source: G. Matheis, Flavoring Pharmaceutical Products, *Dragoco Report*, Vol. 38, pp. 22–33, © Dragoco.

The simplest way of making a dry flavoring is the adsorption of the liquid or viscous precursor on a solid carrier substance. These adsorbates are relatively unstable because the flavoring is unprotected. The majority of dry flavorings are spray-dried products which, at acceptable prices, have adequate storage lives. Carrier substances include plant gums (gum arabic), starch hydrolysates (maltodextrins) and simple sugars (glucose). We shall not consider freeze-drying and microencapsulation here because these processes are of minor importance for the flavor industry (Heath & Reineccius, 1986). Encapsulation with cyclodextrins (β-cyclodextrin in particular) provides ideal inclusion for liquid flavorings (Szejtli, 1982; Reineccius & Risch, 1986).

Table 12–5 Physical State of Flavorings

Physical state	*Typical examples*
Liquid	Essential oils, distillates, compositions[a]
Emulsion	Compositions[a]
Viscous	Extracts, oleoresins, compositions[a]
Solid	Dry flavorings,[b] dry mixtures[c]

[a]Mixtures with or without solvent.

[b]Dry versions of liquid or viscous flavorings, usually with a carrier substance.

[c]Mixtures of two or more dry flavorings.

Source: G. Matheis, Flavoring Pharmaceutical Products, *Dragoco Report*, Vol. 38, pp. 22–33, © Dragoco.

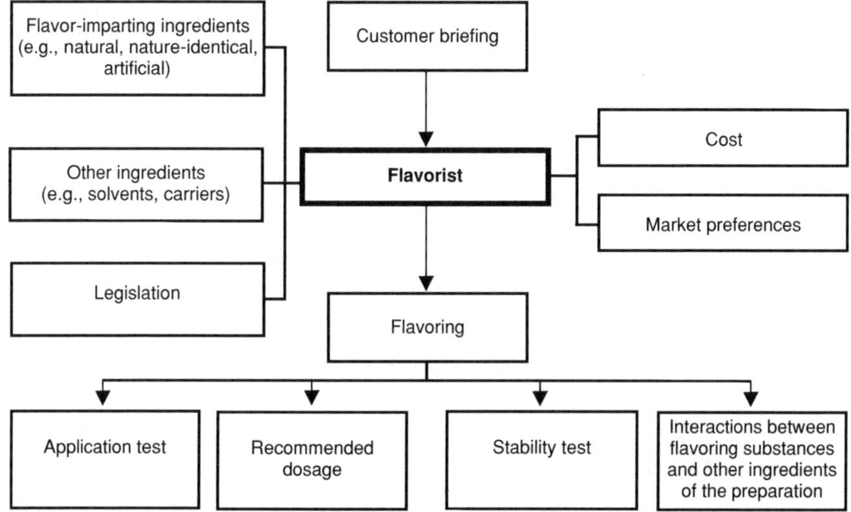

Figure 12–2 Examples of parameters that the flavorist has to take into account when developing a flavoring.

12.3 INTERACTIONS BETWEEN VOLATILE FLAVORING SUBSTANCES AND OTHER CONSTITUENTS OF THE PREPARATIONS

Pharmaceutical preparations are usually complex mixtures of various components that can interact with each other both physically and chemically. This applies particularly to volatile flavoring substances that are affected by fillers and carrier substances and can react with active substances. Volatile flavoring substances must reach the oral and pharyngeal cavity in the gaseous phase before they can be smelled retro-nasally in the olfactory epithelium (Figures 12–3 and 12–4). Chewing and temperature in the oral cavity encourage the release of the volatile flavoring substances.

The amount of volatile flavoring substance that is released in the gas phase depends on its vapor pressure (Overbosch et al., 1991). This, in turn, is affected by various factors, including temperature and possible interactions with ingredients of the preparation. If an ingredient has a negative effect on release, we speak of flavor binding. The term binding is defined here as the enrichment of concentration of the volatile substance within or in the proximity of an ingredient of the preparation, without specifying precisely the type of binding involved (whether it is physical or chemical, and what type of physical or chemical bond is involved).

Table 12–7 gives a summary of the possible interactions between volatile flavoring substances and some fillers, carriers and other substances. The majority of

Table 12–6 Examples of Dry Flavorings (Heath & Reineccius, 1986; Sinki & Schlegel, 1990)

Designation	*Type of distribution of the liquid or paste-like flavoring on or in the carrier substance*	*Brief description of the common manufacturing process*
Adsorbate (adsorbed or plated flavoring)	Surface film	Plating liquid or paste-like flavoring on solid carrier
Spray-dried flavoring	Surface film and inclusion	An oil-soluble liquid or paste flavoring is made into an oil-in-water emulsion, the flavoring being the disperse phase and water being the continuous phase, and the carrier—if suitable—acting as the emulsifier (if the carrier lacks emulsifying properties, an emulsifier may be added). A water-soluble liquid or paste flavoring is made into a solution or slurry with the carrier. These preparations are forced through a nozzle or centrifugal device as a mist into a stream of hot gas (usually air) moving through a chamber. The hot gas causes the mistlike feedstock to evaporate and instantly produce dried particles.
Freeze-dried flavoring	Surface film and inclusion	Similar to spray-dried flavoring except that the preparation is frozen and the water removed by sublimation of ice crystals at low pressure.
Microencapsulated flavoring	Surface film and inclusion	Coacervation or aqueous phase separation processes. Complex coacervation is most important. In this process, a dilute aqueous solution of two colloids with opposite electric charges are mixed with a liquid or paste flavoring. The pH is adjusted to the isoelectric point where coacervation is induced, causing the coating

continues

Table 12–6 continued

Designation	*Type of distribution of the liquid or paste-like flavoring on or in the carrier substance*	*Brief description of the common manufacturing process*
		material to precipitate around the liquid or paste flavoring droplets. The liquid microcapsules are solidified by chilling or by chemical methods.
Encapsulated flavoring	Inclusion	An alcoholic solution of liquid or paste flavoring is added to an aqueous-alcoholic solution of special carrier (e.g., β-cyclodextrin) at 50–55°C. The solution is cooled slowly to room temperature and the carrier/flavoring precipitate is removed by filtration and air dried.
Extrusion flavoring	Surface film and inclusion	A liquid or paste flavoring is incorporated into a molten carrier followed by extrusion and solidification of the mixture.

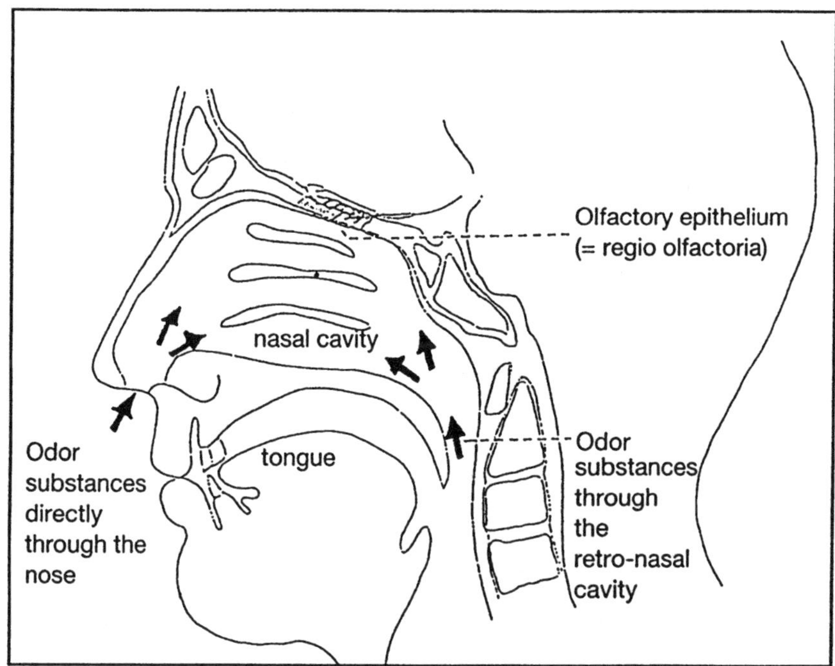

Figure 12–3 Schematic drawing of the human olfactory organ. *Source:* G. Matheis, Taste, Odor, Aroma, and Flavor, *Dragoco Report*, Vol. 39, pp. 50–65, © 1994, Dragoco.

interactions are of a physical nature, and therefore reversible. This means that, under certain conditions (for instance in chewing, mixing with saliva, and warming in the mouth), at least part of the bound flavoring substances are released again and can thus be perceived retro-nasally. However, chemical reactions are also possible; these lead to covalent bonds and are irreversible. Here, the part of the flavoring substance that has reacted is no longer available to create an aroma. An example of this is the reaction of volatile aldehydes with the amino groups of proteins, peptides and amino acids (Figure 12–5), which leads to Schiff bases. These in turn lead to a series of subsequent products. Known reactions of this type are those of hexanal (aldehyde C_6) and vanillin with amino groups in various proteins (Meier, 1985; Dhont, 1975).

12.4 COOPERATION BETWEEN THE PHARMACIST AND THE FLAVORIST

It can be seen from the foregoing that it takes an expert's experience to discover what flavoring is best suited to a specific preparation. It is advisable, there-

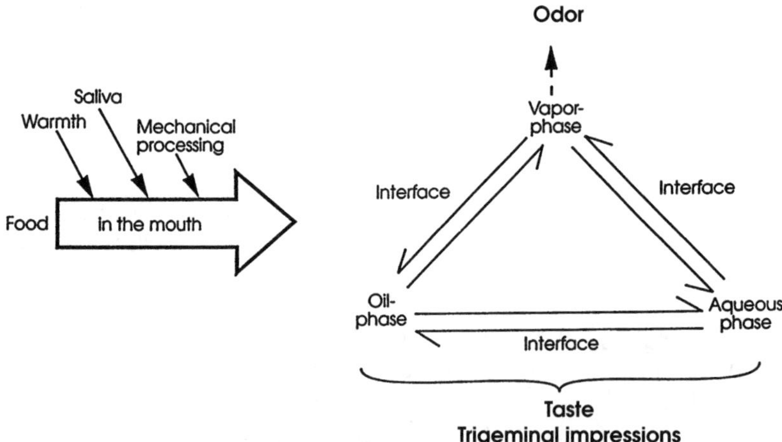

Figure 12–4 Processes in the mouth that lead to the perception of the flavor. *Source:* G. Matheis, Trigeminal Perceptions in the Nasal and Oral Cavities, *Dragoco Report*, Vol. 40, pp. 72–82, © 1995, Dragoco.

fore, for the flavorist to be consulted as soon as possible by the pharmacist. Mutual trust and collaboration are vital, going so far, even to providing the flavorist with not only the rough composition of the preparation, but its precise formulation. The original active ingredients themselves should, where possible, be made available. It is only in this way that one can ensure that interactions between volatile flavoring constituents and the constituents of the preparation can be truly investigated (Figure 12–6). Undesirable reactions between fillers, carriers and active substances on the one hand and volatile flavoring substances on the other must be avoided (of course this also applies to non-volatile flavoring substances). Reversible and irreversible reactions with active ingredients are possible if, for example, the active substance has peptide, amino acid, purine alkaloid or phenolic structures (see Table 12–7).

The compatibility of the flavorings with constituents of the preparation must also be determined by means of a shelf-life test. A rough idea can be obtained through stress tests at increased temperatures (Labuza & Schmidt, 1985).

12.5 EXAMPLES OF FLAVORINGS FOR PHARMACEUTICAL PRODUCTS

Although it has already been mentioned that general rules are very hard to find, we give examples here of some frequently-occurring taste and odor types. In preparations on the German market, the following flavorings are frequently found: orange, lemon, banana, caramel, strawberry, raspberry, cocoa, cherry,

Table 12–7 Possible Interactions between Volatile Flavoring Substances and Other Ingredients of the Preparation

Preparation constituent	*Type of interaction*
Carbohydrates	
• Simple sugars	Adsorption (dry medium) Unknown (aqueous medium)
• Starches, cyclodextrins	Inclusion complexes
• Dextrins, maltodextrins	Adsorption
• Pectin, alginate	Binding to carboxyl groups Chemical reactions
• Cellulose, methyl cellulose	Hydrogen bridges?
• Agar-agar	Unknown
Proteins	Hydrophobic interactions Hydrogen bridges Inclusion complexes? Chemical reactions
Fats	Distribution between hydrophilic and lipophilic phases Adsorption on interfaces Hydrogen bridges?
Peptides and amino acids	Hydrogen bridges Chemical reactions
Phenolic compounds	Hydrogen bridges
Fruit acids	Unknown
Purine alkaloids	Unknown
Inorganic salts	Salting-out effect

Source: G. Matheis, Flavoring Pharmaceutical Products, *Dragoco Report*, Vol. 38, pp. 22–33, © Dragoco.

mandarin, port wine, cream and vanilla. Table 12–8 shows which flavorings, in the experience of our company, are particularly suitable for preparations with a strong inherent taste or odor. Tables 12–9 and 12–10 show our flavoring recommendation for the German and US markets, respectively, based on the type of preparation. We must stress, however, that flavoring combinations are composed for a great many preparations that are not listed in Table 12–9, because they do not represent a defined flavoring direction (for example, lemon/peppermint, cream/caramel, cream/coconut or curacao/orange).

Peppermint flavoring, with its principal constituent menthol, which gives rise to the familiar, pleasantly cooling effect in the mouth, is of particular value. At higher dosage rates the taste nerves are to a certain extent anesthetized, which is helpful in masking a particularly unpleasant or bitter taste.

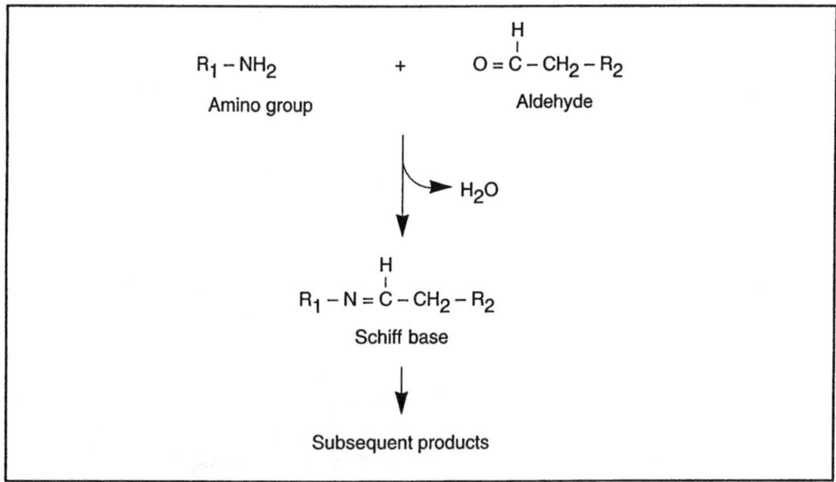

Figure 12–5 Formation of a Schiff base from an aldehyde and an amino group. *Source:* G. Matheis, Flavoring Pharmaceutical Products, *Dragoco Report*, Vol. 38, pp. 22–33, © Dragoco.

Finally, it must be stressed that the flavoring has to be metered and adapted in such a way that a medication remains a medication, and does not give the impression of being a treat. Otherwise it can lead to a craving for sweet things in children, and "addiction" in adults.

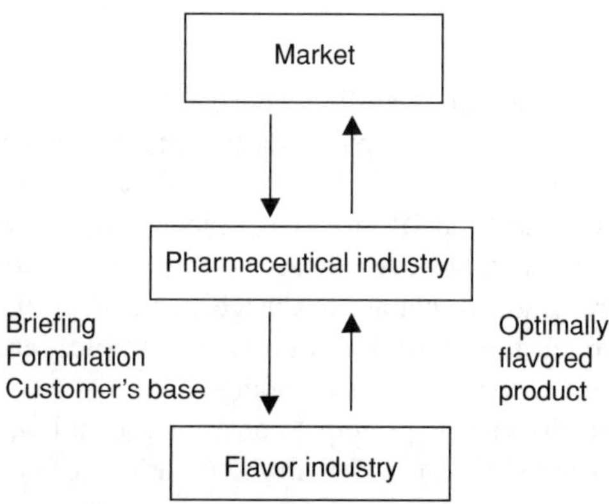

Figure 12–6 Collaboration between the pharmaceutical industry and the flavor industry.

Table 12–8 Suggestions for Flavoring Pharmaceutical Preparations with Strong Inherent Taste or Odor

Inherent taste or odor of the preparation	*Examples of suitable flavorings (Dosage rate: 0.005–0.5%)*
Sweet	Dessert and spice flavorings
Sour	Fruit flavorings
Bitter	Cocoa, wine, liqueur and spice flavorings
Salty	Anise, peppermint and spicy flavorings
Alkaline	Peppermint, anise and banana flavorings
Metallic	Wine, peppermint and fruit flavorings
Burning	Spice flavorings
Unpleasant odor	Peppermint, spearmint, spice and exotic fruit flavorings

Source: G. Matheis, Flavoring Pharmaceutical Products, *Dragoco Report*, Vol. 38, pp. 22–33, © Dragoco.

Table 12–9 Suggestions for Flavoring Pharmaceutical Preparations Based on the Type of Medicament for the German Market

Type of medicament	*Suitable flavorings*
Analgetics and antipyretics	Lemon, orange, mandarin
Antibiotics	Apricot, peppermint, strawberry, raspberry, cherry, nougat, black currant, orange, grapefruit, chocolate, lemon, mandarin
Antihistamines	Coconut, cream, chocolate, raspberry
Antacids	Banana, apricot, spearmint, tutti frutti, cream, caramel
Barbiturates	Lemon, orange, raspberry, port wine, caramel, peppermint, spearmint, cherry
Cough preparations	Raspberry, cherry, banana, orange, cocoa, caramel, apricot, anise
Digestive preparations	Anise, bitters, rum, maraschino, spice flavorings
Iron preparations	Orange, pineapple, nut, brandy, sherry, herb liqueur, raspberry, black currant, cherry, tutti frutti
Laxatives	Banana, anise, orange, cream, caramel
Liver preparations	Apricot, black currant, chocolate
Protein hydrolysates	Spice flavorings
Vitamin preparations	Orange, lemon, cherry, raspberry, honey, strawberry, banana, chocolate
Yeast preparations	Herb flavorings, spice flavorings, cola, orange, lemon

Source: G. Matheis, Flavoring Pharmaceutical Products, *Dragoco Report*, Vol. 38, pp. 22–33, © Dragoco.

Table 12–10 Suggestions for Flavoring Pharmaceutical Preparations Based on the Type of Medicament for the US Market

Type of medicament	*Suitable flavorings*
Analgetics and antipyretics	Pomegranate
Antibiotics	Banana, butterscotch, chocolate, citrus, maple, mint, peach, pineapple, vanilla, wild cherry
Antihistamines	Black currant, cherry, chocolate, citrus, coconut, raspberry
Barbiturates	Black currant, cherry, lime, mint, orange, raspberry, strawberry, vanilla
Cough preparations	Black currant
Digestive preparations	Chocolate, maple, orange, mint, raspberry, vanilla, wild cherry
Iron preparations	Black currant, strawberry
Laxatives	Pineapple, strawberry
Liver preparations	Black currant
Vitamin preparations	Banana, butterscotch, caramel, maple, pineapple, strawberry, vanilla
Yeast preparations	Black currant, strawberry

12.6 FINAL REMARKS

The pharmaceutical industry has long recognized that good flavoring increases the acceptance of their preparations. The number of products that could be used for flavoring is very great. In practice, preliminary classification into the categories of fruit flavorings, spicy flavorings, dessert flavorings and those for preparations containing alcohol has proved useful. The most important fruit flavorings are citrus, red fruits and exotic fruits. Significant spicy products are clove, cinnamon, caraway seed, pepper, onion and garlic flavorings.

The dessert flavorings include vanilla, cocoa, caramel, cream, nut, coffee and honey. Rum, wine, brandy and liqueur flavorings are particularly suitable for preparations containing alcohol.

REFERENCES

Council of the European Communities (CEC). 1988. Council Directive of 22 June 1988 on the Approximation of the Laws of the Member States Relating to Flavourings for Use in Foodstuffs and to Source Materials for Their Production. *Off. J. Eur. Comm.* L184, 61–67.

J.H. Dhont. 1975. Reaction of Vanillin with Albumin. *Proc. Int. Sympos. Aroma Res. Zeist.* 193–194.

H.B. Heath and G. Reineccius. 1986. *Flavor Chemistry and Technology.* Avi Publishing, Westport, CT.

International Organisation of the Flavour Industry (IOFI). 1990. *Code of Practice for the Flavour Industry.* IOFI, Geneva.

T.P. Labuza and M.K. Schmidt. 1985. Accelerated Shelf-life Testing of Foods. *Food Technol.* 39, 57–64, 134.

H.T. Lawless and C.B. Lee. 1993. Common Chemical Sense in Food Flavor. In: *Flavour Science. Sensible Principles and Techniques* (T.E. Acree and R. Teranishi, eds.). American Chemical Society, Washington, D.C., pp. 33–66.

W.E. Lee III and R.M. Pangborn. 1986. Time-Intensity: The Temporal Aspects of Sensory Perception. *Food Technol.* 40(11), 71–78, 82.

H.G. Meier. 1985. Bindung flüchtiger Aromastoffe an Lebensmittelbestandteile. Lecture Given at the 43rd Seminar of the Food Industry Research Group (43. Diskussionstagung des Forschungskreises der Ernährungsindurstrie).

P. Overbosch, W.G.M. Afterof, and P.G.M. Haring. 1991. Flavour Release in the Mouth. *Food Rev. Int.* 7, 137–184.

G.A. Reineccius and S.J. Risch. 1986. Encapsulation of Artificial Flavor by β-Cyclodextrin. *Perfum. Flavor.* 11(1), 3–6.

G.S. Sinki and W.A.F. Schlegel. 1990. Flavoring Agents. In: *Food Additives* (A.L. Branen, P.M. Davidson and S. Salminen, eds.). Dekker, New York, pp. 195–258.

J. Szejtli. 1982. *Cyclodextrins and Their Inclusion Complexes.* Akadeiai Kaido, Budapest.

D.M.H. Thomson. 1986. The Meaning of Flavour. In: *Development in Food Flavours* (G.G. Birch and M.G. Lindley, eds.). Elsevier, London, pp. 1–21.

Chapter 13

Tobacco Flavorings and Their Application

Roger N. Penn

13.1 INTRODUCTION

The use of various natural flavorants added to mixtures of smoking tobaccos has been with humankind for many years, with records (Polunin & Robbins, 1992) listing the use of extracts/tinctures/solid materials by the Aztecs of the Southern Americas through to the Native Americans of the North Americas. These smoking products were within our modern product ranges and similar to snuff, pipe, cigarette or cigar tobacco. Furthermore, in other early civilizations, such as the early Chinese dynasties and early Arabic races, tobacco was consumed in both dry and moist powder forms, as well as in the hookah or water pipe.

These product types gave rise to powdered snuff, and what we know today as moist snuff and chewing tobacco. Products in this physical form have been used for medical, religious and recreational activities for centuries.

Most of the above forms of tobacco, for some reason, need some type of addition of sweetening/flavoring agent to either make the product more enjoyable (e.g., sweeter/fruitier, etc.) or to reduce some of the negative characteristics (e.g., bitterness, lack of continuity of flavor quality of the gathered leaf, etc.). These can be due to crop to crop variation or subsequent to climatic changes (drought/flooding) or soil content variation (nutrient depletion, etc.). This last aspect of crop to crop variation has never left the realms of the vagaries of the tobacco leaf purchasing area of this industry and still demands great attention from the tobacco flavorist to address modern day continuity of the organoleptic characters of tobacco products. The raw material, the tobacco leaf, is still a biological product and therefore subject to the whims of Mother Nature in the tobacco leaf plantations and in recent times to suffer under the influence of El Ninõ!

Nevertheless, with the major agronomic advances in the husbandry of the various types of tobacco varieties around the world, and the subsequent improvements in fertilizer technology and pest control, the industry still has the challenge of attempting to maintain a consistent flavor character for each product type. This is to ensure that the consumer receives, as he or she indeed expects, the continuity of product performance and thus continuing product satisfaction. The use of flavor systems therefore is a crucial part of consumer product acceptability in all its forms, as well as being a marketing/sales challenge in the form of product differentiation and uniqueness. This is the concept of the flavor "signature" in the marketplace and gives to the product, whether a cigarette, cigar or other form of tobacco product, a particular taste or aroma theme which gives the consumer particular brand identity and assurance.

The references above, citing the early origins of use of tobacco products, really are the forerunners of the story of tobacco use (Gabb, 1990), prior to the classic story of the introduction of tobacco into the European arena by Christopher Columbus upon his return from the New World and his first encounters with the American Indians. This arrival of the New World Mariners in Portugal allowed Jean Nicot (the 16th century French Ambassador to Lisbon) to carry the early seed samples to the then French Queen, Catherine de Medicis. Appreciating the inherent interest (and value!) of this new product, the French court began to disseminate both smoking and snuff products to the four corners of the world (Mookherjhee & Wilson, 1990; Dossier of Jardins de France, 1996). Thus, Jean Nicot is forever rewarded with his name being used for the botanical descriptor for tobacco—*Nicotiana* (the species *Nicotiana* is now known to contain over 1,000 subspecies in the genus).

Specifically in 1560 (Dossier of Jardins de France, 1996), it is claimed that Jean Nicot offered to Queen Catherine a sample of dry snuff powder for her to imbibe to reduce her persistent and consistent headaches. The charlatans and sorcerers of the Americas were known to prescribe this product to "clear" the head and better predict the future (Dossier of Jardins de France, 1996). Thus with these positive medical observations, the French court, being good courtesans, adopted this new practice of snuff taking (appearing to imitate the Queen) and thus named the new product "Poudre à la Reine." Thus the age of snuff imbibing was born, with many "new" aromas being added to extend the product line and to ensure the broadest possible customer acceptability!

The world renowned herbalist Gerard in 1597 in *The Herbal* gave the following description of the smoking of pipe tobacco: "the dried leaves are used to be taken in a pipe set on fire and suckt into the stomach, and thrust forth again at the nostrils, against the pains of the head!" Other, more recent herbals by Lewis and Elvin Lewis (1977), also Polunin and Robbins (1992), observe other ways of "drinking" the smoke of cigar tobacco and powdered snuff.

Thus, the *Nicotiana* genus is considered to consist of three major subgenera, *viz.*:

- *N. tabacum*, major smoking tobacco type
- *N. rustica*, some smoking tobacco types
- *N. petunoides*, mainly ornamental tobacco flowers

Almost all of these genera originated in South America, journeying to the north with the nomadic indigenous Indian tribes to Mexico and the southwest USA, via countries such as Argentina, Bolivia, Brazil and Paraguay (Dossier of Jardins de France, 1996). In a later section, we will briefly discuss the specific types of "industrial" tobacco plants grown for factory use, compared with the ornamental styles of *Nicotiana* used purely for the delight of botanists and gardeners alike.

So, the above few words are a small excursion into the realms of the historical botany of the noble tobacco leaf. Let us now briefly consider the historical aspects of some of the traditional flavoring practices in tobacco products. Various published accounts have been made by Triest (1966), describing observations on the new role of additives in tobacco. Additionally, there has been a plethora of articles on the need for new approaches to cigarette tobacco flavoring to address the issues of the development of the relatively new "low delivery" type of product (i.e., lower tar and nicotine content of the mainstream smoke), relative to products on the market before the 1980s. Dietrich (1978) gave perhaps one of the broadest discussions, at that time, on the emergence of the so-called reconstituted and nature identical tobacco flavors for tobacco use, whereas Wakeham (1978) provided a more technological approach by describing some of the existing and potential flavor delivery systems used at that time. Another article discussed the ever increasing use of flavorants in low delivery products, and the relative flavor delivery systems ("A Question of Taste," 1978).

At a later date, discussion centered upon the types of raw materials used in tobacco products (Heyzer, 1988), and thus the relative performance and success of the flavoring system being more a question of raw material selection and innovative use and blending of components, rather than the highest quality or highest priced products.

The above references will help the interested reader to construct a working background knowledge of the historical evolution of tobacco flavors, whereas Gabb (1990) eloquently describes the past and present use and prohibitions of tobacco.

13.2 DEFINITION OF TOBACCO TYPES

For simplification of discussion, the overall consideration of this section will be limited to those general classes of tobaccos used in the regular tobacco industries.

Initial differentiation can be made by the type of treatment or curing the tobacco receives, and the specific dimensions of the leaf, *viz.*:

Virginia (flue cured)

- bright tobacco
- cured by forced hot air heating in barns (thus the term flue-chimney)

	• enzyme activity is stopped, therefore contains high level of sugars • medium to large leaf size
Burley (air cured)	• darker tobacco • cured at ambient air temperature • enzyme activity maintained, therefore sugar levels completely (or almost) negligible • medium sized leaf • can be heat treated later in process (toasted)
Oriental	• dark tobacco, very small leaf size • heat sensitive leaf, but very flavorful character • particular chemistry (cembranes/labdanes)
Fire cured	• normally from Kentucky or Latakia, leaves are suspended over an open wood fire and thus smoke deposition occurs on the surface • gives heavy, smoky, phenolic notes, used in cigarettes and pipe tobacco
Fermented	• leaves are kept in piles or bays in bulk • natural fermentation takes place • gives heavy, dark, acetic notes to smoke

Other types of tobacco exist, e.g., sun-cured, where the leaves are suspended in "hands" by twine tied around the stems, then the leaves are dried in the open in direct, hot sunlight (mainly Africa and Latin America).

As well as the leaf portion (lamina), the veins of the leaf (stem) are also used. They are normally separated upon arrival at the tobacco factory by ripping or a destemming process. The stems can be steam-treated and then rolled or expanded by using liquid carbon dioxide to give a larger volume. Due to their high cellulosic content, stems sometimes receive separate treatment with flavor modifiers to give a more acceptable smoke, with a significantly reduced cellulosic off-taste.

Another form of usage of tobacco materials, the so-called byproducts, is in the form of reconstituted tobacco, whereby a mix of tobacco leaf fragments is made into a slurry and then processed via a paper making process. This tobacco sheet is then cut into strands and appears slightly similar in nature to the original leaf. Flavoring systems are also used in this reconstituted tobacco to develop more acceptable taste characteristics, and again cover the inherent off-tastes. World-wide, there are many more tobacco types for specific applications, but these are too specialized for discussion here.

13.3 TYPES OF FLAVOR SYSTEMS FOR TOBACCO PRODUCTS

The tobacco leaf represents a smokeable product only after it has undergone treatments such as curing, drying, fermentation or ageing. Once the farmer has harvested the leaf and "processed" it, depending upon what type of product it is, the leaf is then normally ready for auction, when it is destined for particular factories. Now the leaf is in the hands of the leaf blender, one of the most skilled people in the industry and certainly one of the main practitioners in the sensory properties, as well as the physical appearance, of the leaf.

The so-called "grading" of tobacco leaf, simply put, is the characterization of the smoking quality and basic chemical indicators (e.g., sugars and nicotine levels and leaf position in the plant plus the leaf color and friability). The position of the leaf on the plant will further dictate the expected smoking characteristics of the individual leaves; this is due to the varying distribution of metabolites (active principles) within the growing plant.

An early article (Enzell, 1986) gives an excellent review of the scientific evidence regarding the role of tobacco leaf precursors and the resulting cigarette mainstream smoke flavor developed from the burning of the tobacco. Enzell (1986), as part of his significant contribution in this area along with Wahlberg & Enzell (1977), gives much insight into the involvement of the isoprenoid flavor components in the leaf, and their formation from the biodegradation of carotenoids (Figure 13–1a) and the deterpenoids of the cembrane and labdane types (Figure 13–1b). Indeed a significant contribution to natural product chemistry, specifically in the area of stereo-isomerism, has been made by the work on the biomimetic routes involving singlet oxygen reactions, epoxidations and their ensuing rearrangements. There has been major chemical interest in the last two decades to unravel the biosynthetic routes occurring during postharvest treatment (*viz.* curing, fermentation, etc.), as well as the subsequent bio-degradation, thermolysis and pyrolysis of these structures, to give rise to the flavor (which is constituted by true taste and aroma) in the mainstream and sidestream smoke.

Schematically, the genesis of the tobacco smoke constituents can be represented as in Figure 13–2.

Thus, from an extremely complex biomass—the tobacco leaf—one can follow, thanks to the hundreds of thousands of hours spent by dedicated scientists on unravelling the unique molecular story, the development of what the final consumer perceives as a total "flavor," or "satisfaction" parameter, of the smoked product.

The resulting complex mixture, whether it be cigarette, cigar or pipe tobacco smoke, is really a whole palette of individual chemical moieties, which, at each of their own specific vapor pressures, contribute their own individual character to the overall profile. Another whole area, too long for discussion here, is that of specific threshold detection limits and the relative levels at which individual flavor components become detectable and furthermore the levels at which they become recognizable (so-called recognition thresholds). This whole area above,

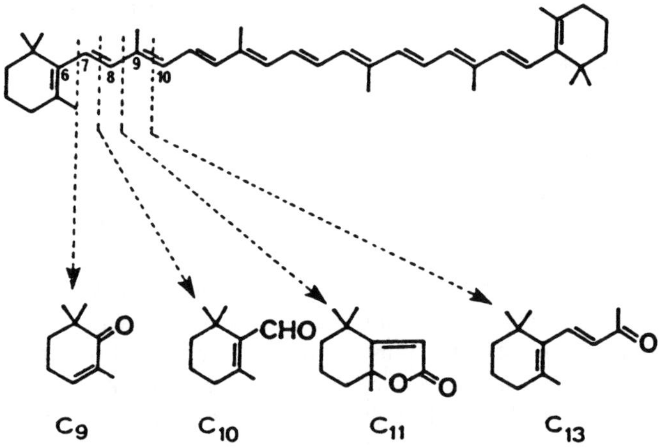

Figure 13–1a,b Isoprenoid flavor components, in the leaf, and their formation from the biodegradation of carotenoids (**A**) and the deterpenoids of the cembrane and labdane types (**B**). *Source:* Reprinted with permission from C.R. Enzell, in *Flavour '81*, P. Schreier, ed., p. 452, © 1981, Walter de Gruyter GmbH & Company (Fig. 13–1a); Reprinted with permission from C.R. Enzell and I. Wahlberg, in *Bio of Flavour '87*, P. Schreier, ed., p. 244, © 1988, Walter de Gruyter GmbH & Company (Figure 13–1b).

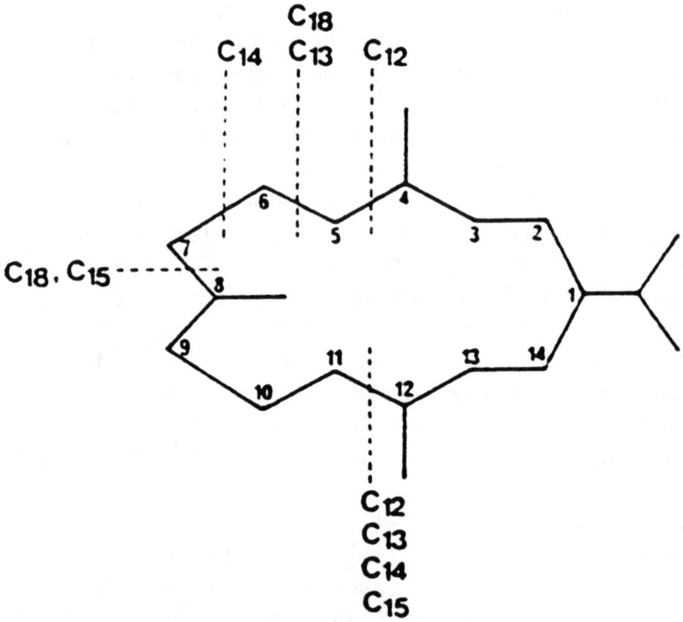

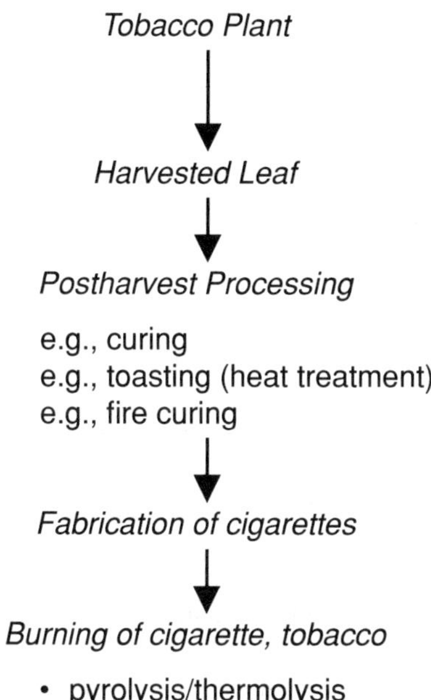

Figure 13–2 Genesis of tobacco flavor volatile systems.

which only constitutes a very small part of sensory assessment, now constitutes another segment of the so-called psycho-sensory effects. It would require a separate chapter to even begin to explore this fascinating subject, both in terms of the creative development of flavor systems as well as the evaluation of the final product profile in terms of consumer acceptability. The relevance of the relative taste and odor thresholds of the single constituents of a complex mixture can be considered as the following scenario: One has present some materials with very low thresholds but in very small amounts (i.e., at the parts per billion range); these materials will have a very much greater contribution to the flavor character than those materials with high detection thresholds but present in higher relative concentrations (i.e., in the parts per million range).

Some of the chemical nature (Mookherjhee & Wilson, 1990) of the aroma of cured tobacco leaf can be divided into the following chemically functional group structures:

acids
alcohols

aldehydes
ethers
hydrocarbons
ketones
lactones
nitrogenous components
phenols
sulfurous components

Many groups of researchers, both from the tobacco industry itself as well as the flavor industry, have carried out analysis (Green & Rodgman, 1996) either by extraction (solvent, liquid CO_2, etc.) and subsequent gas chromatography–mass spectrometry (GC/MS) or by headspace volatile analysis (Wahlberg & Enzell, 1977) of the fresh and cured/processed leaf.

Many classes of components identified by the methods above have been positively identified and then subsequently chemically re-engineered or re-constructed to give the so-called nature identical taste and aroma of the various types of tobacco.

Some "key" single chemicals have been isolated and identified then synthesized and finally used as so-called "character identity compounds" (i.e., one single chemical whose aroma and taste is totally representing a specific, identifiable flavor character). For example, 3-methyl valeric acid and caproic acid mixtures have been used as major contributors for adding typical Oriental tobacco character to base neutral blends; various substituted pyrazines have been used to add roasted, toasted, nutty notes; and finally some sulfur-containing molecules are used, which like most of the "character identity compounds" are obnoxious and unacceptable when concentrated but when suitably diluted give representative and acceptable characters of the required organoleptic targets.

The many scientific endeavors in this area have now given creative flavorists, flavor chemists and flavor technologists, in both the tobacco industry and the flavor industry, many opportunities for the combination of all of these nature identical molecules with many of the natural products that have been traditionally used for many years. The improvements necessary in modern day products and the needs of the modern day consumer have thus been addressed.

Both national and international regulations govern very closely the majority of countries in the world in terms of the type, and in some cases the maximum levels, of flavoring materials added to tobacco products. An outline of a few of the major current regulations applicable to tobacco products is set out in Table 13–1.

These major guidelines have been adopted, both in total or in part, in other areas of the world, whether by private companies or national monopolies. Other countries have adopted various combinations or extracted parts as guidelines for their industries.

Table 13–1 Outline of Regulations Covering Tobacco Products

Country	*Responsible organization*
United Kingdom	Independent Committee on Smoking and Health
USA	Food Extract Manufacturers Assn. (Generally Recognized As Safe) FEMA/GRAS "599" list of ingredients
Europe	Council of Europe list Germany regulations France regulations

At present, the most comprehensive regulatory coverage is considered to be an "umbrella" requirement of FEMA/GRAS (USA), German and UK legislation. Additionally, individual companies will impose their own requirements and exceptions.

13.4 FLAVOR SYSTEM CHARACTERISTICS FOR SPECIFIC TOBACCO PRODUCTS

13.4.1 *Cigarettes*

The largest volume of tobacco is consumed in cigarettes worldwide. In 1996 approximately 5.7 trillion (5.7×10^{12}) cigarettes were produced, with 120 nations producing 7.2 million metric tons of leaf tobacco; the major geographical cigarette manufacturers in 1996 are shown in Table 13–2.

Many different types of cigarette constructions exist around the world, some of which merit separate treatment later in this chapter.

The world's most popular cigarette character is perhaps currently in the direction of so-called American blended type. This is closely followed by Virginia (or so-called English style), modified Virginia, as in the case of the flavored Chinese tobacco, or modified Virginia blends containing other tobaccos. Finally, there are the so-called Oriental type and dark air cured or fermented type. In Table 13–3 an indicative market guide to branded products is given for illustrative purposes.

Table 13–2 Total Volume of Cigarettes Produced By Major Markets

Country	*No. of cigarettes (billions)*	*% World production*
China	1,721	30.1
USA	754	13.2
Japan	271	4.7
Germany	218	3.8
Total		51.8

Table 13–3 Major Cigarette Brands Worldwide

Cigarette type	*Brand*	*Manufacturer*
American blended	Marlboro	P. Morris
	Winston	RJ Reynolds
	Lucky Strike	B.A.T.
Virginia	State Express 555	B.A.T.
	Benson & Hedges	Gallaher
Oriental	Samsun	Tekel
Dark air cured or fermented	Gauloise	Seita
	Ducados	Tabacalera

The type of tobacco leaf in the blend varies within each product type. For example, Table 13–4 exhibits the individual leaf types for each blend, as well as a few comments on some of the leaf chemistry.

Depending upon the type of tobacco, there are other very important stages, once the leaf is harvested, that may be carried out before it reaches the factory for final processing:

- curing/drying
- fermentation/sweating
- ageing

These physical changes result in dramatic chemical changes within the leaf and are also reflected in major modifications of the appearance (color, consistency, etc.) and flavor attributes of the leaf. All these parameters have to be con-

Table 13–4 Basic Smoke Chemistry of Major Tobacco Types

Product type	*Leaf contained in blend*	*Chemistry*
Virginia (flue cured)	100% Virginia	High sugar content High pH in smoke
American blend	Virginia	High sugar content
	Burley (air cured)	Low–no sugar levels Neutral pH smoke
	Oriental	Very high flavor profile Textural character in smoke
Oriental	100% Oriental	High flavor profile
Dark air cured/ fermented	Mixture of Virginia fermented air cured	Very high flavor level Basic pH in smoke

sidered by the leaf blender of the tobacco company when the "grading" process is begun with the leaf, prior to manufacture. This, a complete subject in itself, will not be further discussed here.

We have to concentrate somewhat further on the flavor chemistry of the leaf at this stage. For instance, sugars are present in high levels in both flue cured and Oriental tobacco whereas burley tobacco has very low sugar levels but does possess a high content of proteins, nicotine and total nitrogen. Oriental tobacco contains high levels of essential oils (thus delivering a very aromatic, rich note to the overall flavor profile) as well as high levels of total volatile short-chain organic acids.

Much research work has been carried out to examine the significance of the contribution of the various chemistries and volatile constituents to the total flavor profile. Coleman (1992), for instance, gives a very good overview on the analysis of headspace volatile materials of certain tobaccos and their relation to the leaf chemistry and changes effected by tobacco processing, and Enzell (1981) has demonstrated the chemical changes that occur during the curing of Virginia tobacco.

13.4.2 *Pipe Tobacco*

Historically, there have been fewer types of pipe tobacco compared to the thousands of brands of cigarettes around the world. They are normally prepared using co-called Cavendish types of blend with very heavy use levels of components such as aromatic fruit extracts, molasses, liquorice and cocoa. Perique type of tobacco is also used, incorporating both fire-cured tobacco and heavy fermented or sweated tobacco. This, together with added flavor materials such as those described above, gives a very phenolic, fermented character.

The so-called "twist" tobacco originated from the old practice (Stone, 1978) of sailors on-board ship dipping their tobacco leaves in treacle and rum, twisting the mixture into a roll, wrapping it around with a cord to keep it tight, and then leaving it for a week or two, usually to ferment. This procedure generated very agreeable taste characteristics when the tobacco was smoked and finally resulted in the commercial process for so-called twist tobacco. This was very much in favor at the turn of the 19th century and is still popular in some countries even now. Many other flavorants were and are used in the pipe tobacco industry, usually derived from the practice of using natural extracts and fruit syrups of bygone days, combined with modern flavor enhancers and boosters. Basically, these flavor aromas are also transferred to the ambient smoke, giving the characteristic strong, sometimes sweet/fruity aroma to the smoke, that is so well appreciated in many countries around the world.

13.4.3 *Cigars*

Historically and to the present day, the tobacco leaf used for cigars has very different characteristics to those of cigarette tobacco. There are considered to be three classes of cigars: large, tipped and little. Classical premium cigars are considered to be acceptable if they are from Cuban grown plants. However, with the Cuban embargo and the meteoric rise in consumer demand over the last 3–4 years for cigars in general, there has been a dramatic increase in the cultivation of good cigar leaf especially in places like the Dominican Republic and in some far Eastern countries.

13.4.3.1 *Premium cigars*

Traditionally, these premium cigars are unflavored; they rely heavily on high quality filler tobacco but also on the best (in organoleptic terms) leaf for the outer leaf or wrapper leaf. Furthermore, the premium cigar is usually hand rolled, not machine made. The filler tobacco is normally a carefully blended mixture of air cured and fermented tobaccos.

However, as the current market develops and diversifies, some flavored premium cigars are appearing and are identified in the market as such. Some of the typical (Kucharski, 1998, personal communication) flavor types are vanilla, cherry, chocolate, rum and other cordial types such as creme de menthe, anisette, cognac, etc. Vanilla and rum flavor types are by far the most popular and are applied by spray onto the cut-filler tobacco, at levels from 1 to 5% by weight.

13.4.3.2 *Tipped cigars*

These make up the largest category of flavored cigars. They are machine made with a sheet binder and wrapper. Examples of flavor types used include vanilla, fruit and cordial.

13.4.3.3 *Pipe-tobacco cigars*

Over the last 4–5 years, a new product category has evolved, that of the so-called pipe-tobacco cigars. These cigars use real pipe tobacco blends as a filler, and are heavily cased, sweetened and heavily top flavored, but use a regular cigar leaf wrapper on the outside of the product.

13.4.3.4 *Little cigars*

Machine made, of cigarette size, these have a reconstituted sheet wrapper instead of paper, and only air-cured tobacco as filler. Many of these products are top flavored. The most popular flavors are menthol, aromatic or pipe aroma types. The latter use pipe tobacco but only have low levels of top flavors using the sweet, fruity notes of the well known pipe tobaccos.

13.4.4 *Narghili (Water Pipe) or Hookah*

This form of application where tobaccos are heated or charred by burning charcoal embers has been around for centuries, in the Arabic countries in particular, originating in what is now Turkey and evolving with the growth of the Ottoman Empire (Kiernan, 1991). The use of the water pipe is considered in these parts of the world to be a social activity where a group may sit sharing one water pipe but each person using his or her own mouth piece. Many coffee shops or areas in bazaars are reserved for this sociable practice.

The inclusion of the water "bottle" in the device changes the whole perception of the flavor characteristics of the smoke, compared with pipe smoking. The typical, habitual Narghili smoker inhales deeply and blows smoke out through the nose. This is easier than with other tobacco products due to the fact that the water humidifies the smoke, thus reducing irritation, and also selectively extracts some of the smoke products. Overall the experience is less harsh than a normal pipe, but the main difference is in the style of traditional flavoring of the tobacco.

Normally, tobacco used for the water pipe is "molassed," that is to say, contains a very high level of sugar syrups (up to 60% in some cases) and is therefore a sticky wrapped mass when purchased. Various traditional flavor types are found in these markets, notably apple, strawberry, apricot and rose. Recently though, the tastes are beginning to change and other flavor types are being introduced (e.g., liquorice, coffee, fruit notes, etc.).

The average amount of molassed flavored tobacco per charge of an average water pipe is approximately 7–10 g. A normal smoker or group can keep the charge alight for 5–15 minutes, obviously depending upon rate of draw. Sometimes, a slice of lemon is placed into the water bottle; this will also affect the smoke chemistry due to citric acid present in the lemon. The water in the bottle is normally changed after three charges of tobacco.

Finally, it can be said that although this Narghili smoking practice is steeped in tradition, there is now beginning to develop a larger range of flavored "molasses" to give the product a newer taste and image for current and future consumers.

13.4.5 *Kretek*

The name of the Kretek cigarette is derived from a compilation of Indonesian words for "snap, crackle and pop," which is exactly what the chopped clove buds (contained in the blend) do when they burn in the tobacco blend of this unique type of cigarette. The Kretek is a very traditional cigarette product, originating in, and sold mainly in, Indonesia. Until recent times it was only hand-made as the traditional non-filter, trumpet shaped products. Typically, these cigarettes are made up of chopped clove buds as upwards of 30% by weight of the total product. They also use very high levels of casings and top flavor systems. Both mainstream and sidestream smoke of a Kretek cigarette are very easily recognized due to the highly characteristic aroma of eugenol/clove. Locally grown tobaccos are also used in the blend. They originate in Java, Sumatra and Borneo, and each delivers its own characteristic notes to the blend. The use of specialized casings and "typical" top flavors in this type of product is well appreciated. Many fruity, estery combinations are used together with specialties such as traditional Salak cider, Nangka, Havana and Manila style compounded flavors which further add herbaceous, spicy characteristics. These all blend together with the overall clove/eugenol character.

Today a certain quota of machine-made cigarettes is allowed by the Indonesian government. This method of manufacture is becoming more important as the evolution of medium to low tar products continues in the indigenous markets. It must not be forgotten, however, that the traditional hand-rolling method of production employs tens of thousands of people in Indonesia. To maintain their interests machine production must be carefully controlled.

Today, the Kretek is found in other countries, and an export market may well develop, probably with the availability of reduced tar and nicotine products. However, although the cut clove is used in the blends, the typical clove aromatized sidestream will remain the hallmark of Kretek taste and aroma.

13.4.6 *Bidis*

These are a "hybrid" of a traditional cigarette product, mainly made and consumed in India, but also found in other parts of the subcontinent. They are again hand-made items, more related to a cigar in structure because they are products that are pieces of tobacco wrapped inside a leaf, but more of cigarette size. They have neither cigarette paper nor filtertip.

The wrapper leaf is from an indigenous tree called "Tendu" (Rao, 1997). The filler tobacco is domestic Virginia tobacco, with small amounts of air cured tobacco added. Sometimes, the filler is cased and top flavored with very fragrant natural extracts such as Pan or Pan Masala that are also found in other types of products in the area. These are a type of chewing or moist snuff product, sometimes used wrapped in a betel leaf or as straight moist snuff flavored with typical rose/mint/menthol combinations. Real silver flakes are sometimes in the mix.

In the total Indian cigarette market the classic Bidi cigarette outrivals the production figures of its "white," machine-made counterpart by a ratio of 10:1. Thus the Bidi is the mass market product and is much cheaper than the up-market machine-made alternatives. The Bidi market is huge and remains a source of pleasure for tens of millions of smokers. This characteristic trumpet shaped green leaf product, normally tied with a piece of cotton at the mouth-end (to hold the roll together), is very practical. Its distinctive sidestream aroma is provided by the Tendu wrapper leaf and mainstream character by the flavored cut filler.

13.4.7 *Moist/Dry Snuff*

As alluded to in the opening section of this chapter (Mookherjhee & Wilson, 1990), one of the earliest uses of tobacco was in the form of dry snuff. Although the current-day, world-wide consumption of dry snuff has been largely replaced by cigarettes and cigars, the use of flavored snuffs has not been fully exploited, except perhaps in the form of mint and mentholated products. As described below, there is a resurgent interest in this product range.

On the other hand, the use of moist snuff has seen some resurgence since the 1980s in markets such as the USA and Europe, where dramatic increases in public area smoking bans have been made, together with almost total bans on workplace smoking. These demands have increased the requirements for newer types and flavor ranges of the moist snuff to give products that will be more appealing to both traditional and new users.

More or less 10 years ago moist snuff was a very high moisture (ca. 60%) product consumed in a hand-made pellet form. Typical taste characters were fire cured, phenolic, leathery notes, sometimes with added ammoniacal notes. Again, some use of menthol, liquorice and fruit syrups was made as a flavoring base.

Then came the use of the “tea bag” sachet with high tensile strength and highly porous paper. This form of packaging and product presentation moved the product into a new sphere. It became much easier to use, was more easily packaged, offered a more elegant presentation and thus created more possibilities for product segmentation and identity. The use of different flavor concepts in these new products was launched (e.g., dried fruit types, apricot, whisky, rum, etc.). Recently, the increase in female users of this type of product has led to more feminine tastes being created (e.g., citrus, mint, lighter characters).

Again, the tobacco base that is used for these styles of products differs from country to country and by manufacturer. Some consumers prefer more fire cured phenolic tobacco, others air-cured, Virginia. There are significant levels of salt in the more traditional moist snuffs, whereas in the newer, lighter brands there is lower salt, lower nicotine tobacco and more light, fruity characters in the flavor profile.

13.4.8 *Chewing Tobacco*

This style of tobacco product really only exists in significant manufactured volume in North America. In other parts of the world tobacco is “chewed” but in nothing like the volume that it is used in the USA. Here, the well known portrayal of the baseball player with a cheek bulging waiting for the big throw typifies the product image. In fact, with the banning of tobacco smoking from almost all inside and outside sports events in the USA, a huge new market demand for the chewing tobacco product range has been created.

Normally, very high levels of sugars are applied to the coarse cut/thin leaf blends. Fruit syrups are also used to complement the somewhat lighter, "toppy" Virginia tobacco which is mixed with light air-cured material. Overall, chewing tobacco is much sweeter than the moist snuff product, and normally has a totally different customer profile. Apart from traditional mint and menthol types, there is of course the omnipresent American taste of root beer (methyl salicylate) in several major selling brands.

13.4.9 *Roll Your Own (RYO) Tobacco*

This tobacco segment (Dymond, 1998) has enjoyed a tremendous growth rate over the last 10 years in Europe. This has largely been due to significant tax increases on machine-made cigarettes, whereas the RYO segment has enjoyed a much lower rate of increase, and consumers can decide how many "rollies" they can make from a given amount of tobacco (i.e., thinner cigarettes can make their tobacco last longer!).

The major European markets for RYO are segmented into the following principal geographic areas: Scandinavia, Germany, Holland and the United Kingdom. Other areas of the world contribute some volume but the four areas above are the largest markets.

Scandinavia, Germany and Holland all tend to have heavier, very aromatic RYO tobacco blends with the United Kingdom having a lighter, slightly fire cured character that is more in keeping with UK consumers' liking of Virginia cigarettes. From a flavor point of view they have a burnt, mellow, more acidic character than a typical American blended product containing other tobacco types, such as burley or Oriental.

The Dutch style, highly aromatic RYO tobaccos or "shags" are internationally renowned for their intense and pervading sweet, vanillic aromas. Other varieties are heavily oriented toward a fire cured, phenolic character. Rich dried fruit, alcoholic whisky, cognac and rum notes are all used in these types of products.

Some consumers prefer to "blend" their own mixtures to obtain the correct amounts of tobacco character, sweetness and fruit notes. This activity also occurs widely with connoisseurs of pipe tobacco where various blends and flavored tobaccos are kept in separate humidors to be blended just prior to consumption.

13.5 NEW TOBACCO PRODUCT DEVELOPMENTS AND TRENDS

As the author has noted elsewhere (Penn, 1997, 1998), the future of "traditional" tobacco products is still very much assured in various parts of the world for some considerable time to come. Perhaps this consumption of tobacco may well see several significant changes in how the product is consumed and, more importantly, how customer satisfaction is maintained. On one hand, the move both by industry and legislation in many countries to reduce tar and nicotine consumption has led to less opportunity to continue tobacco products. On the other, an exponential growth in the traditional non-smoking areas has to be considered. Thus, the challenge to the industry is further compounded by the issue of sidestream smoke and its inherent nuisance factor ("passive smoking") to non-smokers and smokers alike.

13.5.1 *Tar and Nicotine Reduction*

The question of the reduction in tar and nicotine delivery has been addressed over the last 20 years, and Durocher (1996) gives an account of the advances made by the tobacco industry, associated organizations and research groups in this area. The basic mechanism for the reduction of both parameters is in part achieved by dilution of the mainstream smoke, using ventilation holes introduced into the filter circumference. This allows the ingress of outside air into the mainstream during inhalation (Figure 13–3).

Product developers, using both filter perforation and high porosity wrapper paper, can in the case of the ultra low delivery products obtain up to 60% dilution or more of the mainstream smoke. This overall dilution results in very "thin" smoke characteristics and certainly lack of flavor quality. Therefore, the need for the overall enhancement of the flavor character is a current challenge for the flavorist and product developer alike and will remain so for some time in the future.

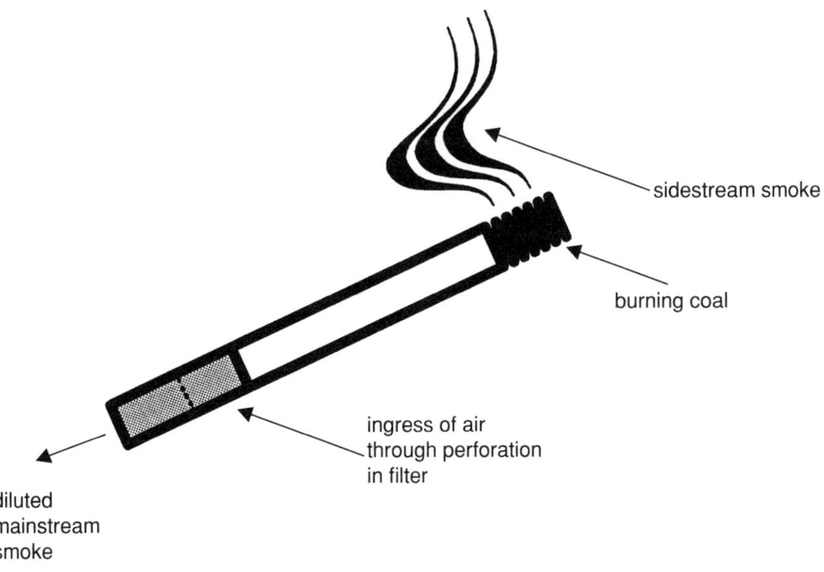

Figure 13–3 Diagrammatic description of cigarette smoke.

13.5.2 *Sidestream Issues*

For some years now the issue of cigarette sidestream smoke has been a major issue in terms of its nuisance factors:

- visibility
- irritation
- odor

Technically, the use of "low visibility sidestream papers" has been an option for the industry and several attempts have been made to launch brands with reduced visible sidestream.

These launches were not generally successful due to the fact that although the cigarettes did have up to 20–30% reduced visible sidestream when compared to controls, other product attributes were rendered more negative (i.e., very flakey white ash associated with chalky, negative aftertaste). Overall consumer response was negative. Several new product designs have been attempted to address the issues of irritation and odor.

13.5.2.1 *Sidestream aroma/irritation*

Several manufacturers have looked at various ways to reduce the above characteristics. Research has shown that the majority of sidestream smoke is formed during the cigarette smoulder phase; various attempts have been made to introduce types of ingredients into the wrapper paper which when heated or burned give rise to a "masking" aroma or agent that perceptibly covers the malodor of sidestream smoke or the sensitized reaction of nose/eyes to the irritation effect. One of the current market successes is that of RJ Reynolds in the Japanese market with their brand "Pianissimo." In this brand they apply so-called precursor materials to the wrapper paper. These are liberated once the cigarette is lit and produce a sweet, light odor in the sidestream, thus "re-aromatizing" the negative, acrid sidestream smoke.

The new product gives greatly reduced sidestream smoke compared to leading regular cigarette brands, and is claimed in product/pack advertising to be 80% less. The tar and nicotine deliveries are significantly reduced too. Improvements continue on the initial lighting procedure which was also problematic when compared to a regular cigarette.

13.5.2.2 *Tobacco that is heated not burned*

This approach has been researched over 10–12 years, originally by RJ Reynolds and more recently by Philip Morris.

RJ Reynolds originally carried out consumer testing in the USA with a product called Premier in the mid 1980s. This product contained a heat source, aerosol generator and flavor source, combined with wrappers of cut tobacco leaf. After the ignition of the graphite heat source, there was an aluminum heat conductor that transferred the heat to granules containing an aerosol generator (glycerine) and flavoring. Thus the tobacco was not pyrolized as in a traditional cigarette, but heated to release volatile flavor components into the aerosol generated by the glycerine and thus inhaled by the consumer.

Very recently, in four countries, USA, Germany, Sweden and Japan, RJ Reynolds has been test marketing a related product design, as a smokeless product ("The Smoke Clears Again," 1997). Known as "Eclipse" in the USA, "HI-Q" in Germany, "Airs" in Japan and "Inside" in Sweden, they all bear a resemblance to their forerunner (Premier) but significant changes have been made to the flavor profile, one of the major weaknesses of the original product.

13.5.2.3 *Tobacco heated in an enclosed space*

In late 1997 ("P. Morris Explores," 1997), Philip Morris announced the results of several years of basic research in which they had been working on, literally, a "black box" approach, where a specially tailored cigarette is locked into position in a small box with the mouthpiece protruding. When the smoker sucks on the mouthpiece, the electronic sensors in the box allow one puff to be taken and then extinguish the rod, which is then ready for the next puff at a later time by the smoker. At the time of this writing, these devices have not yet been released into a "public" test market. It is obvious therefore that there is *no* sidestream smoke, but there is exhaled smoke. The design of this so-called "Accord" cigarette is different from regular cigarettes in terms of its construction and constituents. Thus, the puff cycle is very much an "on-demand" system. The outcome of initial consumer tests of this ultra–high-tech answer to one of society's current obsessions—nuisance removal in the form of sidestream smoke—is awaited with interest.

13.6 SUMMARY

The art and science of tobacco flavoring, as described in this chapter, has encouraged many learned, skilled scientists and intuitive, creative flavorists to respond to the needs and desires of both consumers and controlling bodies alike.

As stated here, the humble work of the Aztecs and North American Indian tribes has been supplanted by the high-tech world of instrumental analysis, biochemical pathways and gene splicing. For the tobacco products of the future, the understanding of what types of flavors will be needed to give sufficient consumer acceptability will be paramount. New flavor delivery systems will have to be conceived, whether of precursor or encapsulated type, whether they have to be heat stable or thermally labile and where they are to be placed—in filter, tobacco or on the wrapper paper. All these questions will require both scientific and creative thought and product development capacities par excellence.

Given the pressures from the various groups, whether they be governmental or private in the tobacco products arena, we can still be sure that the resources and capabilities of the tobacco flavor sector will be required for many years to come!

REFERENCES

W.N. Coleman, *Journal of Chromatographic Science* 30, 159–163 (1992).

P. Dietrich, *Tobacco Reporter,* October, 40–43 (1978).

Dossier of Jardins de France, Paris & L'Institut de Bergerac, France, *Sacre Tabacs* (1996).

D.F. Durocher, Recent Advances in Tobacco Science, 22 (1996).

H.F. Dymond, *Tobacco Journal International,* 64–71 (1998).

C.R. Enzell, Influence of Curing on the Formation of Tobacco Flavour, *Flavor* 81, 449–478, (1981).

C.R. Enzell, Isoprenoid Flavour Components of Tobacco and Their Formation, *Banbury Report* 23, 163–178 (1986).

S. Gabb, Smoking and Its Enemies, *Forest* (1990).

J. Gerard, in *The Herbal or General Historie of Plantes,* John Norton, London (1597).

C.R. Green, A. Rodgman, *Recent Advances in Tobacco Science,* 22 (1996).

E. Heyzer, *World Tobacco,* March (1988).

V.G. Kiernan, *Tobacco: A History,* Hutchinson, London (1991).

W.H. Lewis, M.P.F. Elvin Lewis, in *Medical Botany—Plants Affecting Man's Health,* Wiley Interscience (1977).

B.D. Mookherjhee, R.A. Wilson, Tobacco Constituents—Their Importance in Flavor and Fragrance-Chemistry. *Perfum. Flavour.* 15, 27–49 (1990).

"P. Morris Explores 'New Smoking System,'" *Tobacco International,* December (1997).

R.N. Penn, Tobacco Flavouring: An Overview, *Perfumer & Flavourist* 22 (1997).

R.N. Penn, A Flavour of Tobacco, *World Tobacco,* March (1998).

M. Polunin, C. Robbins, in *The National Pharmacy,* Dowling Kindersley, London (1992).

A Question of Taste, *Tobacco Reporter,* October, 30–34 (1978).

B. Rao, The Ubiquitous Bidi, *Tobacco Asia,* December (1997).

"The Smoke Clears Again," *Tobacco Asia* December (1997).

M. Stone, *Tobacco Reporter,* October, 35–37 (1978).

F.J. Triest, Function of Tobacco Flavor, *Tobacco,* July, 168–170 (1966).

I. Wahlberg, C.R. Enzell, *Phytochemistry,* 16 1217 (1977).

D. Wakeham, *Tobacco Reporter,* October, 67–70 (1978).

FURTHER READING

R.A. Heckam, M.F. Dube, D. Lyons, J.M. Rivers, *Recent Advances in Tobacco Science* 7, 107–153 (1981).

New Role of Additives in Tobacco, *Tobacco Reporter*, July (1967).

J.M. Ockers, *Tobacco International,* September, 9–12 (1978).

F. Robicsek, in *Smoking Gods, Tobacco in Maya Art,* Barley Bros., Oklahoma (1979).

Tobacco Merchants Association, Princeton, NJ, USA (1997).

Appendixes

APPENDIX I: COMPOSITION OF LEMON AND ORANGE OILS

Lemon Oil, Italian

Obtained by pressing peels of *Citrus limon* L., N.L. Burman, a lemon species grown in Italy. Pale yellow to pale greenish-yellow liquid with a characteristic lemon peel odor.

Evaporation residue	1.6–3.6%
Acid number max	1.4%
Carbonyl compounds content, calculated as citral	3.0–5.0%

Main components of lemon peel oils are terpenes and composition depends on variety grown and country of origin, e.g,. in American oils,

(+)-Limonene	65%
beta-Pinene	8–10%
gamma-Terpinene	8–10%

The characteristic odor of lemon oil, which is different from that of other citrus oil, is caused to a large extent by the two citrals:

Neral and geranial	less than 3% (but can be as high as 10%)

The esters neryl and geranyl acetate contribute to the initial aroma impression and changes with increasing alpha-terpineol and terpinen-4-ol content. These alcohols are, in part, artefacts formed under the influence of the citric acid present in the juice during the production of the oil.

Orange Oil, Sweet

Obtained by cold pressing the peels of the orange species, *Citrus* sinesis L. Osbeck, or for Guinea type oils, of the varieties *limoviridis* A. Chevalier and *djalonis* A. Chevalier. Yellow to reddish-yellow liquid with the characteristic odor of orange peel; the oil may become cloudy when chilled. The physical properties of the oil depend on the variety and origin:

	Italy	*Guinea*	*Brazil*
Evaporation residue (%)	1.6–3.5	1.0–3.2	2.0–3.6
Carbonyl compounds content (as decanal) (%)	0.9–2.2	1.8–3.1	1.4–3.1

Sweet orange oil is produced in many countries in combination with orange juice, e.g., United States, Brazil, Israel and Italy. Terpene hydrocarbon content, mainly (+)-limonene is always >90%. Various oils differ in oxygen-containing compounds. The aroma is determined by the aldehydes, mainly octanal, decanal, and both citrals and esters, mainly octyl and neryl acetate. The sesquiterpene aldehydes alpha- and beta-sinensal, which also occur in other citrus oils, although in lower concentrations, contribute particularly to the specific sweet orange aroma.

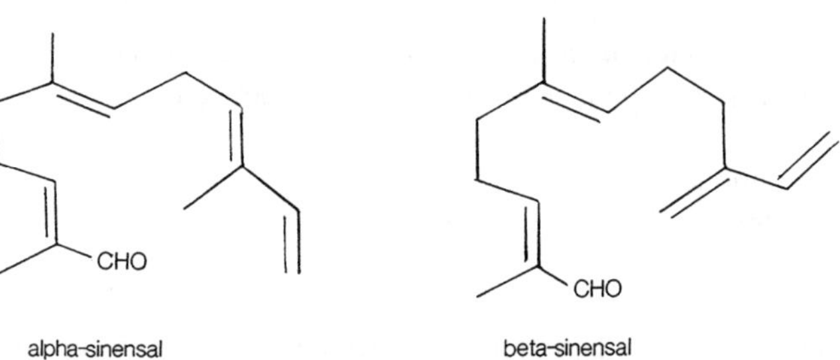

alpha-sinensal beta-sinensal

APPENDIX II: BOTANICAL CLASSIFICATION OF FRUITS

Citrus

Fleshy, edible fruit, segmented into sections and surrounded by a thick skin that contains essential oil characteristic of the fruit: orange, lemon, lime, tangerine/mandarin, grapefruit, citron, bergamot, kumquat.

Berries

Generally small fruits (exception kiwi) with pulpy edible part that has many seeds and generally juicy: blackcurrant, blackberry, blueberry, boysenberry, cranberry, gooseberry, huckleberry, kiwifruit, loganberry, raspberry, redcurrant, strawberry.

Fruits Containing Stones

Contains a single (normally large) stone which is the seed surrounded by the fleshy, edible pericarp: apricot, cherry, damson, peach, plum.

Tropical Fruits

Avocado, banana, date, fig, mango, pineapple, pomegranate.

Pomes

Body of fruit is formed by swelling of receptacle surrounding seed capsule: apple, pear, quince.

Melons

Large fruit with either soft pulpy or cellular flesh containing many seeds: water, honeydew, musk, canteloup.

Grapes

Small, usually very sweet fruits growing in clusters on vines with both seed-containing and seedless varieties. There are many varieties of both red and white.

Index

B

F

N

O

P

Q

R

S

T